Tropical Islands Biodiversity Crisis

The Indo-West Pacific

TOPICS IN BIODIVERSITY AND CONSERVATION

Volume 13

For other titles published in this series, go to www.springer.com/series/7488

Tropical Islands Biodiversity Crisis

The Indo-West Pacific

Edited by

David J.W. Lane

Reprinted from *Biodiversity and Conservation*, volume 19:2 (2010)

Springer

Editor
David J.W. Lane
Universiti Brunei Darussalam
Department of Biology
Jalan Tungku Link 1410
Brunei Darussalam
David.Lane@ubd.edu.bn

ISBN: 978-94-007-0167-0

Springer Dordrecht Heidelberg London New York

Library of Congress Control Number: 2010937225

Printed on acid-free paper

Springer is a part of Springer Science+Business Media (www.springer.com)

Contents

Biodivers Conserv (2010) 19:313–316
DOI 10.1007/s10531-009-9759-3

PREFACE

Tropical islands biodiversity crisis

David J. W. Lane

Published online: 25 December 2009

This Special Issue of *Biodiversity and Conservation* has its inception in a 3-day Conference—"Biodiversity Crisis on Tropical Islands"—organized by the Department of Biology, Universiti Brunei Darussalam and held at the University campus from 11th to 13th June 2007 as part of the 20th Anniversary of the founding of this relatively new University on Borneo Island. With generous university funding, the conference organisers were able to invite several distinguished keynote speakers to deliver plenary addresses on biodiversity issues related to islands in the tropics. These and other presentations, plus additional invited keynote contributions, resulted in a series of manuscripts which are more than a record of the Conference proceedings. Biodiversity depletion is considered to be, or should be, of crucial importance in the wider sphere of human affairs and in this context the invitation to contribute papers was further extended to other biodiversity researchers, including several keynote invitees, who, due to other commitments, were unable to participate in the Conference itself.

Fortuitously, but fittingly, the Conference and the preparation of this Special Issue also coincided with the 150th anniversary of the publication of Charles Darwin's 'Origin of Species'. The very last lines of this book encapsulate the grandeur of the diversity of life on Earth with the oft-quoted words "... from so simple a beginning endless forms most beautiful and most wonderful have been, and are being, evolved". Darwin and his contemporary, Wallace, in the development of their ideas on evolution, drew extensively on the recognition of islands as species 'foundries', yet widespread realization of the vulnerability of island life to extinction has come much later as Malthusian projections of human population pressure on biotic resources have become a reality.

Islands, often with a high proportion of endemic, in effect 'limited-edition', species have accounted for the bulk of the world's recognized extinctions of species in historical times. Additionally, these recognised loses alone have undoubtedly taken with them, to oblivion, a far greater number of obligately associated organisms (including symbionts, parasites and parasitoids), many of which may not even have been named. Much of the Earth's terrestrial biodiversity is located at tropical latitudes, in the rainforests, and the

D. J. W. Lane (✉)
Department of Biology, Universiti Brunei Darussalam,
1410 Jalan Tungku Link, Brunei Darussalam
e-mail: david.lane@ubd.edu.bn

accelerating loss of these forests is currently perceived to be a major threat to the rich biodiversity heritage of countries and islands in the Southeast Asian region, and elsewhere. Island endemics, including those in the tropics, are particularly vulnerable to extinction from any one or a number of causes. The theme of this Conference brought together researchers and conservation workers with common interests in biological diversity and conservation issues and was concerned with patterns of biodiversity loss and strategies for mitigation, principally (but not necessarily exclusively) on tropical 'islands'. In this context the definition of 'islands' encompasses not only land masses isolated by water but also, for example, remnant forest fragments on mainlands, inselbergs, water bodies, wetlands, etc., and even marine domains. The geographical focus was Southeast Asia (but not exclusively so) and the Conference programme allocated the final session to the dramatic habitat and biota losses underway on Borneo in particular.

So, introductions aside, what then of the content and format of this Special Issue? In any analysis of extinctions an appropriate viewpoint, at least for terrestrial vertebrates, might perhaps be the zooarchealogical record, here more than adequately summarized for the late Quarternary megafauna of the Borneo region by the Earl of Cranbrook (pp. 373–391). This extinction perspective, played out over millennia, may implicate early human hunters (they provided the data in cave repositories) but also involved natural climate shifts, associated with ice ages and sea level low-stands, resulting in repeated land connections of mainland Southeast Asia to what is now Borneo. Extinctions, even during the intervening periods between the Earth's mass extinctions, are thus nothing new but current major biota and habitat losses, incurred regionally and globally in just a few decades, are of major conservation concern and are highlighted for Southeast Asia in three opening papers, two of which by respectively Navjot Sodhi and his associates (pp. 317–328), and by Damien Fordham & Barry Brook (pp. 329–342), deal with losses and projected faunal extinctions in archipelagic and mainland Southeast Asia, a region of high species richness and endemicity; the next paper, by Peter Ashton (pp. 343–356), highlights for Borneo the extraordinary species and endemic richness of tree communities in areas, particularly in Northwest Borneo, that are critically endangered by deforestation and land degradation. The current initiatives adopted by (alphabetically) Brunei, Indonesia and Malaysia to include these Bornean botanic hotspots within the 'Heart of Borneo' conservation initiative or in other protected zones, if ultimately realised, are commendable.

Any discussion of extinctions requires an examination of the evidence upon which extinction projections are based and this is tackled, no doubt controversially in the eyes of some, in the subsequent essay by Nigel Stork (pp. 357–371). His analysis finds that recognised extinction rates are lower than expected. 'Recognised extinctions' is perhaps the operative phrase here and part of the controversy since, restricted populations on islands excepted, it is generally hard to confirm global extirpations for any but mega and macro animal and plant life forms. However Nigel's paper does concede that future extinctions are likely to be high and will be driven by the synergism of human-induced threats, including deforestation, with climate shifts.

The overriding or ultimate threat to both climate stasis and biodiversity richness is, as noted in Fordham and Brook's above mentioned paper, the resource and space demands of a burgeoning human population. This pressure is increasing dramatically in Southeast Asia, a region with a high population growth and rapid rates of development that coincide and conflict with the resource and habitat needs of what is essentially the world's highest concentration of terrestrial endemic species. Not surprisingly it is island endemics that are the first to disappear or become critically endangered. Climate change—in particular, CO_2 and other greenhouse-gas induced environmental warming—is now widely accepted as having at least partly anthropogenic causes and potentially catastrophic effects on biodiversity. One of the major contributions to enhanced atmospheric CO_2 is the conversion of Peat Swamp Forests. This is under-recognised even by many in the scientific community and is hardly recognised at all in the wider world and by the global media. Catherine Yule's paper (pp. 393–409) on the severe losses of the extensive Peat Swamp Forests in Southeast Asia puts this to rights, emphasising not only the losses of endemic biota of these specialized, challenging environments but also the stark fact that, area for area, the living biomass plus the deep peat deposits underlying these forests represent carbon sinks that are in the region of an order of magnitude greater than the biomass for dryland forests, even old-growth forests, and that peat forest systems are inadequately protected, vanishing rapidly and are already a major contributor to global CO_2 emissions to the atmosphere.

Deliberate or inadvertent introductions of alien species due to human activity, trade and travel, are shown in a review by Richard Corlett (pp. 411–423) to be another human factor with a potential or actual high impact on native biodiversity, most notably in chronically disturbed areas. The invasibility and vulnerability of island biotas also increases with island remoteness where ecological simplification and the absence of competing native species permits invaders to become dominant. Formulation and enforcement of global quarantine regulations to block or mitigate the transmission of invasives and the setting up of effective eradication programmes may seem a daunting challenge, at least administratively, but this has actually been done on a small scale, as documented by Michael Samways and his co-workers (pp. 425–434) for Cousine Island in the Western Indian Ocean. This restoration success story is perhaps a model for other 'disturbed' and invaded islands and can be seen as a beacon of hope in an otherwise gloomy, biodiversity-challenged scenario. But whether such efforts could ever be scaled up on larger islands and be economically viable is open to question.

The remaining papers cover a wide range of biodiversity topics and conservation challenges, ranging from case studies by Wong Khoon Meng et al. (pp. 435–447) of the vulnerability of endemic plants in small habitat islands in Malaysia, Matt Struebig et al. (pp. 449–469) on the alarming disturbance-related under-representation of forest interior bats in many of Borneo's existing so-called 'protected' forests, and Tigga Kingston (pp. 471–484) who reviews the conservation status and research needs for bats in general in Southeast Asia. Sodhi et al. (pp. 485–490) and McFadden et al. (pp. 491–502) focus respectively on insect extinctions and endangerment of mammalian carnivores on small transformed islands, while the increasing conservation importance of endemic-rich herpetofaunal diversity in an upland forest refuge threatened by agricultural conversion is recorded by Olga Nuneza et al. (pp. 503–518). Forest conversion and isolation of remnant forest fragments is a theme that recurs in several other papers in this Issue, with the general consensus that forest/non-forest or forest/plantation boundaries pose effective dispersal barriers to forest-dependant animal species (as opposed to generalists) thus threatening the viability of small populations isolated within forest remnants (see papers by Carsten Bruehl & Thomas Eltz (pp. 519–529), Anthony Shadbolt & Roselina Ragai (pp. 531–541), and Joe Charles & Bee Biaw Ang (pp. 543–561)). The majority of the papers in this Special Issue relate to status and projections for terrestrial biodiversity but aquatic topics are addressed too, with geographic and genetic differentiation demonstrated for Spanish mackerel stocks that were once considered to be contiguous (Sulaiman & Ovenden, pp. 563–574). Unexpected crustacean diversity and abundances around a small island in the Philippines near the geographic centre of the zone of maximum marine biodiversity, challenges former perceptions of rarity (Mendoza et al., pp. 575–586). Indeed the marine domain and its biodiversity remains under-explored in many tropical and deep water regions, a fact that is also recognized in the present contribution from representatives (Chaîneau et al., pp. 587–600) of a major player in the onshore and offshore oil and gas industry and by the commitment and financial/logistics support of this company in terms of biodiversity research. Apart from macroscopic organisms, the marine sediments, and the terrestrial soils in particular, harbour the most diverse and densely packed species assemblages of micro life on Earth, yet many of these organisms—ranging from bacteria to micro and meso invertebrates and fungi—and their ecological community functioning, are as yet little-studied. Given the global importance of soil biotic cycling processes, an important but largely unresolved issue is the level of functional redundancy in soil communities in the event of species losses.

The issues threatening biodiversity today are multifarious and often interacting. In attempting to compile a cohesive introductory presentation it soon became apparent that many logical sequences for the contributions to this Issue, quite different from the schema summarized above, could have been presented. Readers are certainly not expected to read from cover to cover—to do so would no doubt lead to information overload—but are instead encouraged to dip in and focus on topics of interest and then spread their understanding through informed discussions with others. It is hoped too that media science correspondents may be enlightened on key issues and thereby educate and inform government policy makers involved in bio-resource and land resource decisions.

Lastly it remains for me to thank the many colleagues and students in my Department and other staff in the University who helped to organise the original Conference, the Conference sponsors, the Editorial staff at Springer for assisting in the logistics of putting this Special Issue together, and to thank especially the numerous reviewers who generously contributed their time to criticizing and improving these contributions.

Biodivers Conserv (2010) 19:317–328
DOI 10.1007/s10531-009-9607-5

ORIGINAL PAPER

The state and conservation of Southeast Asian biodiversity

Navjot S. Sodhi · Mary Rose C. Posa · Tien Ming Lee · David Bickford · Lian Pin Koh · Barry W. Brook

Received: 20 March 2008 / Accepted: 19 February 2009 / Published online: 4 March 2009

Abstract Southeast Asia is a region of conservation concern due to heavy losses of its native habitats. In this overview, we highlight the conservation importance of Southeast Asia by comparing its degree of species endemism and endangerment, and its rate of deforestation with other tropical regions (i.e., Meso-America, South America, and Sub-Saharan Africa). Southeast Asia contains the highest mean proportion of country-endemic bird (9%) and mammal species (11%). This region also has the highest proportion of threatened vascular plant, reptile, bird, and mammal species. Furthermore, not only is Southeast Asia's annual deforestation rate the highest in the tropics, but it has also increased between the periods 1990–2000 and 2000–2005. This could result in projected losses of 13–85% of biodiversity in the region by 2100. Secondary habitat restoration, at least in certain countries, would allow for some amelioration of biodiversity loss and thus potentially lower the currently predicted extinction rates. Nonetheless, urgent conservation actions are needed. Conservation initiatives should include public education, sustaining livelihoods, and ways to enhance the sustainability of agriculture and increase the capacity

N. S. Sodhi (✉) · M. R. C. Posa · D. Bickford
Department of Biological Sciences, National University of Singapore, 14 Science Drive 4, Singapore 117543, Republic of Singapore
e-mail: dbsns@nus.edu.sg

T. M. Lee
Ecology, Behavior and Evolution Section, Division of Biological Sciences, University of California, San Diego, 9500 Gilman Drive, 92093-0116 La Jolla, CA, USA

L. P. Koh
Department of Ecology and Evolutionary Biology, Princeton University, 106A Guyot Hall, 08544-1003 Princeton, NJ, USA

L. P. Koh
Institute of Terrestrial Ecosystems, ETH Zürich, CHN G74.2, Universitätstrasse, 16 8092 Zurich, Switzerland

B. W. Brook
Research Institute for Climate Change and Sustainability, School of Earth and Environmental Sciences, University of Adelaide, Adelaide, SA 5005, Australia

of conservation institutions. Furthermore, these actions should be country-specific and not ignore areas heavily populated by humans, as they can also harbour high numbers of threatened species. We urge that cooperative conservation initiatives be undertaken and support (e.g., capacity-building) be given by more developed countries in the region and beyond.

Keywords Extinctions · Endangered species · Deforestation · Habitat loss · Species–area relationship

Introduction

Among the world's tropical regions, Southeast Asia (Brunei, Cambodia, Indonesia, Laos, Malaysia, Myanmar, the Philippines, Singapore, Timor-Leste, Thailand, and Vietnam) is of particular conservation concern because it has the highest rate of habitat loss (Sodhi et al. 2004; Sodhi and Brook 2006). Almost the entire Southeast Asia is considered a biodiversity hotspot because it harbours an exceptionally high number of endemic species that are threatened by the loss of >70% of original habitats (Myers et al. 2000). Further, Southeast Asia is highlighted as an area where past and present human-driven land-use changes are expected to cause extinctions across a wide range of taxa (Brook et al. 2003; Cardillo et al. 2006; Lee and Jetz 2008). However, environmental apathy, corruption, poor natural resource governance, poverty and lack of conservation funding remain formidable challenges for conservationists in Southeast Asia (Sodhi et al. 2004; Sodhi and Brook 2006; Posa et al. 2008).

This overview has three objectives. First, we highlight the conservation importance of Southeast Asia by comparing the degree of species endemism and endangerment, as well as the rate of deforestation in this region with that in other tropical areas. Second, we discuss the likely extent of extinctions in Southeast Asia due to deforestation. There has been a recent debate as to whether regenerating forests can dampen some of the deforestation-driven extinctions (Brook et al. 2006; Wright and Muller-Landau 2006; Gardner et al. 2007; Koh 2007; Laurance 2007). Therefore, we discuss our previously published extinction estimates (Brook et al. 2003; Sodhi and Brook 2006) in light of regenerating forests and other threats to biodiversity such as climate change, invasive species and fire. High human population density also jeopardizes biodiversity through facilitating habitat loss, the spread of invasive species and overharvesting (Cardillo et al. 2004; Sodhi et al. 2004). Thus, we correlate human population density with deforestation and species endangerment to illuminate implications for conservation. We restrict our overview to terrestrial realms but hope to inspire similar analyses for the region's equally threatened freshwater and marine realms (e.g., Halpern et al. 2008).

Southeast Asian biodiversity

The high species richness and endemism in Southeast Asia is linked to its complex geological history (Sodhi et al. 2004). The region's floral and faunal elements have diverse origins, having been brought together by the collision of several tectonic plates at different geologic times (Hall 1998). Shifting paleoclimates caused periodic fluctuations in sea level, in turn enabling migrations from the mainland to the Indo-malayan archipelago and isolating islands (Meijaard 2003). These processes, coupled with a tropical climate,

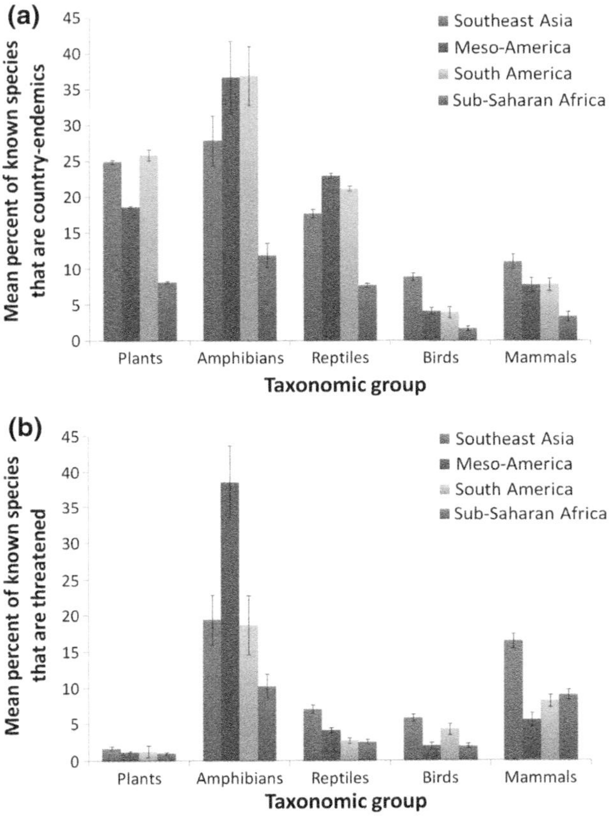

Fig. 1 Degree of species **a** endemism and **b** endangerment for vascular plants, amphibians, reptiles, birds, and mammals for the tropical regions of Southeast Asia (n = 9 countries), Meso-America (n = 13), South America (n = 12), and Sub-Saharan Africa (n = 41) (UNEP–WCMC 2004; IUCN 2006). *Error bars* represent standard errors

facilitated speciation and resulted one of the world's most biodiverse regions. Southeast Asia contains the highest mean proportion of country-endemic bird (9%) and mammal species (11%) and the second-highest proportion of country-endemic vascular plant species (25%) compared to the other tropical regions of Meso-America, South America, and Sub-Saharan Africa (Fig. 1a; UNEP–WCMC 2004). Southeast Asia also has the highest proportion of threatened species across all taxonomic groups (i.e., vascular plants, reptiles, birds and mammals) except amphibians (Fig. 1b; IUCN 2006). The proportion of endemic amphibians (28%) and reptiles (18%) in Southeast Asia is relatively lower than in the new world (Fig. 1a) and it has the lowest proportion of threatened amphibians among the tropical regions (Fig. 1b). However, this likely reflects the paucity of herpetological research in the region (see Sodhi et al. 2007, 2008b) rather than a lack of diversity in these taxa. Indeed, as recently as 2007, the first lungless frog was discovered on Borneo (Bickford et al. 2008a) and there are many amphibian and reptile species awaiting description. Furthermore, amphibians remain the most highly endangered of all terrestrial vertebrates in all areas and the fact that they are not as threatened in Southeast Asia as they are in other areas, makes their plight no less severe in Southeast Asia (see absolute values from Fig. 1). The data above demonstrate the relatively elevated level of threat to Southeast Asia's highly endemic biota.

Deforestation rates in Southeast Asia

Not only is Southeast Asia's annual deforestation rate the highest among all tropical regions, it has also been recently increasing (Fig. 2a; FAO 2006). Less than 10% of Southeast Asia's forests are under some form of protection (World Conservation Union [IUCN] categories I–VI), which suggests that habitat loss in the region likely will continue (Fig. 2b; Iremonger et al. 1997; FAO 2006). Furthermore, many of the region's protected forests are also being degraded and destroyed by illegal logging activities (Curran et al. 2004; DeFries et al. 2005; Bickford et al. 2008b). During the period 1990–2005, many secondary forests in Southeast Asia were also being lost (Fig. 3), which only makes the story grimmer. The deforestation estimates we have used are based on the 2005 Forest Resource Assessment (FRA) of the United Nations Food and Agriculture Organization (FAO) and are undoubtedly prone to errors (see Grainger 2008). Therefore, our results should be interpreted with caution.

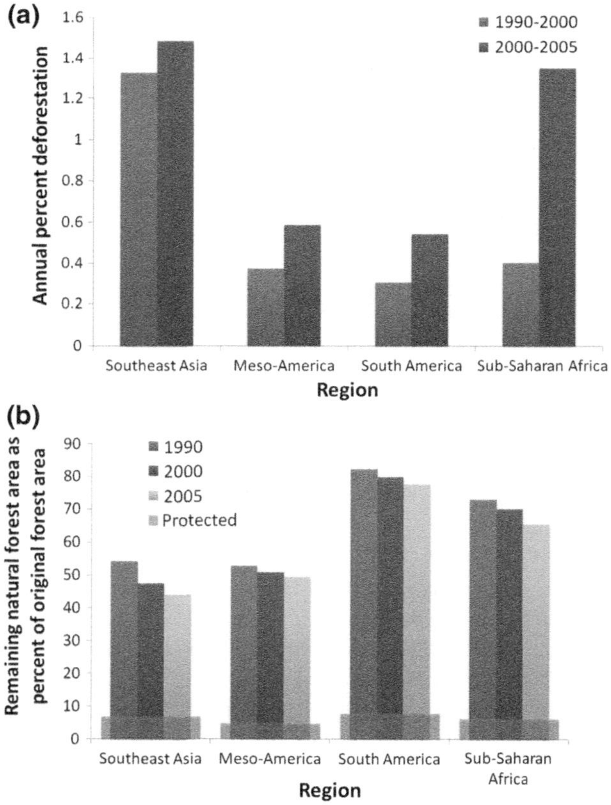

Fig. 2 **a** Annual percent deforestation of natural (primary and secondary) forests for the period 1990–2000 and 2000–2005; and **b** areas of remaining natural forest and protected forest for the tropical regions of Southeast Asia, Meso-America, South America, and Sub-Saharan Africa (Iremonger et al. 1997; FAO 2006)

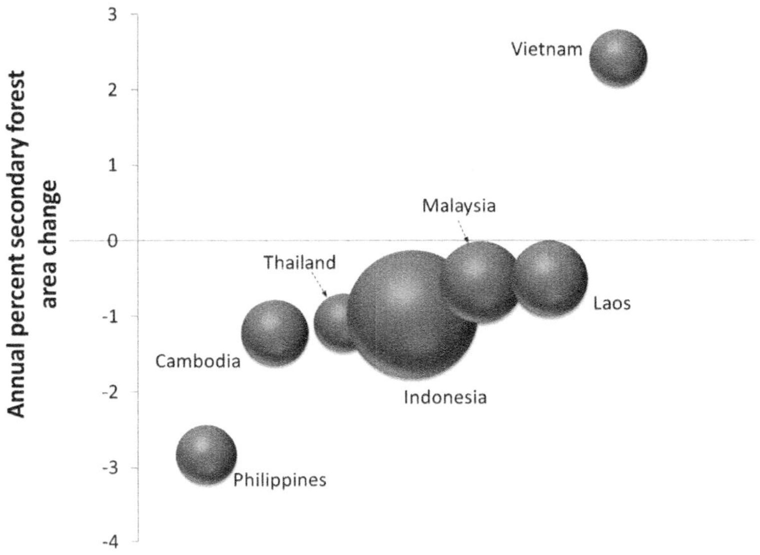

Fig. 3 Annual percent deforestation of secondary forests for the period 1990–2005 for the Southeast Asian countries of Cambodia, Indonesia, Laos, Malaysia, the Philippines, Thailand, and Vietnam (FAO 2006). *Bubble size* reflects the relative area of secondary forest remaining in each country

Projected biotic extinctions from Southeast Asia

The relationship between habitat area and the number of species of a region is typically characterized by the log-linear species–area relationship (SAR; Plotkin et al. 2000). The SAR has been used in a number of studies to predict future extinctions in Southeast Asia as a result of deforestation (Brooks et al. 1997; Brook et al. 2003). The broad regional projections by Brook et al. (2003) predicted that between 13 and 42% of all Southeast Asian species would be lost by 2100 due to the loss of 72–90% of habitat area—with the majority of the species already committed to extinction due to historical clearance of roughly 59% the region's original forest cover (Brook et al. 2006). An elaboration of this method by Sodhi and Brook (2006), using country-specific deforestation rates, implied that even these seemingly high estimates are likely to be conservative, because those countries with the greatest number of endemic species were also those which had experienced greatest historical or current rates of forest loss (e.g., Indonesia, Vietnam and the Philippines). As such, Sodhi and Brook (2006) estimated that 24–63% of Southeast Asian endemic taxa, representing 859–4,815 vertebrate species (including 66% of the region's birds and 85% of its mammals) and 8,343–48,043 species of vascular plants were at risk of extinction due to deforestation this century.

Yet, these SAR-based projections are admittedly a blunt tool for assessing future biodiversity loss because they ignore a number of real-world feedbacks on extinction risk, both positive and negative (see He and Legendre 1996; Rey et al. 1982). For instance, large swathes of tropical forest that were cleared over the last few decades to centuries for swidden agriculture or timber harvesting are now re-establishing as secondary forest. In some areas, the rate of re-growth is sufficiently rapid as to equalize or even overcome the rate of clearance of pristine habitat (Wright and Muller-Landau 2006). Additionally, there is evidence that a good proportion of forest species can survive in secondary forests, selectively logged production forests and even exotic tree plantations (Peh et al. 2005; Barlow et al. 2007; Meijaard and Shiel 2008). Although these areas generally support fewer species than original primary forest, especially in the short term, it is reasonable to expect that secondary habitat restoration would allow for some amelioration of biodiversity loss (Wright and Muller-Landau 2006).

However, despite the above mentioned biases in these SAR-projections they still are likely to underestimate extinctions. This is due to additional impacts of unmodelled factors, such as climate change, invasive competitors and predators, acceleration of forest clearance driven by growing economic demand for mature rain forest timber, and perhaps most insidiously, by the synergistic interactions among these threats (Brook et al. 2008). For instance, deforestation also inevitably causes habitat fragmentation, which hampers dispersal and isolates populations. Logging trails increase access to forest interiors and thereby facilitates access by hunters. Fragmentation, loss of forest cover, and global climate change, could also cause a localised drying around edges of rain forest patches, a regional drop in evapo-transpiration rate, and broad-scale shifts in rainfall due a disruption of the Indian monsoon and El Nino–La Nina cycle, all potentially enhancing fire risk, water loss and thermal stress (Laurance et al. 2006; Brook et al. 2008; Knof and Zickfled 2008). All of these unknowns bias SAR extinction estimates in the same direction—underestimation. So most realistic (but unknown) impacts or synergies will make the region even more threatened than what is already anticipated.

The challenge, if predictions of biotic extinctions in Southeast Asia are to be improved (and ultimately mitigated), is for future research to focus on quantifying these general observations or theoretical expectations. For instance, detailed longitudinal studies across multiple habitat types are needed to assess the extent and length of time for faunal relaxation after disturbance as well as species recovery (or stability of diversity) following habitat restoration. Satellite monitoring of Southeast Asian forests has recently provided, and will continue to deliver, the type of robust spatial data on forest recovery rates needed for more detailed projections of future forest cover. As much as possible, these should be verified with ground truth information. Continual improvement in the down-scaled predictions of atmosphere-ocean coupled global circulation climate models (Intergovernmental Panel on Climate Change 2007) will yield more certainty in estimates of regional rainfall trends and temperature changes that will affect biodiversity. If this information can be adequately linked to vegetation-change and fire models, then improved predictions of extinction rates for Southeast Asia will no doubt emerge. But the reality is that given current limits to knowledge, the disturbingly broad SAR-based estimates of 13–85% extinctions by 2100 are as precise as we can currently derive.

Human density, deforestation and biodiversity endangerment

The growth of the human population and its greater demand for resources are the driving forces behind land transformation and resulting biodiversity losses (Vitousek et al. 1997). The link between high human population densities and risk to biodiversity is mainly indirect, but are nevertheless linked (Cincotta et al. 2000). We correlated human population density with forest cover and threatened species for the countries in Southeast Asia. A negative relationship was found between proportion of remaining forested areas and human population density (Fig. 4a). This correlation suggests that increasing human population density may make forests vulnerable as they are converted to agricultural or urban areas, or logged to provide timber. In addition, forests in heavily populated areas also remain

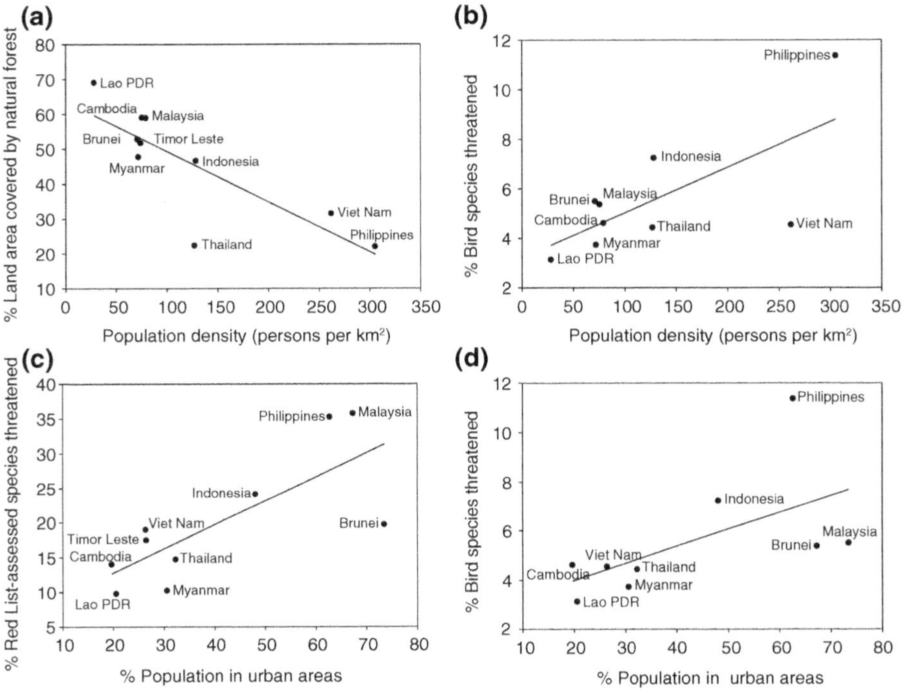

Fig. 4 Scatterplots of correlated biodiversity and socioeconomic indicators, with regression lines shown. Population density correlated with **a** the proportion of land still forested ($r_s = -0.82$) and **b** the proportion of bird species threatened in each country ($r_s = 0.71$). The percentage of the population found in urban areas correlated with **c** the proportion of vascular plant and vertebrate species assessed (see Fig. 1) for the IUCN Red List that are classified as Vulnerable, Endangered and Critically Endangered ($r_s = 0.77$) as well as **d** the proportion of bird species threatened in each country ($r_s = 0.71$). Singapore was excluded from the analysis due to extreme values. Spearman's rank correlations were significant for **a** and **c** at $P = 0.02$. Data sources: US Census Bureau (population density), World Resources Institute (urban population, number of known species) Food and Agriculture Organization of the United Nations Forestry Department (extent of natural forests), and The World Conservation Union Red List 2007 (number of threatened species)

vulnerable to overharvesting (Bulte and Horan 2002). Human population density as a whole in Southeast Asia has been increasing steadily over the past decade at a rate of 1.5% per annum (United Nations Statistics Division 2007), indicating that remaining forests will continue to face high risk of degradation both locally, and over the entire region.

A positive, but non-significant relationship between human population density and proportion of bird species threatened (i.e., in the vulnerable, endangered and critically endangered categories) as assessed by the IUCN was also found (Fig. 4b). Not surprisingly, we also correlated human population density with the proportion of threatened species of other taxa (plants, amphibians, reptiles and mammals) but found no significantly positive correlations, probably because of small sample sizes and poorly assessed status data. Because urbanization will almost certainly greatly expand in the future (Palmer et al. 2004), we determined whether human population in urban areas was correlated with the proportion of threatened species. The proportion of species of all taxa, as well as the subset of bird species that are threatened in each country, were all positively correlated with the percentage of the human population found in urban areas (Fig. 4c, d). However, not all correlations were statistically significant probably due to small sample sizes. These results

Table 1 Major threats and conservation imperatives for terrestrial realms for each country in Southeast Asia

Country	Major threat(s)/opportunities for biodiversity conservation	Conservation imperative(s)
Brunei Darussalam	High human development (Global HDI Rank: 30[a]; GDP per capita Rank: 4[b])	Invest more in the conservation efforts/building capacity in the regions
Cambodia	High rate of deforestation (1.3% natural forest lost annually from 1990–2005[c])	Protection of forests
	Corruption (Global Rank: 162[d])	Increase the capacity of conservation-related institutions
	Poverty (HPI-1 Rank: 85[a])	Integrate sustaining livelihood into conservation regimes
Indonesia	High rate of deforestation (1.7% natural forest lost annually from 1990–2005[c])	Protection of forests/reforestation
	Commercial logging (top tropical log producer from 2004–2007[e])	Reduce tropical timber demand/sustainable logging
	Illegal wildlife trade[f]	Curb illegal wildlife trade
	Oil palm production (12.5% annual growth rate from 1996–2006[g])	Sustainable oil palm production
	Corruption (Global rank: 143[d])	Increase the capacity of conservation-related institutions
Lao PDR	Rapid growth of agricultural production (80% from 1995–2005[g])	Sustainable and efficient agriculture
	Corruption (Global rank: 168[d])	Increase the capacity of conservation-related institutions
	Poverty (HPI–1 Rank: 70[a])	Integrate sustaining livelihood into conservation regimes
Malaysia	Commercial logging (top tropical log producer and exporter from 2004–2007[e])	Reduce tropical timber demand/increase sustainable logging
	Oil palm production (6.6% annual growth rate from 1996–2006[g])	Sustainable oil palm production
Myanmar	Few protected areas (4.63% of surface area[h])	Protection of forests
	High rate of deforestation (1.3% natural forest lost annually from 1990–2005[c])	Protection of forests/reforestation
	Rapid growth of agricultural production (57% from 1995–2005[g])	Sustainable and efficient agriculture
	Corruption (Global Rank: 171[d])	Increase the capacity of conservation-related institutions
Philippines	Dwindling natural forest (22% of land area in 2005[c])	Protection of forests
	High rate of deforestation (1.7% natural forest lost annually from 1990–2005[c])	Reforestation
	Corruption (Global Rank: 131[d])	Increase the capacity of conservation-related institutions
Singapore	Extreme deforestation (3% natural forest remaining in 2005[c])	Reforestation/Reintroductions/Wildlife Corridors
	High population density (6,667 persons per km^2 in 2007[i])	Limit human population density
	Few protected areas (2.24% of surface area[h])	Protection of forests
	High human development (Global HDI Rank: 25[a], GDP per capita Rank: 7[b])	Invest more in conservation efforts/capacity in the region

Table 1 continued

Country	Major threat(s)/opportunities for biodiversity conservation	Conservation imperative(s)
Thailand	Dwindling natural forest (22% of land area in 2005[c])	Protection of forests/Reforestation
	Oil palm production (5.5% annual growth rate from 1996–2006[g])	Sustainable oil palm production
Timor-Leste	Few protected areas (1.22% of surface area[h])	Protection of forests
	High population growth (3.4% annually for 2005–2015[a])	Human population control
	Corruption (Global Rank: 123[d])	Increase the capacity of conservation-related institutions
	Poverty (HPI-1 Rank: 95[a])	Integrate sustaining livelihood into conservation regimes
Viet Nam	Few protected areas (3.62% of surface area[h])	Protection of forests
	Illegal wildlife trade[j]	Curb illegal wildlife trade
	Rapid growth of agricultural production (69% from 1995–2005[g])	Sustainable and efficient agriculture
	Corruption (Global Rank: 123[d])	Increase the capacity of conservation-related institutions

GDP gross domestic product at purchasing power parity; *HDI* human development index; *HPI-1* human poverty index for developing countries

Data Sources: [a] United Nations Development Programme: Human Development Report 2007/2008; [b] The World Bank 2005 International Comparison Program; [c] Food and Agriculture Organization of the United Nations (FAO): Global Forest Resources Assessment 2005 (Forestry Sector); [d] Transparency International: 2007 Corruptions Perceptions Index http://www.transparency.org/; [e] International Tropical Timber Organization: Annual Review and Assessment of the World Timber Situation 2006; [f] Traffic reports http://www.traffic.org; [g] FAO, Selected Indicators of Food and Agricultural Development 1996–2006; [h] United Nations Environment Programme World Conservation Monitoring Centre World Database on Protected Areas http://www.unep-wcmc.org/wdpa/; [i] US Census Bureau: International Database http://www.census.gov/ipc/www/idb/; [j] Lin (2005) suggest that threatened species may remain vulnerable in heavily populated areas (Cincotta et al. 2000; Cardillo et al. 2004) and given the congruence between high population densities and threatened species, conservation should not be ignored in urban areas (see also Miller 2005).

Conclusions and recommendations

All data suggest that Southeast Asian biodiversity remains extremely vulnerable to habitat loss. The protection of the region's remaining forests and biodiversity will require the integration of social issues (e.g., rural employment) into conservation planning (Bickford et al. 2008b; Posa et al. 2008; Sodhi 2008; Sodhi et al. 2008a). Corruption in countries such as Indonesia has resulted in high levels of illegal logging (Sodhi et al. 2007). Therefore, efforts are urgently needed to curb corruption for effective governance of existing "natural capital" of Southeast Asia. Additionally, for tangible conservation, the reliance of rural communities on natural resources should be minimized by creating employment opportunities (e.g., nature guides).

While the highest priority should be given to protecting primary habitats, conservation initiatives should not ignore areas heavily populated by humans. Conservation efforts in human-dominated landscapes must intensify and include public education, sustaining livelihoods and ways to enhance the sustainability of agriculture and strengthening of the capacity of conservation institutions. Additionally, where possible, reforestation, reintroductions and reestablishing severed habitat connections should be attempted in these landscapes.

The region is plagued by a staggering diversity of threats and imperatives (Table 1). Efforts need to be tailored to the conservation priorities in each country and enhanced through cooperative agreements among countries. Even within some countries, there may be multiple priorities for conservation (e.g., excessive bushmeat hunting and deforestation in Sulawesi), so even country-wide efforts should reflect the diversity of threats. We urge the higher income countries (Brunei Darussalam and Singapore) to invest more into conservation initiatives in poorer countries in the region (e.g., Indonesia and the Philippines). For example, Singapore can train biologists from the region and provide expertise in environmental management. Similarly, broader networks can be made with countries such as Japan and Australia to further facilitate capacity building and sharing of biodiversity information. In addition, cooperation within the region and beyond (e.g., China and South Korea) will be critical in curbing the illegal animal and timber trade. Amelioration of massive and ongoing environmental problems (e.g., Lohman et al. 2007) will only be successful with regional and global cooperation and in a region with such tremendous variation in problems and solutions, Southeast Asia faces a challenging task ahead.

Acknowledgments This research was supported by the National University of Singapore (R-154-000-331-112). We thank David J. W. Lane for inviting us to write this article, and two reviewers for comments. NSS thanks the Sarah and Daniel Hrdy Fellowship at Harvard University for support while this manuscript was revised.

References

Barlow J, Gardner TA, Araujo IS, Ávila-Pires TC, Bonaldo AB, Costa JE, Esposito MC, Ferreira LV, Hawes J, Hernandez MIM, Hoogmoed MS, Leite RN, Lo-Man-Hung NF, Malcolm JR, Martins MB, Mestre LAM, Miranda-Santos R, Nunes-Gutjahr AL, Overal WL, Parry L, Peters SL, Ribeiro-Junior MA,

MNFd Silava, CdS Motta, Peres CA (2007) Quantifying the biodiversity value of tropical primary, secondary and plantation forests. Proc Natl Acad Sci USA 104:18555–18560. doi:10.1073/pnas.0703333104

Bickford D, Iskandar DT, Barlian A (2008a) First lungless frog found on Borneo. Curr Biol 18:R374–R375

Bickford D, Supriatna J, Nanda Y, Iskandar D, Evans B, Brown R, Umilaela Townsend T, Azhari D, McGuire J (2008b) Indonesia's protected areas need more protection–suggestions from island examples. In: Sodhi NS, Acciaioli G, Erb M, Tan AK-J (eds) Biodiversity and human livelihoods in protected areas: case studies from the Malay Archipelago. Cambridge University Press, Cambridge, pp 53–77

Brook BW, Sodhi NS, Ng PKL (2003) Catastrophic extinctions follow deforestation in Singapore. Nature 424:420–423. doi:10.1038/nature01795

Brook BW, Bradshaw CJA, Koh LP, Sodhi NS (2006) Momentum drives the crash: mass extinction in the tropics. Biotropica 38:302–305. doi:10.1111/j.1744-7429.2006.00141.x

Brook BW, Sodhi NS, Bradshaw CJA (2008) Synergies among extinction drivers under global change. Trends Ecol Evol 23:453–460. doi:10.1016/j.tree.2008.03.011

Brooks TM, Pimm SL, Collar NJ (1997) Deforestation predicts the number of threatened birds in insular Southeast Asia. Conserv Biol 11:382–394. doi:10.1046/j.1523-1739.1997.95493.x

Bulte EH, Horan RD (2002) Does human population growth increase wildlife harvesting? An economic assessment. J Wildl Manage 66:574–580. doi:10.2307/3803125

Cardillo M, Purvis A, Sechrest W, Gittleman JL, Bielby J, Mace GM (2004) Human population density and extinction risk in the world's carnivores. PLoS Biol 2:e909. doi:10.1371/journal.pbio.0020197

Cardillo M, Mace GM, Gittleman JL, Purvis A (2006) Latent extinction risk and the future battlegrounds of mammal conservation. Proc Natl Acad Sci USA 103:4157–4161. doi:10.1073/pnas.0510541103

Intergovernmental Panel on Climate Change (2007) Climate change 2007: synthesis report. http://www.ipcc.ch/pdf/assessment-report/ar4/syr/ar4_syr.pdf. Accessed 17 March 2008

Cincotta RP, Wisnewski J, Engelman R (2000) Human population in the biodiversity hotspots. Nature 404:990–992. doi:10.1038/35010105

Curran LM, Trigg SN, McDonald AK, Astiani D, Hardiono YM, Siregar P, Caniago I, Kasischke E (2004) Lowland forest loss in protected areas of Indonesian Borneo. Science 303:1000–1003. doi:10.1126/science.1091714

DeFries R, Hansen A, Newton AC, Hansen MC (2005) Increasing isolation of protected areas in tropical forests over the past twenty years. Ecol Appl 15:19–26. doi:10.1890/03-5258

FAO (2006) Global forest resources assessment 2005: progress towards sustainable forest management. Forestry paper 147. http://www.fao.org. Accessed October 2007. FAO (United Nations Food and Agriculture Organization), Rome

Gardner TA, Barlow J, Parry LW, Peres CA (2007) Predicting the uncertain future of tropical forest species in a data vacuum. Biotropica 39:25–30. doi:10.1111/j.1744-7429.2006.00228.x

Grainger A (2008) Difficulties in tracking the long-term global trend in tropical forest area. Proc Natl Acad Sci USA 105:818–823. doi:10.1073/pnas.0703015105

Hall R (1998) The plate tectonics of Cenzoic SE Asia and the distribution of land and sea. In: Hall R, Holloway JD (eds) Biogeography and geological Evolution of SE Asia. Backbuys Publishers, Leiden, pp 99–131

Halpern BS, Walbridge S, Selkoe KA, Kappel CV, Micheli F, D'Agrosa C, Bruno JF, Casey KS, Ebert C, Fox HE, Fujita R, Heinemann D, Lenihan HS, Madin EMP, Perry MT, Selig ER, Spalding M, Steneck R, Watson R (2008) A global map of human impact on marine ecosystems. Science 319:948–952. doi:10.1126/science.1149345

He F, Legendre P (1996) On species-area relations. Am Nat 148:719–737. doi:10.1086/285950

Iremonger S, Ravilious C, Quinton T (1997) A statistical analysis of global forest conservation. In: Iremonger S, Ravilious C, Quinton T (eds) A global overview of forest conservation CD-ROM. World Conservation Monitoring Centre (WCMC) and Centre for International Forestry Research (CIFOR), Cambridge

IUCN (2006) 2006 IUCN Red List of Threatened Species. http://www.redlist.org. Accessed October 2007. IUCN (International Union for Conservation of Nature and Natural Resources), Gland

Knof B, Zickfled K (2008) Stability and sensitivity analysis of the Indian monsoon under climate change. Geophys Res Abs 10:EGU 2008-A-10195

Koh LP (2007) Impending disaster or sliver of hope for Southeast Asian forests? The devil may lie in the details. Biodivers Conserv 16:3935–3938. doi:10.1007/s10531-007-9177-3

Laurance WF (2007) Have we overstated the tropical biodiversity crisis? Trends Ecol Evol 22:65–70. doi:10.1016/j.tree.2006.09.014

Laurance WF, Nascimento HEM, Laurance SG, Andrade A, Ribeiro JELS, Giraldo JB, Lovejoy TE, Condit R, Chave J, Harms KE, D'Angelo S (2006) Rapid decay of tree-community composition in Amazonian forest fragments. Proc Natl Acad Sci USA 103:19010–19023. doi:10.1073/pnas.0609048103

Lee TM, Jetz W (2008) Future battlegrounds for conservation under global change. Proc R Soc Lond B Biol Sci 275:1261–1270. doi:10.1098/rspb.2007.1732

Lin J (2005) Tackling Southeast Asia's illegal wildlife trade. SYBIL 9:191–208

Lohman DJ, Bickford D, Sodhi NS (2007) The burning issue. Science 316:376. doi:10.1126/science.1140278

Meijaard E (2003) Mammals of south-east Asian islands and their Late Pleistocene environments. J Biogeogr 30:1245–1257. doi:10.1046/j.1365-2699.2003.00890.x

Meijaard E, Shiel D (2008) The persistence and conservation of Borneo's mammals in lowland rainforest managed for timber: observations, overviews and opportunities. Ecol Res 23:21–34. doi:10.1007/s11284-007-0342-7

Miller JR (2005) Biodiversity conservation and the extinction of experience. Trends Ecol Evol 20:430–434. doi:10.1016/j.tree.2005.05.013

Myers N, Mittermeier RA, Mittermeier CG, da Fonseca GAB, Kent J (2000) Biodiversity hotspots for conservation priorities. Nature 403:853–858. doi:10.1038/35002501

Palmer M, Bernhardt E, Chornesky E, Collins S, Dobson A, Duke C, Gold B, Jacobson R, Kingsland S, Kranz R, Mappin M, Martinez ML, Micheli F, Morse J, Pace M, Pascual M, Palumbi S, Reichman OJ, Simons A, Townsend A, Turner M (2004) Ecology for a crowded planet. Science 304:1251–1252. doi:10.1126/science.1095780

Peh KS-H, De Jong J, Sodhi NS, Lim SLH, Yap CAM (2005) Lowland rainforest avifauna and human disturbance: persistence of primary forest birds in selectively logged forests and mixed-rural habitats of southern Peninsular Malaysia. Biol Conserv 123:489–505. doi:10.1016/j.biocon.2005.01.010

Plotkin JB, Potts MD, Yu DW, Bunyavejchewin S, Condit R, Foster R, Hubbell S, LaFrankie J, Manokaran N, Lee H-S, Sukumar R, Nowak MA, Ashton PS (2000) Predicting species diversity in tropical forests. Proc Natl Acad Sci USA 97:10850–10854. doi:10.1073/pnas.97.20.10850

Posa MRC, Diesmos AC, Sodhi NS, Brooks TM (2008) Hope for threatened tropical biodiversity: lessons from the Philippines. Bioscience 58:231–240. doi:10.1641/B580309

Rey JR, Strong DR Jr, McCoy ED (1982) On overinterpetation of the species area relationship. Am Nat 119:741–743. doi:10.1086/283949

Sodhi NS (2008) Tropical biodiversity loss and people—a brief review. Basic Appl Ecol 9:93–99. doi:10.1016/j.baae.2007.11.001

Sodhi NS, Brook BW (2006) Southeast Asia biodiversity in crisis. Cambridge University Press, Cambridge

Sodhi NS, Koh LP, Brook BW, Ng PKL (2004) Southeast Asian biodiversity: an impending disaster. Trends Ecol Evol 19:654–660. doi:10.1016/j.tree.2004.09.006

Sodhi NS, Brook BW, Bradshaw CJA (2007) Tropical Conservation Biology. Blackwell, Oxford

Sodhi NS, Acciaioli G, Erb M, Tan AK-J (2008a) Biodiversity and human livelihoods in protected areas: case studies from the Malay Archipelago. Cambridge University Press, Cambridge

Sodhi NS, Bickford D, Diesmos AC, Lee TM, Koh LP, Brook BW, Sekercioglu CH, Bradshaw CJA (2008b) Measuring the meltdown: drivers of global amphibian extinction and decline. PLoS One 3:e1636. doi:10.1371/journal.pone.0001636

UNEP–WCMC (2004) Species data. http://www.unep-wcmc.org. Accessed October 2007. UNEP–WCMC (World Conservation Monitoring Centre of the United Nations Environment Programme), Cambridge

United Nations Statistics Division (2007) Demographic yearbook system. http://unstats.un.org/unsd/demographic/products/dyb/dyb2.htm. Accessed 15 February 2008

Vitousek PM, Mooney HA, Lubchenco J, Melillo JM (1997) Human domination of earth's ecoystems. Science 277:494–499. doi:10.1126/science.277.5325.494

Wright SJ, Muller-Landau HC (2006) The uncertain future of tropical forest species. Biotropica 38:443–445

Biodivers Conserv (2010) 19:329–342
DOI 10.1007/s10531-008-9529-7

ORIGINAL PAPER

Why tropical island endemics are acutely susceptible to global change

Damien A. Fordham · Barry W. Brook

Received: 4 July 2008 / Accepted: 14 November 2008 / Published online: 10 December 2008

Abstract Tropical islands are species foundries, formed either as a by-product of volcanism, when previously submerged seabed is thrust upwards by tectonics, or when a peninsula is isolated by rising sea level. After colonisation, the geographical isolation and niche vacancies provide the competitive impetus for an evolutionary radiation of distinct species-island endemics. Yet the very attributes which promote speciation in evolutionary time also leave island endemics highly vulnerable to recent and rapid impacts by modern people. Indeed, the majority of documented human-driven extinctions have been exacted upon island endemics. The causes include over-exploitation, invasive species brought by people and destruction of island's naturally constrained habitats. Imminent threats include inundation by rising sea levels and other adaptive pressures related to anthropogenic global warming. We review recent work which underscores the susceptibility of island endemics to the drivers of global change, and suggest a methodological framework under which, we argue, the science and mitigation of island extinctions can be most productively advanced.

Keywords Biodiversity · Biogeography · Climate change · Deforestation · Extinction · Over-exploitation · Habitat loss · Invasive species · Southeast Asia · Synergistic human impacts

Introduction

Southeast Asia has the world's greatest concentration of endemic species, reflecting the unique geographical and geological history of the region (Sodhi and Brook 2006). Tectonic plate movements and glacial-interglacial fluctuations in Quaternary sea levels have, at different times in prehistory, created islands, formed and severed land bridges and exposed or inundated large areas of Southeast Asia (Voris 2000), creating a foundry for speciation

D. A. Fordham · B. W. Brook (✉)
Research Institute for Climate Change and Sustainability, School of Earth and Environmental Sciences, University of Adelaide, Adelaide, SA 5005, Australia
e-mail: barry.brook@adelaide.edu.au

due to allopatry and regular environmental change (Mittermeier et al. 1998). Yet the very attributes which promote diversification on evolutionary time-scales also leave island endemics highly vulnerable to recent, rapid human impact (Cronk 1997; Sadler 1999). Indeed, the majority of documented human-driven extinctions, from prehistory through to modern times, have been exacted upon island endemics (Diamond 1989), many of these involving birds (Baillie et al. 2004). Today, over-exploitation, introduced species and habitat destruction continue to threaten island biodiversity—especially birds (Johnson and Stattersfield 1990; Blackburn et al. 2004)—threatening dramatic and permanent shifts in ecosystem function (O'Dowd et al. 2003).

Global extinction rates have soared over the past century, due predominantly to the resource demands of a burgeoning human population (Millennium Ecosystem Assessment, MA, www.millenniumassessment.org, 2005). Shifting land use (e.g. forest clearance for agriculture and logging) and wildlife exploitation (e.g. for market consumption and pet trade) have reduced the range and abundance of many island species in Southeast Asia, directly causing severe biodiversity loss at a local scale (Brook et al. 2003), and indirectly limiting the scope for sufficient ecological and evolutionary adaptation to future climate and land use change across the region (Sodhi et al. 2004a). Mitigation of human-mediated impacts continues to be a clear focus for tropical conservation biology, but with an increasing acknowledgement that these deterministic and stochastic extinction drivers are usually interacting and self-reinforcing (Bradshaw et al. 2009). In particular, the play-offs between these historical threats and an emerging, overarching process—climate change—are becoming increasingly apparent (Lovejoy and Hannah 2005). With this, a greater awareness of the problem of synergistic feedbacks is becoming apparent (Brook et al. 2008).

In the face of mounting evidence of the developing biodiversity crisis in Southeast Asia (Sodhi and Brook 2006), and knowledge that human-induced ecosystem shifts may be rapid, large and sometimes irreversible (i.e. once thresholds are crossed, hysteresis will mean it is impossible to restore former interrelationships; Brook et al. 2008), policy makers and resource managers urgently need new mechanisms for informed decision-making-validated systems which incorporate the complexities and adequately capture the uncertainties underlying the response of biodiversity to human impacts (Zimmer 2007). Behind this backdrop, we argue that islands, as microcosms of continental-scale processes, provide an intriguing opportunity to disentangle environmental and ecological constraints owing to their smaller size, distinctive colonisation and extinction dynamics (MacArthur and Wilson 1967), and simpler food webs (Pimm 1991): enabling a 'whole ecology' model approach.

In this overview of the challenges and opportunities offered by the study of tropical island endemic systems, we: (1) describe the historical factors that have combined to influence the high biodiversity and endemism characteristic of archipelagic Southeast Asia (our focal region); (2) underscore the susceptibility of island endemics to human-caused extinction, highlighting the link between biodiversity and ecosystem services; and (3) provide a methodological framework under which, we argue, the science and management of island extinctions can be most productively advanced.

Island endemics in a biodiversity hotspot

Islands are home to many distinct organisms, yet overall species richness is often low on islands compared to continental areas of the same size; diversity builds as isolation time increases (Heaney 1986). This reflects two important prerequisites for high island endemism; long-term environmental stability and geographic isolation (Jansson 2003; Sadler 1999). The former promotes adaptive radiation; the latter reduces inter-specific competition and immigration rates, with less chance of invasion from non-indigenous species and more scope for fine-scale niche partitioning (Cronk 1997). A substantial proportion of the land area of Southeast Asia consists of islands and island groups. The sheer number of islands, both in the oceanic sense and in the sense of one habitat surrounded by a matrix of different habitat forming a barrier that deters movement, is a particular feature of archipelagic Southeast Asia (Sodhi and Brook 2006). Periodic episodes of rise and falls in sea-levels have at various times isolated or permitted connections, allowing freer movement of species. These episodes of isolation, together with thermal buffering during glacial periods, have driven speciation and high endemism in Southeast Asia (Heaney 1986, 1991).

The Pleistocene epoch of the Quaternary period (1.8 million to 11,400 years ago) saw the development of the dramatic alternation between glacial and interglacial periods, triggered by orbital forcing and amplified by greenhouse-gas and ice-albedo feedbacks (Hansen et al. 2007). Southeast Asia experienced warm wet pluvial periods, alternating with colder dry inter-pluvials (correlating with glacial periods in temperate regions). During dry cool periods, seasonal forests and savannah expanded as areas of rainforest contracted and montane vegetation zones shifted to lower elevations (Heaney 1991). Low sea levels during glacial episodes compensated for expanding drier climates, through a concomitant increase in land area of exposed continental shelves, enabling at least some rainforest species to survive in pockets of moist climates (Gathorne-Hardy et al. 2002). The repeated pattern of glaciations frequently connected islands to other land masses (Voris 2000), facilitating biotic interchange and acted as a 'species pump' (Haffer 1969). As sea levels rose, the colonisation of newly formed oceanic islands, followed by genetic differentiation and long-term persistence, resulted in exceptional high species diversity in Southeast Asia, with endemism being highest on those oceanic island chains that remained continuously isolated from neighbouring islands during the Pleistocene, such as Sulawesi and the Philippines (Heaney et al. 2005).

In a landmark paper by Myers et al. (2000), 25 terrestrial global 'biodiversity hotspots' were identified; regions containing high concentrations of endemic species that were simultaneously suffering substantial habitat degradation and the ongoing threat of future loss. Southeast Asia overlaps with four of these hotspots (Indo-Burma; Sundaland; Wallacea; and the Philippines). During Pleistocene glacial episodes, some temperate species expanded their ranges southwards into Indo-Burma (Jablonski 1993). Episodic sea level changes repeatedly connected the islands of Sundaland to the Asian mainland, facilitating colonisation by mainland biota; rising sea levels caused isolation and promoted speciation (Meijaard 2004). Wallacea derived from land fragments that rifted from Gondwanaland at different geological periods (Audley-Charles 1983). This unique geological history, together with its consistent equatorial-tropical climate and numerous insular refugia, enabled the evolution of highly endemic biotas in Wallacea. For instance Sulawesi has one of the Earth's most distinctive mammalian faunas, with 98% endemism for non-volant mammals (Whitten et al. 1987). The other geologically unique region of Southeast Asia, the Philippines, consists of some 7,000 islands of varying ages and geological histories, with multiple centres of endemism (Mittermeier et al. 1998).

Insularity and susceptibility to human-mediated impacts

Island biotas are a product of dispersal events and in situ diversification. Natural colonisation and extinction rates are well known to be broadly related to the geographical isolation of the island, its area and the ecological attributes of the dispersing taxa (MacArthur and Wilson 1967). After an initial phase of colonisation or competitive redistribution of niches, the geographical isolation of oceanic islands and their virgin ecology provides the impetus for an evolutionary radiation of unique biological forms (*neoendemics*); though endemism can also arise through the extinction of the ancestral and continental taxa (*palaeoendemics*; Whittaker and Fernández-Palacios 2007). Biodiversity on young islands, close to continents, tends to be the result of colonisation, while on old remote islands, speciation occurs faster than colonisation (Heaney 1986); exacerbated by the tendency for species that arrive at islands to undergo a series of evolutionary changes—range contraction, leading to reduced population size, precluding strong competition between species (Ricklefs and Bermingham 2002; Wilson 1961)—which not only allow island populations to diverge from their source population, but also accelerate adaptive change (Gillespie et al. 2008). Environmental stability over geological time encourages the evolution and persistence of island endemics (Jansson 2003). In Southeast Asia, thermal and evapotranspirational buffering of rain forests during glacial periods provided ideal conditions for in situ evolution. However stability and geographical isolation, the very conditions that promoted high species diversity in Southeast Asia, also leave island endemics highly vulnerable to modern human impact.

The human-induced extinction rate is particularly high on oceanic islands (especially for birds; Blackburn et al. 2004) because insular environments typically harbour endemic clades, and island endemics are disproportionately vulnerable to human-driven extinctions (Sadler 1999). Evolutionary processes shaping island communities have provided taxa with behavioural, life-history traits and ecological relationships suited to stable conditions—specialised species with narrow ecological niches which are often naïve towards predators (Cronk 1997). Recent, research has stressed the importance of biological (e.g. life-history traits and ecological relationships) as well as environmental settings (regional threats and stochastic impacts) for predicting human-driven extinction risk (Pimm et al. 2006). For instance some insular species are more susceptible to habitat disturbance due to intrinsic characteristics such as naturally restricted distributions and behavioural or habitat specialisation (Sodhi et al. 2004b) derived from evolution in isolation; yet species most vulnerable to habitat loss can be ecologically and evolutionarily different to species that are threatened by over-exploitation due to the lack, or possession, of fortuitous pre-adaptations (Olden et al. 2007).

Anthropogenic impacts are many and varied, but broadly encompass habitat destruction, over-exploitation and introduced species (Millennium Ecosystem Assessment, MA, www.millenniumassessment.org, 2005). There is growing recognition that these impacts are not mutually exclusive, but rather are interactive and mutually reinforcing (Brook et al. 2008); and that species extinctions or population declines can cause cascading effects, resulting in catastrophic chains of co-extinctions and the disruption of ecosystem services (Koh et al. 2004a). Island biotas are particularly susceptible to invasions by non-native species, because of low species richness and a reduced competitive ability or predator awareness among native species, symptomatic of geographical isolation and evolution under stable conditions (Williamson 1989). In short, low species richness promotes 'open' or flexible ecological spaces and in turn facilitates invasability. A low selective pressure for dispersal among indigenous species compared to non-native species, together with poor evolutionary experience in the midst of competing species, enables invasive species to efficiently colonise open spaces on islands following natural or human-induced environmental perturbations (Denslow 2003). For instance, the degree of niche overlap in native plants has been positively correlated with invasion resistance (Mwangi et al. 2007). Accordingly, the history of regular turnover events and establishment of a highly speciose modern biota in Southeast Asia has buffered, to some degree, this region's island ecosystems from invasive species (Sodhi and Brook 2006). However, the rapid rate at which forests are presently being converted to open and degraded areas is eroding the buffers once provided by rich biodiversity and relatively large land areas (for islands), increasing 'open' ecological space for invasive species.

Although islands in Southeast Asia have remained relatively ecologically undisturbed throughout most of the Anthropocene (Flint 1994), today island communities are experiencing a rapid and extensive rate of species decline (Brook et al. 2003). Deforestation is particularly severe in Southeast Asia, where lowland forests are being destroyed at relative rates that exceed other tropical regions (Achard et al. 2002). Consequently, only a few large islands (e.g. Borneo and Sulawesi) retain large tracts of primary forests capable of supporting viable populations of large-bodied vertebrate species (Sodhi and Brook 2006). Furthermore, over-exploitation, concurrent with habitat loss and fragmentation, accelerates species decline (Clayton et al. 1997; Milner-Gulland et al. 2003). Seven vertebrate species are listed as extinct in Southeast Asia; a further species is extant only in captivity (IUCN 2007). Alone, these figures do not appear alarming. However, they fail to convey the level of endangerment presently faced by extant species (Sodhi et al. 2004a) and the high probability that many extinctions in the region have gone unrecorded (Sodhi and Brook 2006). Because large-scale deforestation is a relatively recent phenomenon in Southeast Asia, many of the native species in the region, especially long-lived ones, may be persisting as 'living dead', doomed to eventual human-induced extinction, regardless of future conservation actions (Brook et al. 2003).

The Philippines and Indonesia illustrate the gravity of the situation in Southeast Asia, where the exploitation of vital habitats has brought these archipelagos to the brink of ecological ruin (Posa et al. 2008). Deforestation in the Philippines in the twenty-first century continues largely unabated, despite already extensive commercial logging during the twentieth century: forested land cover is now less than a quarter of land area (Bankoff 2007), while primary forest is much less (FAO 2005). The archipelago's coral reefs are threatened with over-exploitation, harmful fishing practises and siltation (Gomez et al. 1994). The storey is little different in Indonesia, which has the unenviable distinction of being listed among the top few countries with the most threatened bird and mammal species (IUCN 2007) and highest emissions of greenhouse gases from deforestation sources (Santilli et al. 2005). Because most of these species are endemic to Indonesia, their regional loss would represent global extinctions. From a local perspective, species extirpation and population decline will cause co-extinctions and disrupt ecosystem services, further accelerating extinction rates.

Synergistic impacts of global change

The interaction between different threats to the biodiversity of Southeast Asia remain poorly understood (Sodhi et al. 2004a). However, it is becoming increasingly apparent that in general, it is multiple, rather than single factors, that are primarily responsible for recent population or species loss on tropical islands—occurring often in synergy, thereby accelerating extinctions and potentially leading to ecological collapse (Pimm 1996). For example: (1) current methods of timber extraction increase access to forest interiors, amplifying exploitation and invasability (Milner-Gulland et al. 2003); (2) tropical forest clearance and fragmentation causes localised drying and regional rainfall shifts, enhancing susceptibility to forest fires, causing further forest conversion (Laurance and Williamson 2001); and (3) land cover change can transform El Niño Southern Oscillation (ENSO) events into destructive phenomenon that trigger droughts and wildfires with increasing frequency and intensity (Siegert et al. 2001). Failure to recognise that human impacts on island communities are likely to be interactive and amplifying, runs the risk of making overly optimistic projections of the present and future state of biodiversity, and so could lead to inadequate conservation actions.

Ecosystems most severely affected by habitat loss in Southeast Asia—rainforests, savannas, mangroves and coral reefs—are characterised by highly diverse and complex ecological communities, in which many species are inextricably co-dependent. Thus extinction or species decline can also disrupt ecological processes leading to cascading extinctions (Koh et al. 2004a). For example in Southeast Asia: (1) the extirpation of butterfly species has been linked to the loss or decline in abundance of their host plants (Koh et al. 2004b); (2) the loss of frugivorous birds and mammals, which play an important role in seed dispersal in tropical forests, may impact on forest regeneration (Sodhi and Brook 2006). A decline of forest dwelling pollinators may not only impact plant reproduction in forests, but also influence neighbouring agricultural areas, potentially impacting food production (Bradshaw et al. 2009). Thus, the ramifications of rapid and widespread habitat loss and degradation in Southeast Asia are by no means confined to the direct effects of habitat loss. Forest canopies reduce the force of rainfall striking the soil, while tree roots bind soils, reducing erosion and downstream flooding, minimising siltation of natural waterways and marine environments (Sodhi et al. 2007). Siltation from deforestation and agricultural expansion has been identified as a major destructive factor impacting coral reef ecosystems in the Philippines, together with overexploitation and destructive fishing practises such as dynamiting (Gomez et al. 1994).

Over the next few decades climate change is anticipated to cause a major shift in threatening processes, owing to the unprecedented present rate of global climate change and the fact that landscapes are already heavily modified (Lovejoy and Hannah 2005). Evidence of climatic impacts on the biota of Southeast Asia is currently sparse, although regional observations (e.g. Peh 2007) confirm that human-induced climate change is likely to constitute some of the most pressing conservation issues facing tropical species over the coming centuries (Bradshaw et al. 2009). The modification of forest plant cover and soils which sequester carbon, provides a large source of anthropogenically derived greenhouse gases; second only to fossil fuel emissions (Flint 1994). It is estimated that between 1990 and 2000 forest loss in Southeast Asia accounted for 29% of the global carbon released from deforestation over this same period (Phat et al. 2004), contributing to human-forced climate change. Over the last 40 years in Southeast Asia: (1) the number of warm days and nights have increased; (2) the number of cold days and nights have decreased; (3) annually there are fewer rainy days; and (4) extreme flooding events have become more common (Cruz et al. 2007; Manton et al. 2001). The frequency and magnitude of warm-phase ENSO events have increased, although this may reflect a climatic oscillation and not a direct response to human-induced climate change (Malhi and Wright 2004).

Global warming has, and continues to, invoke latitudinal and altitudinal range shifts, altering ecological interactions and threatening the persistence of range-restricted species (Parmesan 2007). Habitat fragmentation limits dispersal, increasing the risk that species which experience a future loss of suitable local climatic conditions will not be able to move to cope (Williams et al. 2007). Tropical species also appear to be physiologically more susceptible than temperate species to small changes in temperature (Deutsch et al. 2008). Climate change-induced sea level rise will lead to greater agricultural intensification on mid elevation habitats, leading to accelerated habitat loss and greater erosion (Millennium Ecosystem Assessment, MA, www.millenniumassessment.org, 2005). Collectively, these changes can be expected to alter ecosystems, increasing the success and spread of non-native species and so accelerating the homogenisations of regional biota (Brook 2008).

Arguably, the impact of climate-induced change will be greatest on oceanic islands because: (1) indigenous biotas tend to have lower genetic diversity and often little evolutionary experience with new predation or competitors, making them prone to rapid shifts in ecosystem processes (Buckley and Jetz 2007); (2) insular taxa tend to be ill-equipped to disperse in response to climate-induced range shifts—a legacy of evolution in isolation (Gillespie et al. 2008); and (3) for those species with the capabilities to migrate, it may be to no avail, because sea level rise will disproportionately affect insular environments (Mimura et al. 2007), cutting off routes of escape and inundating potential refugia (especially on small islands, and low islands with little relief).

Modelling the dynamics of biota in space and time

Models provide conceptual tools to link demographic traits (population growth, dispersal and trends in vital rates) and ecosystem processes (ecological interactions and feedbacks) to how an organism perceives and uses the landscape's spatial structure, allowing a species' range and persistence to be projected under different human impact scenarios (Akçakaya et al. 2004). Multi-species approaches enable 'whole community' model-based management decisions (Ferrier 2002). Species' ranges are traditionally modelled by linking presence or abundance data directly to environmental variables, using habitat models (Elith and Burgman 2003). The implicit assumption, that current climatic and topographic constraints act to define a species distribution and reflect its environmental preferences, is reasonable as a broad assumption, but simplistic in detail in that it ignores the potentially strong influence of interactions between distribution and the spatial structure of suitable habitat, and how a species' life-history traits interact with shifting ranges (Araújo and Rahbek 2006). Susceptibility to human-mediated environmental disturbances will depend on: (1) life history, morphological and physiological traits (e.g. generation length, dispersal ability, body size and heat exchange); (2) ecosystem processes (including species interactions); (3) landscape spatial structure; and (4) changes in allele frequencies through natural selection (Botkin et al. 2007). Thus habitat and bioclimatic niche models provide, at best, only a first approximation of the potential impact of climate-induced landscape change on biodiversity (Pearson and Dawson 2003).

Persistence in a dynamic landscape depends on the interaction between landscape modification (the pattern, scale, rate and direction of landscape change in size structure and quality) and the species' ecology (its ability to disperse between and grow within habitat patches or utilise the habitat matrix) (Akçakaya and Brook 2008). Most spatial (meta)population models attempt to integrate landscape features with demographic and ecosystem processes, enabling a potentially more realistic forecasting of species' persistence. However, to date, their utility has been restricted largely to solving iconic and economically important problems, because it is often intractable to gather detailed demographic and dispersal data for a whole community of species, and because, with rare exceptions (e.g. Carroll 2007), they tend to ignore species interactions (Burgman et al. 2005). Alternatively, pattern-oriented approaches, by relating landscape structure to environmental variables that promote species diversity, allow conservation decisions to be made in the absence of demographic information for most species (Ferrier 2002). This technique, although generalised, avoids the pitfalls of using focal species to extrapolate conservation decisions (Lindenmayer et al. 2002). The caveat is that species responses to temporal and spatial variability are complex, making it hard to identify a suite of landscape or climatic properties that will favour most, let alone all species in a given region (Manning et al. 2004).

Future research directions

While appreciating that the fundamental threat to biodiversity in Southeast Asia is primarily socio-economic in origin (Sodhi et al. 2004a), we argue that the development of modelling systems which try to capture some of the complexities and uncertainties underlying biological mechanisms that drive species distribution and abundance, will be essential if conservation efforts in Southeast Asia are to be converted into conservation successes. Validation via field data and ongoing, well designed monitoring programmes (e.g. Milner-Gulland and Clayton 2002; Ancrenaz et al. 2005) will be essential to this process. Policy makers and resource managers must be armed with the most ecologically 'realistic' projective artillery if the impending biodiversity crisis is to be averted. The model framework that we suggest, and illustrate in Appendix 1 and 2, should integrate demographic processes (dispersal, population fluctuations and trends in vital rates), landscape spatial structure and ecosystem processes (ecological interactions and non-linear feedback loops), providing tools for exploring the management option trade-offs and sensitivities to assumptions surrounding regional conservation alternatives, enabling a more strategic approach to conservation prioritisation under global change (Zimmer 2007). Island communities tend to have lower functional redundancy and simpler food web structures than continental biota (Denslow 2003; Pimm 1991). Thus they provide ideal candidate systems for developing this 'whole ecology' method.

A hard-nosed, strategic approach to adaptation in the face of global change, which aims to ensure that as many species as possible are able to move into new niches or adapt to human pressures, will require ecological triage (Wilson et al. 2006). Forecasting methods which better identify key communities of future conservation importance will: (1) underpin robust reserve design, by revealing 'pinch point' habitats of high conservation priority in the face of global climate change and interactions with other human-mediated impacts; and (2) inform ecological restoration, by identifying sites and communities that will benefit most in the future from present rehabilitation efforts, and those which may be too difficult to maintain. Collectively, the goal must be to create the conditions which enable a rich and genetically robust biota to survive in ecologically adequate refugial sites, or to have the capacity to move into new areas to adapt to future environmental and landscape change and altered ecological relationships.

Model framework

One likely model architecture to achieve these integrative predictions is demographically structured or individual-based, incorporating explicit and dynamic spatial structure (Akçakaya and Brook 2008). Habitat-vegetation modelling approaches are used to quantify the interaction between landscape and species ranges. Statistical techniques which characterise the background environment and differentially weight variables or alternative model structures in ensemble predictions are used to represent species distributions (Elith et al. 2006) and to develop habitat suitability indices (Akçakaya et al. 2004). Ensembles, by generating probability density functions of range shifts, rather than mean-point projections, explicitly capture prediction uncertainty (Araújo and New 2007). Distribution models are validated by assessing their ability to represent real-world data and their sensitivity to environmental gradients (Botkin et al. 2007). Demographic and ecosystem responses to landscape variation are simulated by developing predictive models which link rates and patterns of human impacts (e.g. decreased habitat connectivity, over-exploitation, predation by invasive species and altered fire regimes) to ecological data (e.g. survival and dispersal rates). Future human impact scenarios are modelled under alternative management actions using the latest climate and land use projections, providing policy makers and resource managers with mechanisms for informed decision-making, for both mitigation and adaptation measures. Illustrative examples are given in Appendix 1 and 2.

Summary

The tropical island chains of Southeast Asia have an extraordinarily rich biodiversity and endemism, reflecting the exceptionality of the geographical and geological history of the region. Recently human-driven changes, such as rapid and extensive destruction of habitat and over-exploitation, have become serious threats to the native biota, by shifting ecological baselines, eroding resilience and undermining the scope for sufficient ecological and evolutionary adaptation to anthropogenic climate change and impacts by invasive species. We argue that development of integrated modelling systems, which capture the complexities and adequately represent the uncertainties underlying biological mechanisms that drive species distribution and abundance under the drivers of global change, will need to be an essential component of any multi-disciplinary approach which aims to avert, or at least allay, the looming biodiversity crisis in Southeast Asia.

Acknowledgments We are grateful to E.J. Milner-Gulland, Benoit Goossens and Marc Ancrenaz for their help in developing the species specific conceptual models outlined in Appendix 1 and 2. Suggestions from two anonymous referees and the special issue editor helped improve the manuscript.

Appendix 1: Modelling the coupled effects of pig exploitation and landscape change in Sulawesi

Babirusa (*Babyrousa babyrussa*) and wild pig (*Sus celebensis*) are endemic to the large island of Sulawesi. Concern has been raised about the threat of continued illegal trade of *B. babyrussa* and the high volume in trade of the unprotected, albeit more resilient species, *S. celebensis* (Milner-Gulland and Clayton 2002). Both pig species are hunted together for the same end market; consequently, *B. babyrussa* (the rarer species) are hunted to levels below that at which it is profitable to hunt it alone, threatening extirpation over much of its range (Clayton et al. 1997). *B. babyrussa* is also sensitive to other human threats (being a forest specialist with a 'slow life history'), while *S. celebenis* is hardier (it readily utilises the modified habitat matrix and has a 'faster life history'). Lattice type spatial models have been used to explore the influences of distances from the end market and road condition on hunting pressure (and local persistence) of *B. babyrussa* and *S. celebenis* (Clayton et al. 1997). Adopting a coupled population viability approach would allow the influence of landscape change—primary forests in Sulawesi have been reduced to small isolates (Lee et al. 2007)—and its interaction with exploitation and harvest dynamics to be examined among two sympatric species with different demographic strategies and ecological requirements, by assessing changes in vital life-history traits in response to global change.

Model framework

Following our conceptual method, models are parameterised using life history traits from published field and captive studies and, where necessary, congeneric species (e.g. *S. barbatus*) and expert opinion (Akçakaya and Brook 2008). Habitat characteristics are linked to a function of habitat suitability, enabling spatial use of the habitat matrix to be simulated (Akçakaya et al. 2004). Market survey data and records from hunters are used to assess exploitation rates (Milner-Gulland and Clayton 2002). The relationship between infrastructure development (e.g. roads, villages, logging camps), habitat fragmentation and pig exploitation, is forecast by developing a predictive model which couples rates and patterns of habitat loss with ecological data (e.g. dispersal, population fluctuations and trends in vital rates). The human-mediated negative relationship between *S. celebenis* abundance and *B. babyrussa* survival (due to harvest dynamics of one species mediating hunting effort for the other) is estimated using harvest and survey data (Clayton et al. 1997) and modelled in a dynamic spatial landscape. Future trends in anthropogenic impacts (e.g. habitat loss, exploitation, climate change) are used to forecast population persistence under different management scenarios. Challenges include developing a bio-economic model to predict changes in forest cover/land conversion and exploitation rates in space and time.

Appendix 2: Modelling the impact of landscape change on orang-utans in Borneo

During the Pleistocene, orang-utans (*Pongo* spp.) were widely dispersed through Southeast Asia (Koeningswald Von 1982). However, today their distribution is restricted to the large islands of Borneo (*P. pygmaeus*) and Sumatra (*P. abelii*), where they persist in patches of remaining lowland dipterocarp forests and low-lying freshwater and peat swamps. Previous efforts to model population viability (Singleton et al. 2004) may be overly optimistic because the projections failed to account for the impact of future climate change (e.g. ENSO warming events, sea level rise) and shifts in land use on habitat suitability. For instance: (1) there is a mutually reinforcing association between deforestation, climate (ENSO) and fire (Laurance and Williamson 2001; Siegert et al. 2001), which negatively impacts dipterocarp forests (Curran et al. 2004); (2) the synergistic interaction between deforestation and sea level rise will directly impact orang-utan habitat availability by inundating low-lying freshwater swamps, and indirectly by orchestrating greater agricultural intensification on mid elevation habitats (Millennium Ecosystem Assessment, MA, www.millenniumassessment.org, 2005; Mimura et al. 2007). The Kinabatagan Orang-utan Conservation Project in Sabah Malaysia (Ancrenaz et al. 2007) provides an opportunity to parameterise demographically structured, spatially dynamic, population models for orang-utans (Ancrenaz et al. 2005; Goossens et al. 2006); potentially giving rise to a more 'ecologically realistic' framework for assessing the population persistence of orang-utans at local and regional scales under competing management scenarios. With a total population size of about 11,000 individuals in 16 major populations, Sabah is a stronghold for *P. pygmaeus* (Ancrenaz et al. 2005). More than 60% of *P. pygmaeus* in the state live outside protected areas, in production forests that have been, or continue to, experience selective logging (Ancrenaz et al. 2005).

Model framework

The movement from traditional population viability approaches (e.g. Singleton et al. 2004) to demographically structured, spatially dynamic, coupled model architectures, which capture many of the ecological complexities and uncertainties impacting species abundance and range, will provide more 'realistic' predictive mechanisms to underpin informed orang-utan conservation decision-making, i.e. with the foresight that the biosphere in

Southeast Asia is rapidly changing. Conceptually this approach involves (1) extracting vital rates from the published literature and expert opinion (e.g. Wich et al. 2004); (2) drawing on aerial nest surveys (Ancrenaz et al. 2005) to model the relationship between abundance and landscape properties—habitat suitability; (3) using patterns of genetic diversity (Goossens et al. 2006) to integrate dispersal rates between habitat patches; and (4) using paired correlates of human impacts and ecological data to model demographic responses. Future landscape change is simulated by modelling habitat loss or expansion in response to global change.

References

Achard F, Eva HD, Stibig HJ et al (2002) Determination of deforestation rates of the world's humid tropical forests. Science 297:999–1002. doi:10.1126/science.1070656

Akçakaya HR, Brook BW (2008) Methods for determining viability of wildlife populations in large landscapes. In: Millspaugh J, Thompson F (eds) Models for planning wildlife conservation in large landscapes. Elsevier, New York, pp 449–472

Akçakaya HR, Radeloff VC, Mlandenoff DJ et al (2004) Integrating landscape and metapopulation modeling approaches: viability of the sharp-tailed grouse in a dynamic landscape. Conserv Biol 18:526–537. doi:10.1111/j.1523-1739.2004.00520.x

Ancrenaz M, Gimenez O, Ambu L et al (2005) Aerial surveys give new estimates for orangutans in Sabah, Malaysia. PLoS Biol 3:30–37. doi:10.1371/journal.pbio.0030003

Ancrenaz M, Dabek L, O'Neil S (2007) The costs of exclusion: recognizing a role for local communities in biodiversity conservation. PLoS Biol 5:2443–2448. doi:10.1371/journal.pbio.0050289

Araújo MB, New M (2007) Ensemble forecasting of species distributions. Trends Ecol Evol 22:42–47. doi:10.1016/j.tree.2006.09.010

Araújo MB, Rahbek C (2006) How does climate change affect biodiversity? Science 313:1396–1397. doi:10.1126/science.1131758

Audley-Charles MG (1983) Reconstruction of Eastern Gondwanaland. Nature 306:48–50. doi:10.1038/306048a0

Bankoff G (2007) One island too many: reappraising the extent of deforestation in the Philippines prior to 1946. J Hist Geogr 33:314–334. doi:10.1016/j.jhg.2006.06.021

Blackburn TM, Cassey P, Duncan RP et al (2004) Avian extinction and mammalian introductions on oceanic islands. Science 305:1955–1958. doi:10.1126/science.1101617

Baillie J, Hilton-Taylor C, Stuart S et al (2004) 2004 IUCN Red list of threatened species. A global species assessment. IUCN, Gland and Cambridge

Botkin DB, Saxe H, Araújo MB et al (2007) Forecasting the effects of global warming on biodiversity. Bioscience 57:227–236. doi:10.1641/B570306

Bradshaw CJA, Sodhi NS, Brook BW (2009) Tropical turmoil: a biodiversity tragedy in progress. Front Ecol Environ (in press). doi:10.1890/070193

Brook BW (2008) Synergies between climate change, extinctions and invasive vertebrates. Wildl Res 35:249–252. doi:10.1071/WR07116

Brook BW, Sodhi NS, Ng PKL (2003) Catastrophic extinctions follow deforestation in Singapore. Nature 424:420–423. doi:10.1038/nature01795

Brook BW, Sodhi NS, Bradshaw CJA (2008) Synergies among extinction drivers under global change. Trends Ecol Evol 23:453–460. doi:10.1016/j.tree.2008.03.011

Buckley LB, Jetz W (2007) Insularity and the determinants of lizard population density. Ecol Lett 10:481–489. doi:10.1111/j.1461-0248.2007.01042.x

Burgman MA, Lindenmayer DB, Elith J (2005) Managing landscapes for conservation under uncertainty. Ecology 86:2007–2017. doi:10.1890/04-0906

Carroll C (2007) Interacting effects of climate change, landscape conversion, and harvest on carnivore populations at the range margin: marten and lynx in the Northern Appalachians. Conserv Biol 21:1092–1104. doi:10.1111/j.1523-1739.2007.00719.x

Clayton L, Keeling M, Milner-Gulland EJ (1997) Bringing home the bacon: a spatial model of wild pig hunting in Sulawesi, Indonesia. Ecol Appl 7:642–652. doi:10.1890/1051-0761(1997)007[0642:BHTBAS]2.0.CO;2

Cronk QCB (1997) Islands: stability, diversity, conservation. Biodivers Conserv 6:477–493. doi:10.1023/A:1018372910025

Cruz R, Harasawa H, Lal M et al (2007) Asia. In: Parry M, Canziani O, Palutikof J et al (eds) Climate change 2007: impacts, adaptation and vulnerability. Contribution of working group II to the fourth assessment report of the IPCC. Cambridge University Press, Cambridge, pp 469–506

Curran LM, Trigg SN, McDonald AK et al (2004) Lowland forest loss in protected areas of Indonesian Borneo. Science 303:1000–1003. doi:10.1126/science.1091714

Denslow JS (2003) Weeds in paradise: thoughts on the invasability of tropical islands. Ann Mo Bot Gard 90:119–127. doi:10.2307/3298531

Deutsch CA, Tewksbury JJ, Huey RB et al (2008) Impacts of climate warming on terrestrial ectotherms across latitude. Proc Natl Acad Sci USA 105:6668–6672. doi:10.1073/pnas.0709472105

Diamond JM (1989) The present, past and future of human-caused extinctions. Philos Trans R Soc Lond Ser B 325:469–477. doi:10.1098/rstb.1989.0100

Elith J, Burgman MA (2003) Habitat models for population viability analysis. In: Brigham CA, Schwartz MW (eds) Population viability in plants. Springer-Verlag, Heidelberg, pp 203–238

Elith J, Graham CH, Anderson RP et al (2006) Novel methods improve prediction of species' distributions from occurrence data. Ecography 29:129–151. doi:10.1111/j.2006.0906-7590.04596.x

Ferrier S (2002) Mapping spatial pattern in biodiversity for regional conservation planning: where to from here? Syst Biol 51:331–363. doi:10.1080/10635150252899806

Flint EP (1994) Changes in land-use in South and Southeast-Asia from 1880 to 1980—a data-base prepared as part of a coordinated research-program on carbon fluxes in the tropics. Chemosphere 29:1015–1062. doi:10.1016/0045-6535(94)90166-X

FAO (Food and Agricultural Organization of the United Nations) (2005) Global forest resource assessment 2005. Forestry department FAO country report 202: Philippines. United Nations, Rome

Gathorne-Hardy FJ, Syaukani , Davies RG et al (2002) Quaternary rainforest refugia in South-east Asia: using termites (Isoptera) as indicators. Biol J Linn Soc Lond 75:453–466. doi:10.1046/j.1095-8312.2002.00031.x

Gillespie RG, Claridge EM, Roderick GK (2008) Biodiversity dynamics in isolated island communities: interaction between natural and human-mediated processes. Mol Ecol 17:45–57. doi:10.1111/j.1365-294X.2007.03466.x

Gomez ED, Alino PM, Yap HT et al (1994) A review of the status of Philippine reefs. Mar Pollut Bull 29: 62–68. doi:10.1016/0025-326X(94)90427-8

Goossens B, Chikhi L, Ancrenaz M et al (2006) Genetic signature of anthropogenic population collapse in orang-utans. PLoS Biol 4:285–291. doi:10.1371/journal.pbio.0040025

Haffer J (1969) Speciation in Amazonian birds. Science 165:131–137. doi:10.1126/science.165.3889.131

Hansen J, Sato M, Kharecha P et al (2007) Climate change and trace gases. Philos Trans R Soc Lond A 365:1925–1954. doi:10.1098/rsta.2007.2052

Heaney LR (1986) Biogeography of mammals in SE Asia—estimates of rates of colonization, extinction and speciation. Biol J Linn Soc Lond 28:127–165. doi:10.1111/j.1095-8312.1986.tb01752.x

Heaney LR (1991) A synopsis of climatic and vegetational change in Southeast-Asia. Clim Change 19:53–61. doi:10.1007/BF00142213

Heaney LR, Walsh JS, Peterson AT (2005) The roles of geological history and colonization abilities in genetic differentiation between mammalian populations in the Philippine archipelago. J Biogeogr 32: 229–247. doi:10.1111/j.1365-2699.2004.01120.x

IUCN (2007) IUCN Red list of threatened species. World Conservation Union, Gland

Jablonski NG (1993) Quaternary environments and the evolution of primates in East-Asia, with notes on 2 new specimens of fossil Cercopithecidae from China. Folia Primatol (Basel) 60:118–132. doi:10.1159/000156681

Jansson R (2003) Global patterns in endemism explained by past climatic change. Proc R Soc Lond B Biol Sci 270:583–590. doi:10.1098/rspb.2002.2283

Johnson TH, Stattersfield AJ (1990) A global review of island endemic birds. Ibis 132:167–180. doi:10.1111/j.1474-919X.1990.tb01036.x

Koeningswald Von GHR (1982) Distribution and evolution of the orang utan, *Pongo pygmaeus* (Hoppius). In: De Boer L (ed) The orang utan. Its biology and conservation. Dr. W. Junk publishers, Hague, pp 1–15

Koh LP, Dunn RR, Sodhi NS et al (2004a) Species coextinctions and the biodiversity crisis. Science 305:1632–1634. doi:10.1126/science.1101101

Koh LP, Sodhi NS, Brook BW (2004b) Co-extinctions of tropical butterflies and their host plants. Biotropica 36:272–274

Laurance WF, Williamson GB (2001) Positive feedbacks among forest fragmentation, drought, and climate change in the Amazon. Conserv Biol 15:1529–1535. doi:10.1046/j.1523-1739.2001.01093.x

Lee TM, Sodhi NS, Prawiradilaga DM (2007) The importance of protected areas for the forest and endemic avifauna of Sulawesi (Indonesia). Ecol Appl 17:1727–1741. doi:10.1890/06-1256.1

Lindenmayer DB, Manning AD, Smith PL et al (2002) The focal-species approach and landscape restoration: a critique. Conserv Biol 16:338–345. doi:10.1046/j.1523-1739.2002.00450.x
Lovejoy TE, Hannah L (2005) Climate change and biodiversity. Yale University Press, New Haven
MacArthur R, Wilson E (1967) The Theory of island biogeography. Princeton University Press, Princeton, NJ
Malhi Y, Wright J (2004) Spatial patterns and recent trends in the climate of tropical rainforest regions. Philos Trans R Soc Lond Ser B 359:311–329. doi:10.1098/rstb.2003.1433
Manning AD, Lindenmayer DB, Nix HA (2004) Continua and Umwelt: novel perspectives on viewing landscapes. Oikos 104:621–628. doi:10.1111/j.0030-1299.2004.12813.x
Manton MJ, Della-Marta PM, Haylock MR et al (2001) Trends in extreme daily rainfall and temperature in Southeast Asia and the South Pacific: 1961–1998. Int J Climatol 21:269–284. doi:10.1002/joc.610
Meijaard E (2004) Biogeographic history of the Javan leopard *Panthera pardus* based on a craniometric analysis. J Mammal 85:302–310. doi:10.1644/BER-010
Milner-Gulland EJ, Bennett EL, The SCB 2002 Annual Meeting Wild Meat Group (2003) Wild meat: the bigger picture. Trends Ecol Evol 18:351–357. doi:10.1016/S0169-5347(03)00123-X
Milner-Gulland EJ, Clayton L (2002) The trade in babirusa and wild pigs in North Sulawesi, Indonesia. Ecol Econ 42:165–183. doi:10.1016/S0921-8009(02)00047-2
Mimura N, Nurse L, McLean RF et al (2007) Small Islands. In: Parry ML, Canziani OF, Palutikof JP et al (eds) Climate change 2007: impacts, adaptation and vulnerability. Contributions of working group II to the fourth assessment report of the IPCC. Cambridge University Press, Cambridge, pp 687–712
Mittermeier RA, Myers N, Thomsen JB et al (1998) Biodiversity hotspots and major tropical wilderness areas: approaches to setting conservation priorities. Conserv Biol 12:516–520. doi:10.1046/j.1523-1739.1998.012003516.x
Mwangi PN, Schmitz M, Scherber C et al (2007) Niche pre-emption increases with species richness in experimental plant communities. J Ecol 95:65–78. doi:10.1111/j.1365-2745.2006.01189.x
Myers N, Mittermeier RA, Mittermeier CG et al (2000) Biodiversity hotspots for conservation priorities. Nature 403:853–858. doi:10.1038/35002501
O'Dowd DJ, Green PT, Lake PS (2003) Invasional 'meltdown' on an oceanic island. Ecol Lett 6:812–817. doi:10.1046/j.1461-0248.2003.00512.x
Olden JD, Hogan ZS, Zanden MJV (2007) Small fish, big fish, red fish, blue fish: size-biased extinction risk of the world's freshwater and marine fishes. Glob Ecol Biogeogr 16:694–701. doi:10.1111/j.1466-8238.2007.00337.x
Parmesan C (2007) Ecological and evolutionary response to recent climate change. Annu Rev Ecol Evol Syst 37:637–669. doi:10.1146/annurev.ecolsys.37.091305.110100
Pearson RG, Dawson TP (2003) Predicting the impacts of climate change on the distribution of species: are bioclimate envelope models useful? Glob Ecol Biogeogr 12:361–371. doi:10.1046/j.1466-822X.2003.00042.x
Peh KSH (2007) Potential effects of climate change on elevational distributions of tropical birds in Southeast Asia. Condor 109:437–441. doi:10.1650/0010-5422(2007)109[437:PEOCCO]2.0.CO;2
Phat NK, Knorr W, Kim S (2004) Appropriate measures for conservation of terrestrial carbon stocks—analysis of trends of forest management in Southeast Asia. For Ecol Manag 191:283–299. doi:10.1016/j.foreco.2003.12.019
Pimm S (1991) The balance of nature: ecological issues in the conservation of species and communities. University of Chicago Press, Chicago
Pimm S (1996) Lessons from a kill. Biodivers Conserv 5:1059–1067. doi:10.1007/BF00052716
Pimm S, Raven P, Peterson A et al (2006) Human impacts on the rates of recent, present, and future bird extinctions. Proc Natl Acad Sci USA 103:10941–10946. doi:10.1073/pnas.0604181103
Posa MRC, Diesmos AC, Sodhi NS et al (2008) Hope for threatened tropical biodiversity: lessons from the Philippines. Bioscience 58:231–239. doi:10.1641/B580309
Ricklefs RE, Bermingham E (2002) The concept of the taxon cycle in biogeography. Glob Ecol Biogeogr 11:353–361. doi:10.1046/j.1466-822x.2002.00300.x
Sadler JP (1999) Biodiversity on oceanic islands: a palaeoecological assessment. J Biogeogr 26:75–87. doi:10.1046/j.1365-2699.1999.00285.x
Santilli M, Moutinho P, Schwartzman S et al (2005) Tropical deforestation and the Kyoto protocol. Clim Change 71:267–276. doi:10.1007/s10584-005-8074-6
Siegert F, Ruecker G, Hinrichs A et al (2001) Increased damage from fires in logged forests during droughts caused by El Nino. Nature 414:437–440. doi:10.1038/35106547
Singleton I, Wich SA, Husson S et al (2004) Orangutan population and habitat viability assessment: final report. IUCN/SSC Conservation Breeding Specialist Group, Apple Valley
Sodhi NS, Brook BW (2006) Southeast Asian biodiversity in crisis. Cambridge University Press, London

Sodhi NS, Koh LP, Brook BW et al (2004a) Southeast Asian biodiversity: an impending disaster. Trends Ecol Evol 19:654–660. doi:10.1016/j.tree.2004.09.006
Sodhi NS, Liow LH, Bazzaz FA (2004b) Avian extinctions from tropical and subtropical forests. Annu Rev Ecol Evol Syst 35:323–345. doi:10.1146/annurev.ecolsys.35.112202.130209
Sodhi NS, Brook BW, Bradshaw CJA (2007) Tropical conservation biology. Blackwell Publishing, Oxford
Voris HK (2000) Maps of pleistocene sea levels in Southeast Asia: shorelines, river systems and time durations. J Biogeogr 27:1153–1167. doi:10.1046/j.1365-2699.2000.00489.x
Whittaker RJ, Fernández-Palacios JM (2007) Island biogeography: ecology, evolution, and conservation, 2nd edn. Oxford University Press, Oxford
Whitten AJ, Mustafa M, Henderson GS (1987) The ecology of Sulawesi. Gadjah Mada University Press, Yogyakarta, Indonesia
Wich SA, Utami-Atmoko SS, Setia TM et al (2004) Life history of wild Sumatran orangutans (*Pongo abelii*). J Hum Evol 47:385–398. doi:10.1016/j.jhevol.2004.08.006
Williams JW, Jackson ST, Kutzbacht JE (2007) Projected distributions of novel and disappearing climates by 2100 AD. Proc Natl Acad Sci USA 104:5738–5742. doi:10.1073/pnas.0606292104
Williamson M (1989) Natural extinction on islands. Philos Trans R Soc Lond Ser B 325:457–468. doi:10.1098/rstb.1989.0099
Wilson E (1961) The nature of the taxon cycle in the Melanesian ant fauna. Am Nat 95:169–193. doi:10.1086/282174
Wilson KA, McBride MF, Bode M et al (2006) Prioritizing global conservation efforts. Nature 440:337–340. doi:10.1038/nature04366
Zimmer C (2007) Predicting oblivion: are existing models up to the task? Science 317:892–893. doi:10.1126/science.317.5840.892

Biodivers Conserv (2010) 19:343–356
DOI 10.1007/s10531-009-9717-0

ORIGINAL PAPER

Conservation of Borneo biodiversity: do small lowland parks have a role, or are big inland sanctuaries sufficient? Brunei as an example

Peter S. Ashton

Received: 25 February 2008 / Accepted: 22 September 2009 / Published online: 27 October 2009

Abstract Borneo, the World's fourth largest island, represents a continental mass sandwiched between volcanic arcs. Patterns of tree species richness, here proposed as proxy for overall biodiversity, and endemism within the island are described and interpreted with special reference to Dipterocarpaceae. Most of Borneo's surface is sedimentary shales, mostly inland, and sandstones, mostly coastal, deposited in past epicontinental seas. The shales mainly support the regional West Malesian (Sundaland) tree flora, while the Neogene coastal hills, especially in the north-west, support the richest communities which are also richest in endemics. The forests of these hills, which include the richest tree communities in the Old World and are therefore of global significance, are accessible and critically endangered. Priorities for their future management are outlined.

Keywords Borneo · Trees · Conservation · Diversity · Ecological ranges · Endemism · Dipterocarpaceae

Introduction

The term biodiversity has been used to describe many different aspects of biological diversity, including structural, functional and phylogenetic diversity. Here, I discuss species diversity in the context of a region, the Malesian island of Borneo, which shares a common biogeographic history, in which diversification into different inland habitats has been achieved from a shared set of immigrating species. By biodiversity I mean, fundamentally, the total diversity of gene sequences at a site; that is, of all organisms represented at it. This, of course, is too enormous to be practically measured. Instead, it is argued that biodiversity can reasonably be approximated by the number of tree species at a site: that is the number of tree metapopulations genetically isolated in nature from all others, yet with potential for interbreeding at least over several generations.

P. S. Ashton (✉)
Organismic and Evolutionary Biology, Harvard University, Cambridge, MA 02138, USA
e-mail: PAshton@oeb.harvard.edu

The indigenous biological cover of Borneo is entirely rain forest. Biodiversity thus defined is overwhelmingly comprised in rain forest of arthropods, particularly insects, and also microorganisms particularly pathogens. Trees contribute more than three quarters of the living biomass and are therefore the primary productive resource for biodiversity. Trees have evolved an extraordinary diversity of chemical, besides less diverse physical, defenses against the battalions of organisms which besiege them. In a climate which varies seasonally and from year to year less than any other, it is this panoply of chemical interactions which is the primary mediator of both the evolution and maintenance of the extraordinary biodiversity of tropical rain forest. Tree species richness is therefore likely to correlate with overall species diversity of rain forest ecosystems, as has been shown for insects (Novotny et al. 2006), and inferred for pathogens (Gilbert and Webb 2007). Fortunately, the tree flora of Borneo, particularly northern Borneo, is well known relative to most other regions of perhumid Malesia. I will therefore use tree species diversity as a proxy for overall biodiversity.

Here I will focus on richness, that is the number of species estimated in a particular floristically defined forest community, or landscape; and endemism, the degree to which tree species are unique to a more or less restricted region, as my measures of diversity. This does not take account of phylogenetic diversity, that is genetic distance between these species, which could be assessed at least at familial and higher levels using current vascular plant molecular phylogenies. This would require added analyses beyond the scope of this paper. Besides, trial analyses of phylogenetic diversity by my associates have yielded results heavily influenced by the small number of gymnospermous and other isolated or putatively archaic tree taxa. Very few of these are either local endemics or even restricted to Borneo. Compared with other equatorial continental regions, and even in comparison with Peninsular Malaysia, Borneo tree taxon richness is expressed at the species rather than at higher levels, and variation among communities is also mostly at this level (Ashton 1964, 2004; Potts et al. 2002; Rees 2009). Thus, in 50 ha at Pasoh, Peninsular Malaysia, 814 species were recorded in 290 genera in the first census, yet the 52 ha at Lambir, Sarawak included a remarkable 1,192 species nevertheless in only 288 genera. Most notably, variation in species richness in Borneo is most reflected by variation in the number of species within coexisting congeneric series, in several genera and families (Ashton 1969).

Brunei Darussalam has the benefit of a comprehensive Checklist of its trees (Coode et al. 1996), thereby enabling comparison of distributions, therefore endemicity (Fig. 1). Over 40% are endemic to Borneo, a third of which are endemic to the north and west (Ashton et al. 2003); almost three quarters are endemic to perhumid West Malesia comprising Peninsular Malaysia, Sumatra, Borneo, West Java and intervening islands. That leaves a mere 5% extending to the seasonal tropics of continental Asia of which only 2% extend to South Asia, 3% east of Wallace's Line, and a surprisingly low 19% to the Philippines where many are nevertheless represented by sister species.

Geological history can explain much of this pattern (Ali and Aitchison 2008; Hall 2002; Morley 1999, 2007). Much of Borneo's tree flora originates from Africa by way of the South Asian Gondwanan Noah's Ark, which docked across the Tethys Sea with the southern Laurasian coastline in the Eocene, when that coastline was equatorial. Subsequent early spread likely occurred because the Himalaya, arising as the bow-wave of the collision, may only have risen to great height in the Oligocene, 20–25 mya (though there remains controversy), thereby initiating the single wet and dry monsoon which so few Borneo species can endure. Nevertheless, the existence of sister species in wet South Asia, especially perhumid SW Sri Lanka, and the Far East implies periodic migration

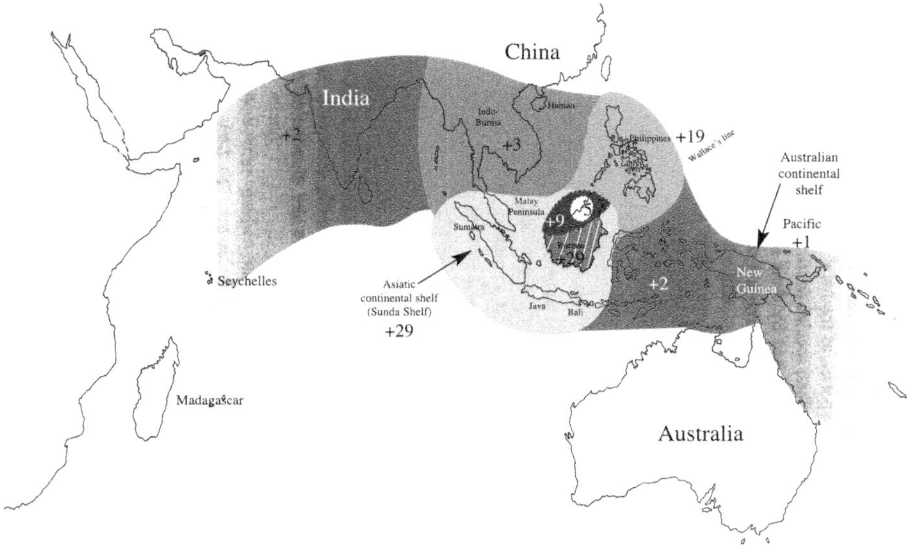

Fig. 1 Geographical range of tree species in Brunei. Each number indicates what percentage of them is accounted for by adding that region (from Ashton 2003)

opportunities, at least for species of weakly seasonal wet climates. The Philippines, an oceanic volcanic island arc, have made contact with NE Borneo since the upper Miocene, c. 5–10 mya, but this has apparently been intermittent. Palawan, which is on the Bornean continental shelf, has surprisingly few species in common: The two largest dipterocarp genera in Borneo, *Shorea* with 129 species and *Hopea* with 42, have yet to be recorded there. This is likely due to extreme seasonality during Pleistocene low sea levels. There appears to have been an overland bridge across Wallace's Line between SW Sulawesi and SE Borneo which share Miocene ophiolites, during Miocene periods of low sea level (Hutchinson 1989). There is little tree biogeographic support for that connection, perhaps because ophiolitic ultramafics support only a restricted set of the Borneo flora, or because the climate of that area was seasonal at the time. By contrast, interpluvial sea levels as much as 120 m below present would have given periodic opportunity for migration throughout West Malesia, although more seasonal rainfall regimes (as opposed to reduced annual rainfall without seasonality) may have restricted such opportunity as is suggested by recent phylogenetic data for vertebrate taxa (Gorog et al. 2004; Harrison et al. 2006; Wilting et al. 2007). In addition, the existence of a river of Congo proportions, the Proto-Mekong, between Borneo on one side and Java, Sumatra and Peninsular Malaysia on the other may explain why those latter regions have more in common floristically than they do with Borneo. As I will show though, this may partially also be explained by their different geology, Borneo being dominated by Cainozoic sediments deposited in epicontinental seas around this easternmost Asian continental outlier.

Concentrations of tree species richness

It is an extraordinary fact that the richest tree communities so far recorded in Borneo are on soils of relatively low nutrient content (Fig. 2), which are on ecological islands of

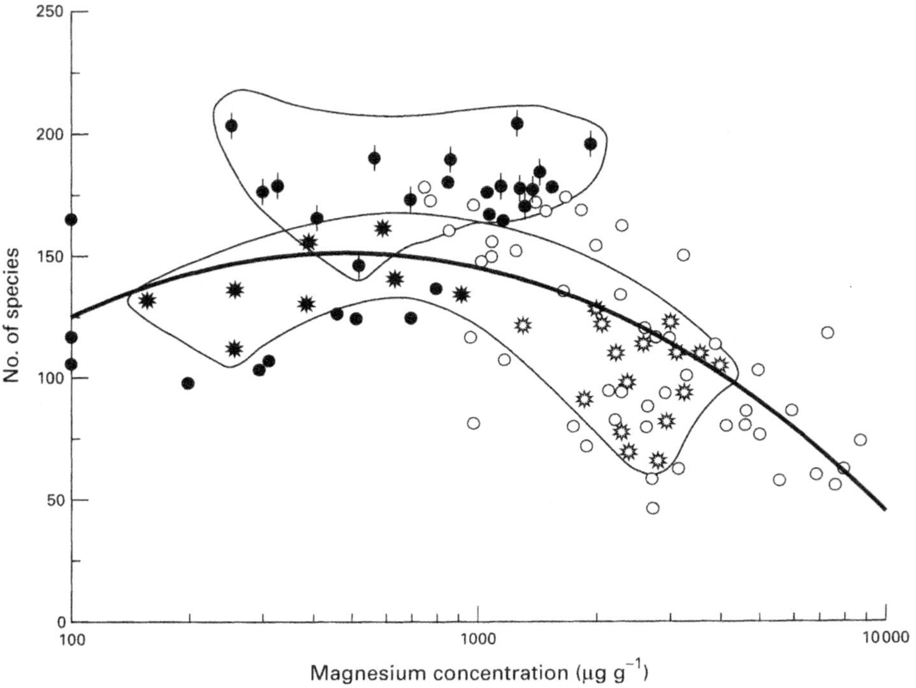

Fig. 2 Pattern of species richness among 105, 0.6 ha plots in mixed dipterocarp forests throughout Sarawak, NW Borneo, in relation to HCl extractable soil magnesium as a proxy for soil fertility. *Black symbols*: lower nutrient sandy soil sites with surface raw humus; *open symbols*: higher nutrient clay-rich soils lacking surface raw humus. *Encircled black disks with vertical line*: rhyolite at Iju Hill, Balingian. *Encircled black and white stars*: respectively sandstone and shale at Lambir National Park, Miri (from Ashton 1998)

surprisingly small area surrounded by other soils which harbour only one-third of their species in common (Ashton 1964; Davies et al. 2005; see Andulau in Turner 2001). The richest of 105, 0.6 ha plots spread throughout Sarawak were on a 25 × 2 km rhyolite extrusion at Iju Hill, on low nutrient yellow soils surrounded by the higher nutrient clay loams over alluvium and shale. The very high tree species richness of a 52 ha plot in Lambir National Park in NE Sarawak is partially due to similar soils and correlated floristic diversity.

Understanding how these centres of tree species arise, or at least diagnosing what environmental factors correlate with them, would provide a base for predicting their distribution thereby aiding rational identification of priority sites for conservation.

It is vital to recognize that, though these Malesian sites of tree species hyperdiversity share a flora in common, tree floras of other habitats—therefore by implication other assemblages of overall biodiversity—equally merit conservation. Only some of these, however are as severely threatened.

Comparative data on tree species composition and richness of upland forest in different regions and sites are only available on an adequate scale in all Malesia in Brunei and Sarawak. It is confined to lowland forests, that is forests below 900 m, though a series of plots exist for Mount Kinabalu (Kitayama 1992), and Mount Mulu in NE Sarawak (Martin 1977). The three widespread forest formations of mixed dipterocarp forest (Ashton 1964,

1998, 2005; Ashton and Hall 1992; Potts et al. 2002), heath forest (*kerangas*) (Brunig 1974), and peat swamp (Anderson 1961) are sufficiently represented but not other habitats with distinct flora, notably limestone and ultramafic substrates. Their comparative general richness can reasonably be judged by our field knowledge and from collections.

These data indicate the following:

- The richest tree communities are mixed dipterocarp forest on soils of relatively low nutrient status, and which are deep. These occur over both volcanic and sedimentary rocks. Soils of comparable nutrient status and physical characteristics are widespread in Borneo, especially along the many ridges which are upheld by more resistant sandstone strata in the widespread matrix of shale. Deep soils of this fertility are confined, however, to relatively poorly consolidated strata which, in the case of the more common sandstones, are confined to Neogene rocks below 400 m and near the coast or immediately behind the swamps. A similar tree flora exists on the Triassic and older coastal hills of Peninsular Malaysia, but is not exceptionally rich there. Comparisons of soils are not available.
- Greatest tree species richness is correlated with greatest structural heterogeneity. Deep sandy soils support very tall forests with individual emergents to 70 m (Ashton 1964, Fig. 25). Periodic severe water stress and lightning lead to predominantly standing emergent mortality of individuals and small groups, leaving small to medium-sized gaps supporting a rich main canopy stratum. Frequent low-level water stress and/or low nutrient levels are associated with diffuse canopy leafage, creating a heterogeneous yet relatively high light understorey climate which is associated with high richness of both understorey species and canopy juveniles (Ashton 2008). Shallow freely draining soils, irrespective of substrate and nutrient status and including most *kerangas* and forest on ultramafic and karst substrates, support shorter forests in which the emergent canopy is short, hardly distinguished from the main canopy, and less rich. There is evidence there for occasional catastrophic canopy mortality, presumably associated with extreme droughts (e.g. Ashton and Hall 1992). The most fertile equatorial soils in perhumid W. Malesia are clay loams with excellent crumb structure and such water storage capacity that droughts may never be experienced; rooting is deep. There, the tall emergent canopy becomes locally continuous, with frequently medium to large gaps occasioned by windthrow. The forest floor light climate is one of pervasive deep shade with contrasting full sun in gaps. Richness, perhaps for this reason, is depressed in all strata but especially in the patchy main canopy which in Borneo forests is the richest stratum (Ashton in 2008). Steep, landslip-prone sites with such soils, and more compact clay soils of intermediate fertility and shallow rooting, bear less continuous emergent canopies and may also be more species rich, though on present evidence less so than on deep yellow sands.
- No comprehensive plot data exist to test whether tree species community richness changes with altitude. Floristic data, and plot sets from selected altitudes from Mount Kinabalu (Kitayama 1992; Beaman et al. 1989–2001; Ashton 2003) and field observations from elsewhere indicate that floristic richness declines precipitously above c. 1,200 m which marks the upper limit of lowland associations, and declines again above the lower–upper montane ecotone. Distribution data for individual families indicate a first decline in species at c. 400 m correlated with coastal hill heights, a second at 600 m correlated with the general height of inland ridges, then continuing decline of lowland species with a few extending as rarities beyond 1,200 m (Ashton 1964, Table 14).

- There is as yet insufficient evidence to test whether analogous floristic associations, or associations in similar habitats, vary in richness through perhumid West Malesia, which might be expected if some regions acted as perhumid forest refuges whereas others experienced Pleistocene intrusion of rainfall seasonality or reduction (e.g. see Turner 2001). The exceptional richness of a mixed dipterocarp forest in Sumatra (Reynolls and Laumonnier in Turner 2001) appears inconsistent with the lower number of species recorded from that island compared with Peninsular Malaysia in many tree families revised in Flora Malesiana. There are several possible explanations for this. First, community samples from particular habitats at the scale documented (usually 1 ha or less) may assemble comparable richness, within certain limits, whatever the richness at landscape scale. Clearly this is not so when, for instance, communities in the much less rich tree flora of perhumid Sri Lanka are compared with those of W. Malesia. Second, different workers may under- or over-estimate the number of species they identify in a sample. In a striking case, Poulsen (cited in Turner 2001) increased by almost half the number of species in an area earlier sampled by me at Kuala Belalong, Brunei (Ashton 1964), probably more correctly as those data had been gathered during my first experience of rain forest census! But frequently also, taxonomists unfamiliar with plot census and particularly with identification of sterile juveniles, overestimate the number of species. Thirdly, our knowledge of different floras differs, Sumatra being less intensively collected than Borneo and particularly Peninsular Malaysia. There is therefore as yet no convincing evidence of regional variation in the species richness of the core mixed dipterocarp forests on clay and sandy clay soils through W. Malesia. The extreme hyperdiversity of mixed dipterocarp forests on deep yellow sands does, however, appear to be unique to NW Borneo, samples from coastal hill dipterocarp forest in Singapore (LaFrankie et al. 2005), and Peninsular hill dipterocarp forest (in Turner 2001), both of which share widespread members of the same flora, being substantially poorer. Many of the species of this habitat are endemic to NW Borneo, as I shall now show.

Endemism

Borneo rain forests, like vegetation throughout the World and most particularly the humid regions, differ in species composition according to their habitat. Mixed dipterocarp forest communities each include a roughly estimated 600–850 tree species, yet those on yellow sands share but a third of those on yellow clays (Ashton 1964; Davies et al. 2005). My estimates of the number of tree species in the different forest types of Brunei (Table 1) total 4,800 species, yet the total number of tree species in Brunei is 2,500–3,000. The difference is because each major forest type differs from others in more than half its species. Within a region it is vital to ensure, therefore, that adequate areas of all distinct habitats which differentiate these forest types are included within the area conserved, notwithstanding their differing species richness and irrespective of how small their area.

All Borneo experiences a wet climate in which drought frequency and intensity, although varying regionally, is nowhere annual and predictable. Astride the equator, there is negligible seasonal variation in temperature. There is no clear evidence of the impact of present climate on lowland species' distributions, though past climates may explain some, as shall be discussed. Floristic associations are therefore correlated with altitude and by inference with the direct or indirect effect of temperature, and with geology. Lowland forest associations are gradually replaced by lower montane forest associations between

Table 1 Estimated number of tree species in Brunei Darussalam, according to their forest types or habitats. (From Ashton 2003)

Forest type	Number of species	Percentage Borneo endemism
Sea shore	25	0
Mangrove	33	0
Back-mangrove, banks of brackish rivers	50	5
Rivers of the plains, alluvium river banks	200	15
White water and rocky river banks	200	35
Floodplains (freshwater swamp forest, empran)	600	25
Peat swamp forest	200	25
Kerangas	300	50
MDF, sandy, sandy clay soils	1,200	50
MDF, clay soils	900	35
Upper dipterocarp forest	600	40
Lower montane forest	500	45
Upper montane forest	100	50

900 and 1,300 m. Upper montane 'cloud forest' associations occur wherever diurnal fog penetrates the canopy, down to 800 m on coastal and isolated peaks but ubiquitous only above 1,800–2,200 m. At all altitudes, floristic associations differ according to substrate: The primary difference is between forests on low nutrient sandy soils on which oligotrophic raw humus (*mor*) accumulates, and mull soils in which there are sufficient nutrients derived from their clay for litter breakdown to be rapid and raw humus more or less absent (Ashton 1964; Potts et al. 2002). Although there are floristic and richness gradients on mulls, such that high nutrient basalt-derived soils carry forests which differ somewhat from sediment-derived mulls, most floristic differentiation occurs on the mor-bearing soils (Potts et al. 2002). *Kerangas* and peat swamp forests are less rich than mixed dipterocarp forest but differ from it in more than half their species. Forests on the acid organic soils of limestone karst summits, and ultramafic rocks, bear tree floras similar to *kerangas* but each with a compliment of characteristic species.

Similar soils-correlated floristic associations occur in the lower montane forests of Borneo, less so elsewhere. On Kinabalu, ultramafic exposures in the upper montane zone are associated with the major concentration of endemic tree species on the mountain.

The dominant roles of altitude, drainage and surface geology in differentiating the floristics of Borneo forests thus provide an easy means for identifying areas of potential conservation priority, though these must subsequently be tested by field investigation.

Endemism varies with habitat and community, and also between and within islands for apparently historic reasons including degree of isolation and apparent extinctions on account of Pleistocene climate change. On the latter account, it is impossible to interpret current regions of exceptional regional or habitat richness, or endemicity, as the cradle of origin of taxa without rigorous phylogeographic analysis. No such regional analyses exist for trees, though Kwek et al. (2007) have provided a convincing case that northern Borneo was a centre of origin for *Crematogaster* ants, symbionts of the pioneer tree genus *Macaranga*. The Philippines manifest high archipelago-wide tree species endemicity, especially in families with poor seed dispersal including Dipterocarpaceae (Ashton 1982) and Fagaceae (Soepadmo et al. 2000), but low island-specific endemism except, apparently (more collection is needed) in upper montane forest, and on ultramafic exposures particularly on Sibuyan, Palawan and the perhumid south-east. Basilan, a basaltic island off SW Mindanao, is also rich in endemic plant species.

The perhumid continental core of W. Malesia—Sumatra, Peninsular Malaysia, Borneo and intervening islands—share a core lowland forest formation, mixed dipterocarp forest on yellow–red clay and sandy clay loams, within which other communities occur as ecological islands and along the more or less fragmented coastal hill ranges. Species of these island habitats have higher levels of endemicity than the core formation as has been shown for Dipterocarpaceae in mixed dipterocarp forest on yellow sands and *kerangas* in Brunei (Fig. 3, Ashton 1964, 1984, 1992). The same is the case on the Sabah ultramafics, especially at Bukit Tawai, Telupid, astonishing for the number of its point endemics, and on the limestone karst but here more particularly for shrubs and herbs. These ecological islands, now critically endangered, must therefore have highest priority in any conservation plan (Fig. 4).

The regional patterns of richness in these habitat islands may reflect intrusion of periodically dry climates during the Pleistocene. One indicator of past extinctions (or failure of immigration) is coincidence of species absences from a shared habitat in certain

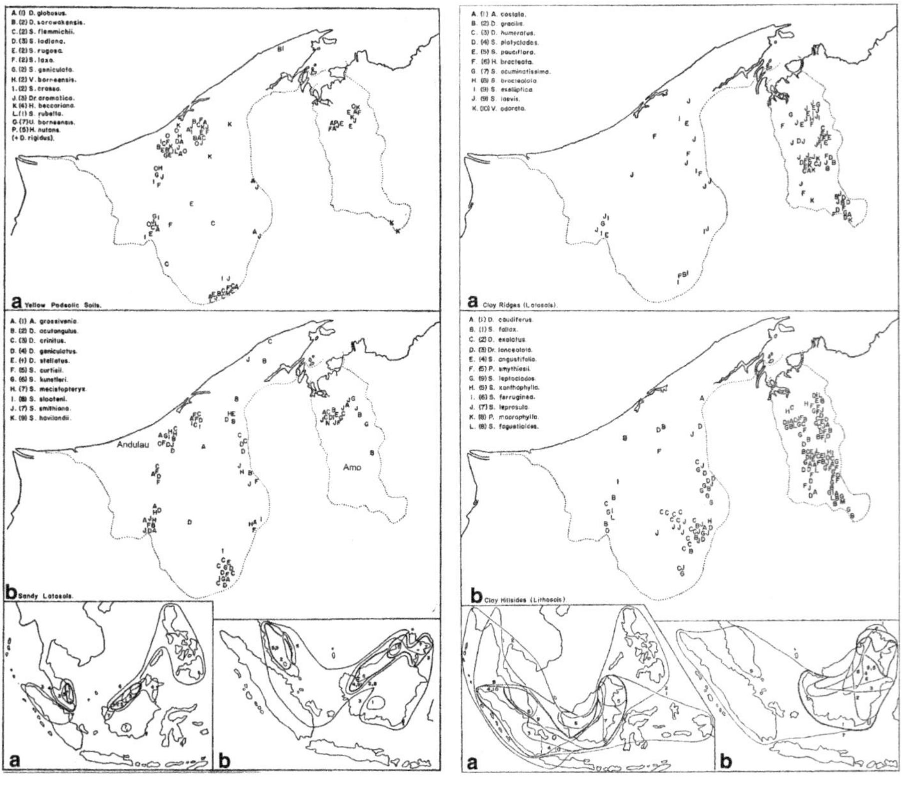

Fig. 3 Recorded distribution within Brunei (*above*) and total geographical range (*below*) of dipterocarp species occurring in Brunei Darussalam. *Left*: species of leached humic sandy soils; *right*: species of clay mull soils (from Ashton 1964). The *letters*, specific to each map, indicate the locations recorded (as collections or my field observations). The *numbers* after each letter refer to the numbered ranges in the relevant map below

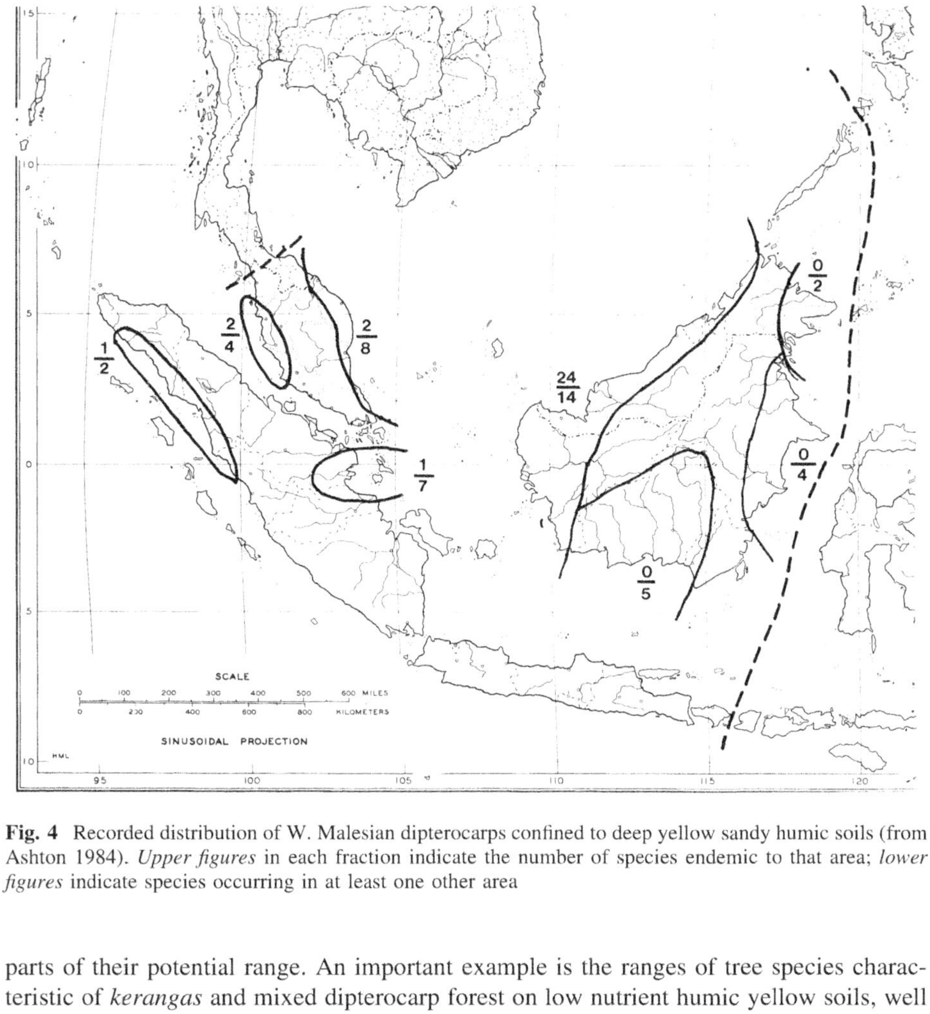

Fig. 4 Recorded distribution of W. Malesian dipterocarps confined to deep yellow sandy humic soils (from Ashton 1984). *Upper figures* in each fraction indicate the number of species endemic to that area; *lower figures* indicate species occurring in at least one other area

parts of their potential range. An important example is the ranges of tree species characteristic of *kerangas* and mixed dipterocarp forest on low nutrient humic yellow soils, well exemplified by Dipterocarpaceae, their seeds being poorly dispersed (Fig. 5). This distribution represents the extension of that which Corner (1960) described as the Riau Pocket. The enclaves illustrated are also rich in endemics in other habitats: peat swamp species in eastern coastal Peninsular Malaysia (Lower Perak) and West Sarawak with North-West Kalimantan; mixed dipterocarp species on clay soils in West Sarawak and eastern Sabah.

We infer from these distributions that northern Borneo was the major Pleistocene refuge for the flora of perhumid equatorial climates, which may have had intermittent connection with smaller refuges in eastern Peninsular Malaysia and the Malacca Strait. The existence of distinct geographical subspecies of three dipterocarps in Lower Perak, two of swamp and one of humic yellow sand in mixed dipterocarp forest, implies survival of suitable habitats there since at least the mid-Pleistocene.

The mountains of Peninsular Malaysia, which may have remained uplifted since the Jurassic, support a higher proportion of point endemic species than those of Borneo where, other than on Kinabalu's ultramafics, endemic species mostly extend over groups of mountains. Notable among these are Kinabalu with the northern Crocker Range, the

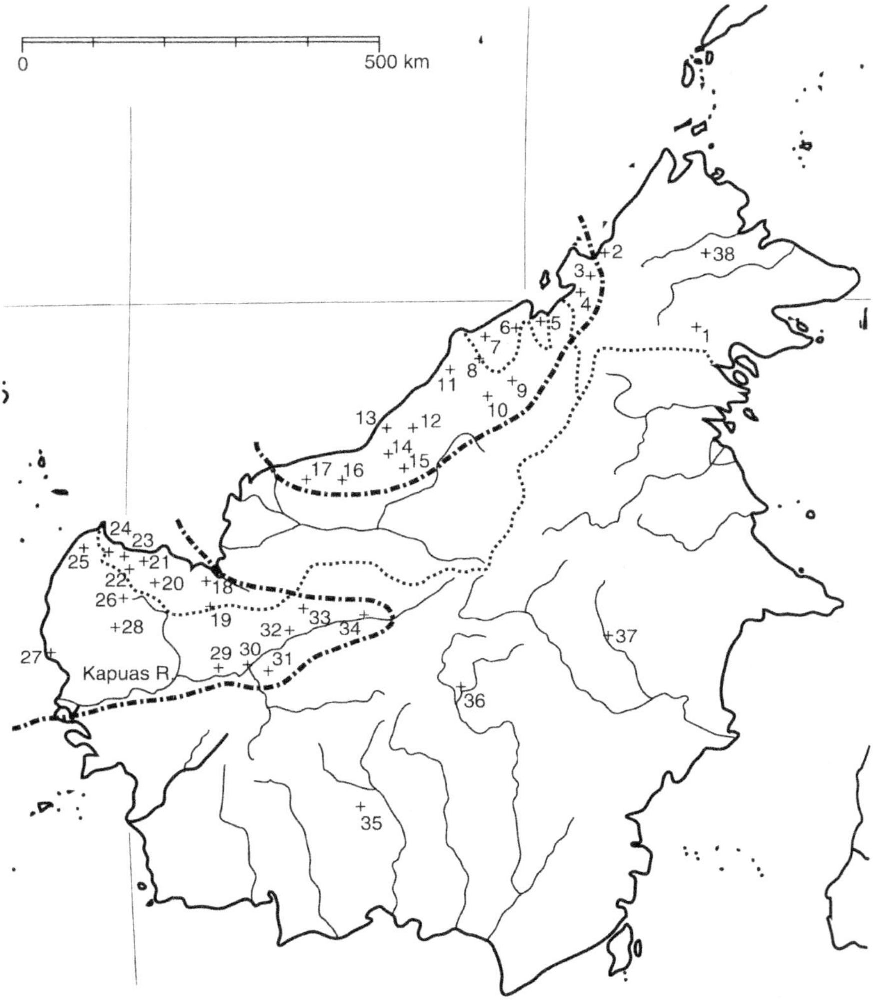

Fig. 5 Recorded localities of dipterocarp species associated with leached humic sandy and sandy clay soils in Borneo; only localities with at least 3 records indicated. Riau Pocket localities (delineated) include the most species (38 of which 24 endemic to Borneo), with only 12 species elsewhere, of which none endemic. The richest individual localities, each with >28 species, are Andulau Forest Reserve (site 7), Lambir National Park (11) and Semengoh F.R. (20) (all fragments of former more extensive forests). Sites numbered as follows: 1. Meliau basin (intact); 2. Papar (degraded); 3. Beaufort Hill (degraded); 4. Mengalong F.R., Sipitang (logged, largely converted); 5. Biang Hill (intact); 6. Northern Ladan Hills (partially intact); 7. Andulau F.R. (mostly logged); 8. Labi Hills (mostly intact), Bukit Teraja (partially intact), Bukit Puan (degraded); 9. Mulu N.P. (intact but potentially hyperdiverse habitat unexplored); 10. Batu Belah (logged, probably converted); 11. Lambir Hills N.P. (intact); 12. Similajau F.R. (converted); 13. Nyabau F.R. (converted); 14. Segan F.R. (converted); 15. Ulu Minah, Kakus (logged); 16. Iju Hill (deforested); 17. Ulu Kenyana, Mukah (converted); 18. Ulu Sebuyau (remnants); 19. Klingkang range, incl. Gunung Gaharu (remnants); 20. Semengoh F.R. (small fragment remaining); 21. Kubah N.P. (Mt. Matang) (partially intact); 22. Undan Hill, Bau (deforested); 23. Meroyong Hill (partially intact); 24. Sebandar Hill, Tamin Hills, Lundu (remnants); 25. Paloh, Sambas; 26. Sanggau; 27. Mempawah; 28. Ngabang, Landak; 29. Sekadau; 30. Sintang; 31. Catit, Melawi; 32. Semitau; 33. Kapuas Lakes; 34. Putussibau; 35. Lebak, Sampit; 36. Ulu Barito; 37. Long Bleh, Kahayan (25–37), Indonesian sites, status uncertain but at best remnants surviving; 38. Tawai Hill and Telupid ultramafics (unconserved state land forest, partially degraded by burning) (from Ashton 2005)

southern Crocker Range with the NE Sarawak mountains, and Sarawak and Kalimantan mountains W. of Batang Lupar and north of Batang Kapuas.

Implications for conservation of biodiversity

It has to be admitted that the areas predicted here to be the richest in overall species diversity of organisms are, by virtue of their relatively low nutrient resources, low in vertebrate and probably other animal biomass—though that does not necessarily imply that their species diversity is low (e.g. Soepadmo et al. 1984). More critical at present, these hyperdiverse and high endemicity sites always existed as ecological islands, but now exacerbated by their lowland coastal geography and consequent endangerment from conversion to non-forest at worst, or isolation by conversion of adjacent forest with proneness to illegal intrusion and poaching of both plants and animals at best.

Already, the most tree species rich of these communities appear to be reduced to three, all others having apparently been degraded or destroyed (Fig. 5). Of these three, the most complete and extensive is Lambir National Park, Miri, Sarawak. Its condition is symptomatic of others. Still experiencing illegal rattan and periodic timber poaching and, more seriously, a recent intrusion with intent to establish a plantation, the vertebrate fauna has experienced catastrophic reduction in the last decade (Shanahan and Debski 2002) and hunting continues.

Within Brunei Darussalam the second surviving area, Andulau Forest Reserve is a fragment comparable in global importance to Lambir N.P. Although logged except for two compartments and a natural arboretum totaling c. 250 ha, the logging was carried out carefully over many years and the regenerating forest, albeit as yet unsampled, still appears rich in climax tree species. It is reported that hornbills have once again been sighted in Andulau. Strict conservation of what remains of the Forest Reserve is of the highest priority. Botanical exploration of the eastern, Belait slopes of the frontier hills south of Bukit Teraja may yield further, probably smaller, tracts of similar floristic associations to those surviving in Andulau, and if so would also merit protection. These recommendations should not be seen as in any way weakening the case for the superb conservation efforts in Brunei's southern Amo District, which serve as an example for the whole region. Their relatively intact fauna and rich flora, both of the clay loams that prevail there and the montane forests of the far south, is as pristine as anywhere in Borneo. Rather, the southern Amo forests do not reflect the whole of Brunei Darussalam's biodiversity, indeed probably not her most biodiverse ecosystem. That effort is not therefore sufficient.

This same case must apply to the Heart of Borneo Project. This has established international accord for the conservation of the inner mountains which run along the Indonesia–Malaysia frontier, with extensions to the Crocker Range and Kinabalu, the Danum lowlands in Sabah, and similar extensions down the Kalimantan mountains. The Amo conservation area is included. These will serve as corridors and sufficient area for conservation of the wider ranging and larger vertebrates notably the orang hutan, clouded leopard, Sabah elephant, larger hornbills, and possibly rhinoceros. Most of this land is above 500 m however, and clay soils with their tree flora of relatively low lowland endemicity predominate. The montane forests, though less rich than the lowland, retain high levels of regional and island-wide endemicity. Yet the yellow sands, lowland *kerangas* and ultramafics are absent, and little limestone karst included except in the Mulu National Park. These areas nevertheless eminently merit conservation on arboreal, and therefore inferred overall biodiversity flora criteria—but again, they are far from sufficient.

The third surviving fragment of hyperdiverse rain forest is what remains of Semengoh Forest Reserve and natural arboretum, south of Kuching, in West Sarawak. It differs from the others. The Batang Lupar is a major phytogeographic boundary (Ashton 1972), Semengoh is to its west, and its tree flora includes c. 15% of species confined to western Sarawak and adjacent Kalimantan. Also, the soils at Semengoh are leached clay, only patchily sandy, on that account supporting a distinctly different albeit comparably rich flora to the deep sands of Lambir and Andulau.

If the immense biodiversity of these few remaining fragments, including those on podsols, ultramafics, limestone and peat are to be sustained in the medium to long term, and the animals that serve as dispersers are to be restored, investment must be made in active management. Obvious needs are investment in adjacent communities aimed at their active and rewarding participation in the conservation programme; and plantation of fruit and other trees in a buffer zone, in order to ensure that food for vertebrate herbivores is available year round at levels sufficient to sustain their populations. What current value has this biodiversity to the forest owners, who are the citizen's of Malaysia and Sarawak? Supporting few large animals even when well protected, these sites attract limited international tourism although they can become, as at Lambir, major venues for local recreation. Certainly, the value of such biodiverse fragments cannot compete with alternative land use, particularly oil palm monoculture. Their future must be uncertain unless the principle beneficiaries of biodiversity conservation, the citizens and industry of the industrialized countries, cease to be free riders and pay fair rental.

Conclusion

The mixed dipterocarp forests of relatively infertile yellow sandy soils, confined to the lowlands not far from present or past coastlines in north and west Borneo, are likely the most biodiverse terrestrial ecosystems in the Old World, but are now reduced to a few fragments themselves experiencing illegal poaching of useful plants and animals, and even local conversion on occasion. Their conservation, and that of other lowland ecological islands rich in endemics, is a critical and global priority, not as an alternative to the Heart of Borneo initiative but in addition to it.

In Brunei Darussalam, Andulau Forest Reserve is the most extensive known surviving fragment of this hyperdiverse ecosystem, although the Belait slopes north of Bukit Teraja may represent an extension of the fragment to its south-east, as yet unexplored and still unlogged. These merit the highest priority for botanical exploration and biodiversity assessment. Only two other comparably rich surviving fragments survive elsewhere in Borneo, at Lambir and Semengoh.

All these fragments, now isolated by forest conversion to agriculture and other uses, will increasingly require active management, in particular to ensure survival of animals, including vertebrates, critical to dispersal of seed and pollen.

References

Ali JR, Aitchison JC (2008) Gondwana to Asia: plate tectonics, paleogeography and the biological connectivity of the Indian sub-continent from the Middle Jurassic through the latest Eocene (166–35 Ma). Earth Sci Rev 88:145–166

Anderson JAR (1961) The ecology and forest types of the peat swamps of Sarawak and Brunei in relation to their silviculture. Doctoral dissertation, University of Edinburgh

Ashton PS (1964) Ecological studies in the mixed Dipterocarp forests of Brunei state. Oxford For Mem 25. Clarendon Press, Oxford

Ashton PS (1969) Speciation among tropical forest trees: some deductions in the light of recent evidence. Biol J Linn Soc 1:155–196

Ashton PS (1972) The quaternary geological history of western Malesia and lowland forest phytogeography. In: Ashton PS, Ashton HM (eds) The quaternary era in Malesia. Trans. 2nd Aberdeen-Hull Symp. Malesian Ecol. Hull Geog Dept Misc Series 13, pp 35–49

Ashton PS (1982) Dipterocarpaceae. In: van Steenis CGGJ (ed) Flora Malesiana, Ser. 1, 9:237–552

Ashton PS (1984) Biosystematics of tropical forest plants: a problem of rare species. In: Grant WF (ed) Plant biosystematics. Academic Press, Toronto, USA, pp 497–518

Ashton PS (1992) Plant conservation in the Malaysian region. In: Yap SK, Lee SW (eds) In harmony with nature. Proceedings of the international conference on the conservation of tropical biodiversity, Malayan Nature Soc, Kuala Lumpur, pp 86–93

Ashton PS (1998) Niche specificity among tropical trees: a question of scales. In: Newbery DM, Prins HTT, Brown N (eds) Dynamics of tropical communities. 37th Symposium of the British Ecology Society, RoyalHolloway College, University of London. Blackwell Science, Oxford, pp 491–514

Ashton PS (2003) Floristic zonation of tree communities on wet tropical mountains revisited. Perspect Plant Ecol, Evol Syst 6/1(2):87–104

Ashton PS (2004) Dipterocarpaceae. In: Soepadmo E, Saw LG, Chung RCK (eds) Tree flora of Sabah and Sarawak, vol 5. pp 63–388

Ashton PS (2005) Lambir forest: the world's most diverse known tree assemblage? In: Roubik DW, Sakai S, Hamid Karim AA (eds) Pollination ecology and the rain forest. Sarawak studies. Springer, New York, pp 191–216

Ashton PS (2008) Changing values of Malaysian forests: the challenge of biodiversity and its sustainable management. J Trop For Sci 20:282–291

Ashton PS, Hall P (1992) Comparisons of structure and dynamics among mixed dipterocarp forests of northwest Borneo. J Ecol 80:459–481

Ashton PS, Kamariah AS, Said IM (2003) A field guide to the forest trees of Brunei Darussalam. Universiti Brunei Darussalam, Brunei Darussalam

Beaman JH and others (1989–2001) The plants of Mount Kinabalu, 4 vols. Natural History Publications (Borneo), Kota Kinabalu

Brunig EF (1974) Ecological Studies in the Kerangas forests of Sarawak and Brunei. Borneo Literature Bureau, Kuching, Sarawak

Coode MJE, Dransfield J, Forman LL, Kirkup DW, Said IM (1996) A checklist of the flowering plants and gymnosperms of Brunei Darussalam. Ministry of Primary Resources and Industry, Brunei Darussalam

Corner EJH (1960) The Malayan flora. In: Purchon RD (ed) Proceedings of the centenary and bicentenary congress of biology, Singapore, pp 21–24

Davies SJ, Tan S, LaFrankie JV, Potts MD (2005) Soil-related floristic variation in a hyperdiverse dipterocarp forest. In: Roubik DW, Sakai S, Hamid Karim AA (eds) Pollination ecology and the rain forest. Sarawak studies. Springer, New York, pp 22–34

Gilbert GS, Webb C (2007) Phylogenetic signal in plant pathogen range. Proc Natl Acad Sci USA 104:4979–4983

Gorog AJ, Sinaga MH, Engstrom D (2004) Vicariance or dispersal? Historical biogeography of three Sunda shelf murine rodents (*Maxomys surifer*, *Leopoldamys sabanus* and *Maxomys whiteheadi*). Biol J Linn Soc 81:91–109

Hall R (2002) Cenozoic geological and plate tectonic evolution of SE Asia and the SW Pacific: computer based reconstructions, model and animations. J Asian Earth Sci 20:353–431

Harrison T, Krigbaum J, Manser J (2006) Primate biogeography and ecology on the Sunda Shelf Islands: a paleontological and zooarchaeological perspective. In: Lehman SM, Fleagle JG (eds) Primate biogeography: progress and prospects. Springer, New York, pp 331–372

Hutchinson CS (1989) Geological evolution of south-east Asia. Oxford University Press, Oxford

Kitayama K (1992) An altitudinal transect study of the vegetation on Mt. Kinabalu, Borneo. Vegetation 102:149–171

Kwek SP, Davies SJ, Ashton PS, Itino T, Pierce NE (2007) The geography of diversification in mutualistic ants: a gene's-eye view into the Neogene history of Sundaland rain forests. Mol Ecol 16:2045–2062

LaFrankie JV, Davies SJ, Wang LK, Lee SK, Lum SKY (2005) Forest trees of Bukit Timah. Population ecology in a tropical forest fragment. Simply Green, Singapore

Martin PJ (1977) The altitudinal zonation of forests along the west ridge of Gunong Mulu. Unpublished report, Sarawak Forest Dept

Morley RJ (1999) Origin and evolution of tropical rain forests. Wiley, Chichester

Morley RJ (2007) Cretaceous and tertiary climate change and the past distribution of megathermal rainforests. In: Flenley JR, Bush MB (eds) Tropical rainforest responses to climate change. Springer, New York, pp 1–31

Novotny V, Drozd P, Miller SE, Kulfan M, Janda M, Basset Y, Weiblen GD (2006) Why are there so many species of herbivorous insects in tropical rain forests? Science 313:1115–1118

Potts MD, Ashton PS, Kaufman LS, Plotkin JB (2002) Habitat patterns in tropical rain forests: a comparison of 105 plots in northwest Borneo. Ecology 83:2782–2797

Rees N (2009) Borneo. A quantitative analysis of botanical richness, endemicity and floristic regions based on herbarium records. National Herbarium Nederland, Leiden

Shanahan M, Debski I (2002) Vertebrates of Lambir National Park, Sarawak, Malaysia. Malayan Nat J 56:103–118

Soepadmo E, Fong FW, Kiew BH, Zakaria MY et al (1984) An ecological survey of Lambir National Park, Sarawak. University of Malaya, Kuala Lumpur

Soepadmo E, Julia S, Go R (2000) Fagaceae. In: Soepadmo E, Saw LG (eds) Tree flora of Sabah and Sarawak, vol 2. Sabah Forest Department, Forest Research Institute, Sarawak Forest Department, Kuala Lumpur, pp 1–117

Turner IM (2001) An overview of the plant diversity of South-East Asia. Asian J Trop Biol 4:1–16

Wilting A, Buckley-Beason VA, Feldhaar H, Gadau J, O'Brien SJ, Linsenmair KE (2007) Clouded leopard phylogeny revisited: support for species recognition and population division between Borneo and Sumatra. Front Zool 4:15

Biodivers Conserv (2010) 19:357–371
DOI 10.1007/s10531-009-9761-9

ORIGINAL PAPER

Re-assessing current extinction rates

Nigel E. Stork

Received: 26 September 2009 / Accepted: 5 December 2009 / Published online: 24 December 2009

Abstract There is a widespread belief that we are experiencing a mass extinction event similar in severity to previous mass extinction events in the last 600 million years where up to 95% of species disappeared. This paper reviews evidence for current extinctions and different methods of assessing extinction rates including species–area relationships and loss of tropical forests, changing threat status of species, co-extinction rates and modelling the impact of climate change. For 30 years some have suggested that extinctions through tropical forest loss are occurring at a rate of up to 100 species a day and yet less than 1,200 extinctions have been recorded in the last 400 years. Reasons for low number of identified global extinctions are suggested here and include success in protecting many endangered species, poor monitoring of most of the rest of species and their level of threat, extinction debt where forests have been lost but species still survive, that regrowth forests may be important in retaining 'old growth' species, fewer co-extinctions of species than expected, and large differences in the vulnerability of different taxa to extinction threats. More recently, others have suggested similar rates of extinction to earlier estimates but with the key cause of extinction being climate change, and in particular rising temperatures, rather than deforestation alone. Here I suggest that climate change, rather than deforestation is likely to bring about such high levels of extinction since the impacts of climate change are local to global and that climate change is acting synergistically with a range of other threats to biodiversity including deforestation.

Keywords Extinction · Extinction crisis · Extinction rates · Red Data lists · Threatened species

Introduction

That there is an on-going extinction crisis is widely accepted by scientists, government and non-governmental organisations, the public and the media. Many suggest that the current

N. E. Stork (✉)
Department of Resource Management and Geography, Melbourne School of Land and Environment, University of Melbourne, 500 Yarra Boulevard, Richmond, VIC 3121, Australia
e-mail: nstork@unimelb.edu.au

extinction phase, frequently referred to as the 'sixth mass extinction event', is as extensive and rapid as the five previous mass extinction events in the last 600 million years when 65–95% of marine animals previously known from fossil records disappeared (Raup 1986). A survey in 1998 suggested that 70% of biologists believe that during the next 30 years as many as one-fifth of all species alive today are committed to extinction, and a third think as many as half the species on Earth will die out in that time (AMNH 1998). There have been a few who have challenged this view (Simon 1995; Lomborg 2001) and their arguments, quite justifiably, have been dismissed as highly selective or unscientific (Pimm et al. 1995; Pimm and Harvey 2001).

The media have a large influence on public opinion, particularly those outlets that are respected for their journalism. For example, the influential magazine, *Time*, ran an article in January 2000 entitled 'Death Row' which noted that surprisingly none of the 25 species of primate that are on the verge of extinction had actually gone extinct in the last century. It stated '*As far as we know, no primate became extinct during the 20th century. That's an impressive record, since the world loses about 100 species a day*'. Almost certainly the latter statement reflects estimates of extinctions which derive from the species–area relationship and loss of tropical forests (e.g. Myers 1988; Wilson 1988) and these are discussed later. In this paper I examine whether figures for extinctions reflect current evidence. I do this by examining the data for recent extinctions and data on the number of all species on Earth, different approaches to estimating extinction rates, and the evidence for differences in the vulnerability of different groups to extinction. In this paper when I use the term *extinction* I am specifically referring to *global extinction* unless otherwise stated. In this respect, I argue that *global extinction* is currently less frequent than many reports indicate and a greater problem is the loss of species at the local and regional level. Finally, I reflect on current evidence that climate change and its interactions with other threats, rather than forest loss alone, may indeed result in the levels of species loss this century that some had previously predicted.

Current data for extinctions and the number of species on Earth

The actual number of species that are officially recognised as extinct since 1600, according to IUCN (IUCN 2009) and other sources, is surprisingly low—around 1,200 species. It is also important to recognise that the most recent 2009 IUCN list of extinct species is in large part a list for 1959 since a species has traditionally only been recognised as 'extinct' when no living individual has been seen for 50 years. This has been changed with a small number of species ($n = 37$) now recognised in a relatively new category of 'extinct in the wild'. There have been remarkable changes in the global environment over the last 50 years with large increases in rates of desertification, the loss of dry and moist tropical forests, and increases in pollution (MEA 2005). Also the threat of extinction for whole faunas and floras has only been measured for a few areas of the world, such as parts of Europe, or globally for only a few groups of organisms such as terrestrial vertebrates, swallowtail butterflies and other well-known animals and plants. Similarly, it is very difficult to determine when the last individual of a species has died (Diamond 1987; Ehrlich and Wilson 1991; Mawdsley and Stork 1995) and figures for recent extinctions are inaccurate. Biologists are still discovering many new species, not all of which are small and cryptic. For example, the large and distinctive Wollemi pine was found 100 km from downtown Sydney (Jones et al. 1995). Even some large vertebrates remain to be described: van Roosmalen et al. (2002) described the 37th and 38th new species of primates discovered since 1990. We should therefore not be surprised that the last remaining individuals of many threatened species of these vertebrates are often impossible to observe and monitor (Chades et al. 2008). For small vertebrates and invertebrates it is virtually impossible to monitor the fate of such threatened species except where their distribution is small and well-known.

A critical component in understanding the scale and impact of the 'extinction crisis' is the knowledge of at least a ball park estimate of how many species there are on Earth and how these species are distributed, group by group, and area by area. Only with that knowledge can we start to understand the scale and extent of species extinctions and translate extinction rates into real numbers of species extinctions and vice versa. Estimates of the number of species on Earth escalated in the 1980s from 1–2 million to 30–100 million due to large increases in the number of species of several key taxa—principally insects and their other arthropod relatives, and to a lesser extent, fungi and nematodes (Erwin 1982; May 1988; Stork 1988). Even the number of species of vertebrates and higher plants is uncertain since the use of DNA techniques suggest that the number of recognised species might increase significantly. However, earlier estimates which suggest up to a hundred-fold increase in the total number of species of insects (Erwin 1988), fungi (Hawksworth 1991) or marine invertebrates (Grassle and Maciolek 1992) are disputed (Stork 1988; May 1992) and have been replaced by estimates of 5–15 million species (Hammond 1995; Stork 1997, 1999; May 1998; Novotny et al. 2002; Dirzo and Raven 2003). Recent use of uncertainty modelling with two models (Hamilton et al., submitted) predict medians of 6.3 and 8.6 million arthropod species globally, with 90% confidence intervals of 1.6–16.0 and 2.7–19.7 million. This analysis confirms the very low probability of there being 30 million species of arthropods. The key implications of such estimates with respect to the extinction debate arguably are (i) that arthropods may comprise 80–90% of all species of terrestrial macro-organisms and, (ii) that 85–95% of arthropods, invertebrates and micro-organisms have yet to be named and described, and (iii) that most of these are found in tropical forests. Finally, if the world is losing about 100 species a day, or roughly 36,500 a year, the vast majority of these losses must come from the plants, fungi or invertebrates, since there are arguably only about 45,000 vertebrate species on Earth (Hammond 1992).

Estimates of extinction rates

The approaches to estimating extinction rates (Reid 1992; Stork 1997) can be grouped into five categories: species–area based estimates, estimates from empirical data of known extinctions, estimates from the use of Red Data lists, co-extinctions and lastly models of the impact of climate change. These are discussed in more detail below and in Table 1.

Species area estimates

The earliest estimates of extinction rates variably suggesting the loss of 5–30% of all species on Earth per decade (Lovejoy 1980; Ehrlich and Ehrlich 1981; Raven 1987; Myers 1988, 1993; Wilson 1988) were based on predictions of species loss from the clearing of tropical forests calculated using species–area relationships. Where these estimates differed is in the value of a key constant z (Reid 1992; Simberloff 1992). If some of these higher estimates were true, then we should have already witnessed the extinction of up to 50% of all species on Earth in the last 30 years. Wilson (1988) for example, extrapolating from

Table 1 Estimated rates of extinction (after Reid 1992; Stork 1997, 1999)

Estimate	Taxon (*all taxa* unless otherwise stated)	% global loss per decade	Method of estimation	Reference
(1) Species-area estimates				
One million species between 1975 and 2000		4.00	Extrapolation of past exponentially increasing trend	Myers (1979)
15–20% of species between 1980 and 2000		8–11	Estimated species–area curve; forest loss based on Global 2000 projections	Lovejoy (1980)
50% loss of species by 2000 or soon after; 100% by 2010–2025		20–30	Various assumptions	Ehrlich and Ehrlich (1981)
9% extinction by 2000		7–8	Based on Lovejoy's calculations using updated forest loss estimates	Lugo (1988)
12% of plant species in neotropics, 15% of bird species in Amazon basin	Plants and birds	–	Species–area curve ($z = 0.25$)	Simberloff (1986)
2000 plant species loss per year in tropics and subtropics; 25% loss of species in 1985–2015	Plants	8–9	Loss of half the species in area likely to be deforested by 2015	Raven (1987, 1988)
At least 7% loss of plant species	Plants	7	Half of species lost over next decade in 10 'hot-spots' covering 3.5% of forest area	Myers (1988)
0.2–0.3% species loss per year		2–6	Half of rainforest species assumed lost in tropical rainforests to be local endemics and becoming extinct with forest loss	Wilson (1988, 1989)
2–13% loss between 1990 and 2015		1–5	Species–area curve ($0.15 < z < 0.35$); range includes current rate of forest loss and 50% increase	Reid (1992)
37–50% extinction rate for 5,308 Amazonian plant species by 2020	Amazonian plant species	14.6–19.7	Estimated number of Amazonian species lost by Laurance et al. (2001) predicted forest loss	Hubbell et al. (2008)
5–9% extinction for Amazonian plants by 2050	Amazonian plant species	1.2–2.2	Revision of above estimate reviewing potential habitat loss and species distributions	Feeley and Silman (2009)
Predict: in next 25 years rate of net tropical deforestation to slow on all continents; net increase in forest area in Latin America and Asia at least within 50 years, and in Africa within 100 years			Slowing shifts of people from tropical rural to urban; suggested reduced deforestation, increased secondary forests succession, both reducing extinction pressures	Wright and Muller-Landau (2006a, b)

Table 1 continued

Estimate	Taxon (*all taxa* unless otherwise stated)	% global loss per decade	Method of estimation	Reference
(2) Estimates from empirical data				
Past loss of 50–90% Pacific Island birds	Pacific island birds		Evidence from recent fossil evidence	Pimm et al. (1995), Steadman (1995)
Current extinction rates 100 times background rates and will increase 15 fold by 2100	All bird species		Range of threats to bird species: habitat/forest loss, invasive species and global change	Pimm et al. (2006)
(3) Estimates from Red lists				
50% extinction in 50–100 years (palms and 300–400 years (birds and mammals)	Selected taxa including palms, birds and mammals	1–10	Extrapolating current recorded extinction rates and by dynamics of threatened status	Smith et al. (1993)
6–50% extinction of selected vertebrates in next 100 years	Some vertebrate taxa including reptile, mammal and bird groups	0.6–5	Fitting of exponential extinction functions based on IUCN categories of threat	Mace (1994)
100,000–500,000 insect species extinctions in next 300 years	Insects	0.04–0.21	Extrapolating from British Red Data Books and assuming 8 million species of insects worldwide	Mawdsley and Stork (1995)
(4) Co-extinctions				
3–5% extinction of helminth species parasitising vertebrates in next 50–100 years	Parasitic helminths	0.15–0.5	Loss of parasites due to loss of hosts; developing co-extinction theory (Stork and Lyal 1993; Koh et al. 2004)	Dobson et al. (2008)
213,830–547,500 insect herbivores committed to extinction in 34 biodiversity hot-spots	Herbivorous insects		Assumes that 150,371 endemic plant species have an average of 5.3–10.6 monophagous insect herbivores	Fonseca (2009)
(5) Models of climate change				
15–37% loss of species by 2050	1,103 species including vertebrates & invertebrates	3.2–7.8	Modelled species distributions against IPCC climate change scenarios	Thomas et al. (2004)

Note that many of these estimated rates include species 'committed' to extinction. Estimates of % global loss per decade are from Reid (1992) or for later estimates are calculated by the current author based on global estimates for all species on Earth of 10 million and for all insects of 8 million species area estimates based on logging and forest loss, suggested that 17,500 species are being lost each year based on a total of 5 million species of organisms. Many of these early authors still consider that such extinction rate estimates are valid (e.g. Raven and McNeely 1998; Pimm and Raven 2000; Myers 2003; Dirzo and Raven 2003).

Further concern for the loss of species due to forest loss came from an analysis of the impact of expansion of road networks and other infrastructure in the Amazon that suggested as little as 40% might be unconverted forest by 2020 (Laurance et al. 2001) (but note even the areas termed heavily degraded or deforested are still mostly covered with forests—see also Chazdon et al. 2009 for discussion of regrowth forests). Hubbell et al. (2008) estimated the number, relative abundance and range size of tree species in the Brazilian Amazon using Hubbell's neutral theory (Hubbell 2001) and then examined the optimistic and pessimistic scenarios of Laurance and co-workers to estimate what proportion of species may become extinct. They estimated that this area of the Amazon might contain 11,210 species with 5,308 of these having less than 10,000 individuals each and as such this group of species might suffer a 50% extinction rate under the pessimistic scenario and as high as 37% under the optimistic scenario. These estimates have been criticised (Feeley and Silman 2008) and an improved assessment of the extinction risks of Amazonian plant species suggests that by 2050 land use change may result in 12–24% reduction of forest habitat and that this may result in 5–9% species being committed to extinction (Feeley and Silman 2009).

In contrast, Wright and Muller-Landau (2006a, b), examined United Nations human population data to demonstrate slowing in population growth and population shifts from rural to urban in tropical countries, globally. They suggested that deforestation pressures might reduce and that secondary forests succession might speed up, both reducing the extinction pressures on tropical forests. Their novel approach has reignited what had been a stale argument and sparked new interest in the fate of tropical forest biodiversity (Brook et al. 2006; Sloan 2007; Laurance 2007).

An inspection of Table 1 shows that over time, as data have become more available and as methods of analyses have improved, the species–area based estimates have reduced very considerably. Indeed if the Wright and Muller-Landau calculations are correct then there may even be a recovery of some species rather than extinctions in the long-term. However, the impacts of climate change are likely to more than counteract this trend as is discussed below.

Estimating from empirical data of actual extinctions

The most compelling data for extinctions relate to those provided by Pimm et al. (2006) and others for birds from oceanic islands and in particular the Hawaiian islands. Earlier Pimm et al. (1995) suggested that 50% or more of bird species on many of the Pacific islands including the Hawaiian islands, S.E. Polynesia, the Marianas, and New Caledonia are missing, endangered, extinct or known from bones only. For the Hawaiian islands this figure is closer to 90%. Much of the evidence for such high levels of extinction is from recent fossil evidence which suggests that on these islands the wave of extinction has largely taken place for birds, and in particular flightless birds, after the arrival of man over the last 1,000 years. Steadman (1995) suggested that the Polynesians eliminated 2,000 species of flightless rails alone (see also Curnutt and Pimm 2001). Subsequently, Pimm et al. (2006) suggested that, globally, recent extinction rates for birds are 100 times higher than background rates and that by the end of the twenty-first century they will be 1,500 times higher than background rates. In practice, the extinction phase for birds of the Pacific

Islands has largely taken place, mostly perhaps several hundred years ago (but note there are still species of bird there on the brink of extinction). It is difficult to argue that such extinctions from islands will be replicated on continental landmasses since bird communities there are of different taxa and vary considerably in their extinction vulnerability. In remote island situations, whole biotas which have evolved from a low genetic stock without invasive species for hundreds of thousands if not millions of years are extremely vulnerable to human invasions and the threats that they bring.

Estimating from Red Lists of threatened species

A third approach to estimating extinctions rates by examining the changing threatened status of a number of groups of animals and plants (Smith et al. 1993; Mace 1994; Mawdsley and Stork 1995) provided an alternative to the species–area model that had previously dominated the extinction rate debate (see Table 1). Smith et al. (1993) summed the number of species added to the list of extinctions (as recorded on the World Conservation Monitoring Centre's database) for a 4-year period (1986–1990) for animals, and 2-year period (1990–1992) for plants. With these data and the estimated number of species worldwide for different taxa, they were able to calculate that the time required for 50% of species to go extinct was 1,500 years for birds and 6,500 years for mammals. They also estimated the net changes in the Red Data Book status of species (from rare, vulnerable, endangered to probably extinct) and again estimated 50% extinction values with resulting values of 350, 250 and 70 years for birds, mammals and palms—much lower than those from their first method. A further method of estimating extinction rates by Mace (1994) also used IUCN red list categories but involved species-by-species assessments of extinction probabilities over time (Mace 1994). Her calculations, based on expected extinction times for selected vertebrate taxa, suggested average extinction times of 100–1,000 years or characteristically 300–400 years for 50% extinction of mammals and birds. Although the way such lists are compiled means that their use in such extrapolations may be questionable (Possingham et al. 2002) this work provides further demonstrations that extinction rates vary considerably between different groups.

Smith and colleagues were unable to produce extinction estimates for less well-known groups such as insects and other invertebrates because of the lack of adequate information on their threatened status. However, Mawdsley and Stork (1995) suggested that one such source of information was the Red Data Book status of British invertebrates (e.g. Shirt 1987; Batten and 1990; Bratton 1991). The British biota has been so well studied for so many years that the threatened status of most groups is very well known. Mawdsley and Stork noted that the percentages of threatened insects, molluscs and spiders were remarkably similar and that these groups were 7.1 times less threatened than birds in Britain. They argued that if this reflected a true difference in the vulnerability of birds to invertebrates, we might be able to use this 'relative extinction rate' to determine how many invertebrate species have gone globally extinct. Since 1% of birds and mammals have become extinct since 1600, $1/7.1 = 0.14\%$ of insects also may have become extinct in the same period of time—equivalent to 7,000–14,000 species of insect, assuming a figures of 5–10 million insect species on Earth. Using the estimated 12–55 fold increase in extinction rates of birds over the next 300 years (Smith et al. 1993) a further 100,000–500,000 species of insect would also become extinct. Mawdsley and Stork (1995) also showed that since 0.7% of British insects have become extinct since 1600, a relative extinction rate with birds of 7.1 would suggest that 5% or 11 British breeding birds should also have been lost. This concurs with Sharrock (1974) who states that 11 species of bird have become extinct in

Britain and Ireland with two later recolonizing. Attempts to re-examine the viability of the 'relative extinction rate' for other parts of the world have been thwarted by a lack of consistency in the assessment of the threatened status of insects and other invertebrates in different countries.

Estimating co-extinctions

The biologies of some species of organisms are so intimately connected with other host species that the extinction of their hosts may lead to their own extinction. Such 'co-extinctions' (Stork and Lyal 1993) are most likely to occur when species are highly host-specific. Although no one has yet estimated precisely how many species are dependent on other species, I would guess this to be possibly 30–40% of all species on Earth (but see discussion by Fonseca, 2009). A number of authors have attempted to estimate the likelihood of extinction based on estimates of host-specificity variously focussing on insects specific to plants and parasites of vertebrates (Koh et al. 2004; Dobson et al. 2008). Dobson et al. (2008) estimate that there are 75,000–300,000 helminth species parasitising vertebrates and that 3–5% is threatened with extinction in the next 50–100 years because of their intimate relationships with threatened vertebrate species. These estimates suggest that co-extinctions are likely to be in the tens of thousands in the next 100 years. Most recently Fonseca (2009) has estimated that the 150,371 endemic plant species in the world's 34 biodiversity hotspots may have an average of 5.3–10.6 monophagous insect hervbivore species feeding on them or a total of 795,971–1,603,423 insect species. He estimates that 213,830–547,500 of these insect species are committed to extinction as their hosts become extinct due to habitat loss. To some extent the co-extinction problem reflects the connectedness of species and is a side issue from the main debate about the importance of the key drivers of extinction—habitat loss and climate change.

Models of the impact of climate change

Another recent set of models have been developed to predict the impact of rising temperatures on organisms (Williams et al. 2003; Thomas et al. 2004). Thomas et al. (2004) looked at six biologically diverse regions around the world covering about 20% of the Earth's land area and modelled the future distributions of 1,103 species including plants, mammals, birds, reptiles, frogs, butterflies and other invertebrates. Using three different IPCC climate change scenarios they found that the 15–37% of these species could become extinct by 2050. Their model is also based on the species–area relationship and the use of this relationship is further critiqued by Lewis (2006) who suggests that these estimates should be considered 'first pass' estimates requiring much greater refinement. The Thomas et al. model and other models have been analysed by Botkin et al. (2008) who suggest how to make such predictions more accurate and reliable. They and others (Willis et al. 2004; Stork et al. 2009) also suggest that we should be making much greater use of the Quaternary record where very few species appeared to go extinct as a consequence of large changes in temperature. In summary, climate envelope models provide an upper limit on the number of extinctions that might be possible rather than the number that might be expected.

There has been considerable attention on the impact of rising temperatures on tropical forest species particularly as poleward range shifts of equatorial species due to global warming are unlikely given the prevailing narrow tropical temperature gradient today (Wright et al. 2009). Upslope elevational range shifts in the tropics are perhaps more likely

(Colwell et al. 2008) but some higher elevation species are likely to disappear as suitable climatic conditions disappear (Williams et al. 2003). Poleward and elevational shifts of tropical species may be hampered in anthropogenically modified and fragmented landscapes and by the lack of some key seed-dispersing animals previously lost through overhunting.

Why predicted extinction rates do not match empirical data

Conservation efforts to protect species

There is no doubt that the massive conservation efforts in the last 50 years by governments and non-governmental organisations to create protected areas and efforts to protect individual species that are threatened has had a great impact on reducing what might have been almost certain extinctions in many instances. The IUCN Red lists and Global Biodiversity Hotspots, for example, are testament to these efforts.

Extinction debt–relaxation times

Many have observed that loss of forest on continental landmasses does not lead to automatic widespread global extinction of species as predicted by the species–area relationship (Heywood and Stuart 1992, but see Raven and McNeely 1998). For example, there is little evidence so far for large scale extinctions following the massive forest loss in the Atlantic forests in north-east Brazil (Brown and Brown 1992). This lack of extinction in the Atlantic forests and elsewhere where there has been extensive forest conversion may be due both to the often fragmented nature of this conversion, that it can take many years for extinction inducing factors to take effect (Diamond 1972; Tilman et al. 1994; Brooks and Balmford 1996; Brooks et al. 1999), that biodiversity hotspots too have their own hotspots (Hugall et al. 2002; Carnaval et al. 2005) and that hotspots of endemism do not necessarily coincide with hotspots of threat (Orme et al. 2005). This phenomenon of higher species diversity for an area than one would predict following a reduction in that area has been termed ‘relaxation time’ (Diamond 1972) or ‘extinction debt’ (Tilman et al. 1994), as the species diversity returns to an equilibrium level that reflects the balance between immigration rates and extinction rates. How long extinction debt may exist after logging and fragmentation is still uncertain but empirical data from Borneo show that regenerating previously selectively logged forests can remain habitat for a range of vertebrate species that might have been considered highly threatened by logging (Johns 1997; Meijaard et al. 2005).

The issue of how important secondary regrowth forests are in maintaining biodiversity and species survival has become increasingly more important as half of all tropical forests now fit that category (Chazdon et al. 2009). With the exception of a few comparative studies (e.g. Lawton et al. 1998; Dunn 2004; Meijaard et al. 2005; Barlow et al. 2007), we still have a poor understanding of what happens to biodiversity (insects, fungi and other microorganisms, as well as birds and mammals) when areas are logged or how biodiversity changes as forests regenerate after logging. A key question is just what proportion of ‘old growth’ species is found in regrowth forests (Chazdon et al. 2009). Interpretation of the results of existing studies have also been complicated because too often the different or compounding effects of aspects of forest fragmentation (e.g. edge effects, size of fragments, degree of isolation, length of time of isolation) or forest disturbance (e.g. amount of loss of vegetation cover, degree of soil compaction, degree of soil erosion, hydrological effects, etc.) have not been measured, frequently because it has been difficult to do so.

Species are not equally prone to extinction

Some assume global extinction rates for birds and large mammals can be directly translated to other organisms (e.g. Pimm and Brooks 2000; Pimm and Raven 2000) and yet there is now a vast body of literature and other evidence from the IUCN lists of threatened species which demonstrate that species and different groups of organisms are not all equally prone to extinction. This also includes analysis of variously combined paleontological, phylogenetic, historical and ecological data (McKinney 1997; Purvis et al. 2008; Stork et al. 2009). Why some species are more vulnerable than others has been the subject of many studies for different taxa (e.g. mammals and birds—Laurance 1991; Newmark 1991, insects—Mawdsley and Stork 1995, and plants—Bradshaw et al. 2008; Sodhi et al. 2008). In addition, the biotas of different parts of the Earth also vary in their vulnerability—a key thesis of the biodiversity hotspots theory (Myers 1988; Myers et al. 2000; but see Manne et al. 1999). As discussed previously, the majority of all recorded extinctions are from islands where their biotas have often evolved from a small gene pool and have suffered from human-related impacts such as hunting or over-extraction, land-use change or invasive organisms. Such impacts on these narrowly distributed species have resulted in the loss of all populations simultaneously with the consequent loss of such species. Species on continental islands, such as mountain tops, are similarly prone to such threats because of their small size (Manne et al. 1999).

Concluding remarks

So what can we conclude about extinction rates? First, less than 1% of all organisms are recorded to have become extinct in the last few centuries and there are almost no empirical data to support estimates of current extinctions of 100 or even one species a day. Second, the most frequently used predictions for global extinction rates are still largely based on the species–area relationship and the fact that large areas of forests (in particular) are being converted. As Lewis (2006) suggested these and the first models of the impacts of climate change on biodiversity, are first passes with subsequent more sophisticated analyses frequently reducing first estimates of extinction rates (see Table 1). With the increasing evidence that some species appear to survive in regrowth forests (Chazdon et al. 2009) the key question is how long are the time lags to extinction for those remaining species which are unable to survive in regrowth or fragmented forests? Third, the evidence is so overwhelming that extinction threats vary for different groups of organisms and different faunas and floras that it is surprising that there are still some who seek to draw conclusions on global extinction rates for all organisms based on the knowledge of just a few very highly threatened groups.

Are estimates of extinction rates useful? Yes I believe so but we should consider some of the highest and ‘first pass’ estimates as upper boundaries rather than real estimates since the assumptions that are made are often unrealistic and require much further modification. However, in spite of the ‘first pass’ nature of the climate change models and their frequent dependence on the assumptions of the species–area relationship, the pervading global nature of such temperature increases seems more threatening than any previous threat. There is also growing evidence that many threats are acting synergistically (Brook et al. 2008; Stork et al. 2009, Supplementary data). Most notably synergistic interactions have been recognised between climate change and all other identified threats (Stork et al. 2009). How such synergies may accelerate species loss is as yet largely unknown although there is some experimental evidence indicating rapid acceleration. Habitat fragmentation and overharvesting combined with experimental warming in rotifer zooplankton resulted in population declines up to 50 times greater combined than when acting on their own (Mora et al. 2007). As Brook (2008) has noted, the potential synergies between species invasion and climate change is one of the prime examples of the type of non-linear, mutually reinforcing ecological responses driving ecosystem change and there are few case studies upon which to base projections for the future.

In contrast to the lack of evidence for mass global extinctions, there is considerable evidence for widespread loss of species at the local and regional level and I suggest that and the consequences of such losses on ecosystem function should be key foci for future research. The loss of this biodiversity, and in particular the loss of the tropical forest megafauna, affects the future regeneration of some key tropical forest plant species. As a consequence, we are now witnessing a massive change in floral composition and the dynamics of many tropical forests around the world (Wright et al. 2007; Wang et al. 2007; Muller-Landau 2007). Corlett (2007), for example, has found that massive range contractions of many of the typical megafauna species in South-East Asian forests including, for example, the Asian elephant, the tiger and the orangutans, mean that less than 2% of this region has an intact or near complete megafauna. Virtually all of these species are still globally extant but their regional contribution to ecosystems has been vastly reduced because of local extinctions.

What can be done to improve our monitoring of extinction beyond the work of IUCN and the Red Data lists? Museums, botanical gardens and other repositories of natural history can play a vital role since their collections provide a historical record of what was found at a particular location. Theoretically we should be able to resurvey some of these locations to determine what has not disappeared and therefore to measure local extinction. One example is the assessments of the species that have gone extinct in Singapore (Brook et al. 2003; Lane et al. 2006). A rapid survey of the largest museums and botanical gardens to identify those species that are most likely to be threatened or are already extinct would be relatively easy to achieve and yet would advance our knowledge of the threatened status of life on Earth enormously.

Acknowledgements This paper is based on the Opening Plenary for the 2007 Biodiversity in Crisis International Conference at the Universiti Brunei Darussalam, and I thank Dr David Lane and his co-organisers, for the invitation to present. I also thank Mark Burgman, Nicholas Williams, Barry Brook and S Joe Wright for their comments on earlier versions of this paper.

References

AMNH (1998) National survey reveals biodiversity crisis—scientific experts believe we are in midst of fastest mass extinction in Earth's history. http://www.amnh.org/museum/press/feature/biofact.html

Barlow J, Gardner TA, Araujo IS et al (2007) Quantifying the biodiversity value of tropical primary, secondary and plantation forests. PNAS 47:18555–18560

Batten LA et al (1990) Red Data birds in Britain: action for rare, threatened and important species. NCC and RSPB, T. and A.D. Poyser, London

Botkin DB, Saxe H, Araujo MB et al (2008) Forecasting the effects of global warming on biodiversity. Bioscience 57:227–236
Bradshaw CJA, Giam X, Tan HTW et al (2008) Threat or invasive status in legumes is related to opposite extremes of the same ecological and life-history attributes. J Ecol 96:869–883
Bratton JH (1991) British Red Data Books. 3. Invertebrates other than insects. Joint Nature Conservation Committee, London
Brook BW (2008) Synergies between climate change, extinction and invasive vertebrates. Wildl Res 35:249–252
Brook BW, Sodhi NS, Ng PKL (2003) Catastrophic extinctions follow deforestation in Singapore. Nature 424:420–423
Brook BW, Bradshaw CJA, Koh LP et al (2006) Momentum drives the crash: mass extinction in the tropics. Biotropica 38:302–305
Brook BW, Sodhi NS, Bradshaw CJA (2008) Synergies among extinction drivers under global change. TREE 23:453–460
Brooks TM, Balmford A (1996) Atlantic forest extinctions. Nature 380:115
Brooks TM, Pimm SL, Oyugi JO (1999) Time lag between deforestation and bird extinction in tropical forest fragments. Conserv Biol 15:1140–1150
Brown KS Jr, Brown GG (1992) Habitat alteration and species loss in Brazilian forests. In: Whitmore TC, Sayer JA (eds) Tropical deforestation and species extinction. Chapman & Hall, London, pp 119–142
Carnaval AC, Hickerson MJ, Haddad CFB et al (2005) Stability predicts genetic diversity in the Brazilian Atlantic forest hotspot. Science 323:785–789
Chades I, McDonald-Madden E, McCarthy MA et al (2008) When to stop managing or surveying cryptic threatened species. PNAS 105:13936–13940
Chazdon RL, Peres CA, Dent D et al (2009) Where are the wild things? Assessing the potential for species conservation in tropical secondary forests. Conserv Biol 23:1406–1407
Colwell RK, Brehm G, Cardelús C et al (2008) Global warming, elevational range shifts, and lowland biotic attrition in the wet tropics. Science 322:258–261
Corlett RT (2007) The impact of hunting on the mammalian fauna of tropical Asian forests. Biotropica 39:292–303
Curnutt J, Pimm SL (2001) How many bird species in Hawai'i and the Central Pacific before first contact? In: Scott JM, Conant S, van Riper III S (eds) Evolution, ecology, conservation, and management of Hawaiian birds: a vanishing avifauna. Stud Avian Biol 22:15–30
Diamond JM (1972) Biogeographic kinetics: estimation of relaxation times for avifaunas of southwest Pacific Islands. PNAS 69:3199–3204
Diamond JM (1987) Extant unless proven extinct? Or extinct unless proven extant? Conserv Biol 1:77–79
Dirzo R, Raven PH (2003) Global state of biodiversity and loss. Annu Rev Environ Resour 28:137–167
Dobson A, Lafferty KD, Kuris AM et al (2008) Homage to Linnaeus: how many parasites? How many hosts? PNAS 105:11482–11489
Dunn RR (2004) Recovery of faunal communities during tropical forest regeneration. Conserv Biol 18:302–309
Ehrlich PR, Ehrlich AH (1981) Extinction: the causes of the disappearance of species. Random House, New York
Ehrlich PR, Wilson EO (1991) Biodiversity studies: science and policy. Science 253:758–762
Erwin TL (1982) Tropical forests: their richness in Coleoptera and other arthropod species. Coleopt Bull 36:74–75
Erwin TL (1988) The tropical forest canopy: the heart of biotic diversity. In: Wilson EO, Peters FM (eds) Biodiversity. National Academy Press, Washington, DC, pp 123–129
Feeley KJ, Silman MR (2008) Unrealistic assumptions invalidate extinction estimates. PNAS 105:E121
Feeley KJ, Silman MR (2009) Extinction risks of Amazonian plant species. PNAS 106:12382–12387
Fonseca CR (2009) The silent mass extinction of insect herbivores in biodiversity hotspots. Conserv Biol. doi:10.1111/j.1523-1739.2009.01327.x
Grassle JF, Maciolek NJ (1992) Deep-sea species richness: regional and local diversity estimates from quantitative bottom samples. Am Nat 139:313–341
Hamilton A, Basset Y, Benke KK et al (submitted) Quantifying uncertainty in estimating global arthropod species richness. Am Nat
Hammond PM (1992) Species inventory. In: Groombridge B (ed) Global biodiversity, status of the Earth's living resources. Chapman and Hall, London, pp 17–39
Hammond PM (1995) Described and estimated species numbers: an objective assessment of current knowledge. In: Allsopp D, Colwell RR, Hawksworth DL (eds) Microbial diversity and ecosystem function. CAB International, Wallingford, pp 29–71

Hawksworth DL (1991) The fungal dimension of biodiversity: magnitude, significance and conservation. Mycol Res 95:641–655
Heywood VH, Stuart SN (1992) Species extinctions in tropical forests. In: Whitmore TC, Sayer JA (eds) Tropical deforestation and species extinction. Chapman and Hall, London, pp 91–117
Hubbell SP (2001) The unified neutral theory of biodiversity and biogeography. Princeton University Press, Princeton
Hubbell SP, He F, Condit R et al (2008) How many tree species are there in the Amazon and how many of them will go extinct? PNAS 105:11498–11504
Hugall A, Moritz C, Mousali A, Stanisic J (2002) Reconciling paleodistribution models and comparative phylogeography in the Wet Tropics rainforest land snail *Gnarosophia bellendenkerensis* (Brazier 1875). PNAS 99:6112–6117
IUCN (2009) http://www.iucnredlist.org/. Accessed 4 Sep 2009
Johns AG (1997) Timber production and biodiversity conservation in tropical rainforests. Cambridge University Press, Cambridge
Jones WG, Hill KD, Allen JM (1995) *Wollemia nobilis*, a new living Australian genus and species in the Araucariaceae. Telopea 6:173–176
Koh LP, Dunn RR, Sodhi NS et al (2004) Species co-extinctions and the biodiversity crisis. Science 305:1632–1634
Lane DJW, Kingston T, Lee BPY-H (2006) Dramatic decline in bat diversity in Singapore with implications for Southeast Asia. Biol Conserv 131:584–593
Laurance WF (1991) Ecological correlates of extinction proneness in Australian tropical rain forest mammals. Conserv Biol 5:79–89
Laurance WF (2007) Have we overstated the biodiversity crisis? TREE 22:65–70
Laurance WF, Cochrane MA, Bergen S et al (2001) The future of the Brazilian Amazon. Science 291:438–439
Lawton JH, Bignell DE, Bolton B et al (1998) Biodiversity inventories, indicator taxa and effects of habitat modification in tropical forest. Nature 391:72–76
Lewis OT (2006) Climate change, species–area curves and the extinction crisis. Philos Trans R Soc B 361:163–171
Lomborg B (2001) The sceptical environmentalist: measuring the real state of the world. Cambridge University Press, Cambridge
Lovejoy TE (1980) A projection of species extinctions. In: Council on Environmental Quality (CEQ), The Global 2000 Report to the President, vol CEQ. Washington, DC, pp 328–331
Lugo AE (1988) Estimating reductions in the diversity of tropical forest species. In: Wilson EO, Peter FM (eds) Biodiversity. National Academy of Sciences, Washington, DC, pp 58–70
Mace GM (1994) Classifying threatened species: means and ends. Philos Trans R Soc Lond B 344:91–97
Manne LL, Brooks TM, Pimm SL (1999) Relative risk of extinction of passerine birds on continents and islands. Nature 399:258–261
Mawdsley NA, Stork NE (1995) Species extinctions in insects: ecological and biogeographical considerations. In: Harrington R, Stork NE (eds) Insects in a changing environment. Academic Press, London, pp 322–369
May RM (1988) How many species are there on Earth? Science 241:1441–1449
May RM (1992) Bottoms up for the oceans. Nature 357:278–279
May RM (1998) The dimensions of life on Earth. In: Raven PH (ed) Nature and human society: the quest for a sustainable world. National Academy Press, Washington, DC, pp 30–45
McKinney ML (1997) Extinction vulnerability and selectivity: combining ecological and paleontological views. Ann Rev Ecol Syst 28:495–516
Meijaard E, Sheil D, Nasi R et al (2005) Life after logging: reconciling wildlife conservation and production forestry in Indonesian Borneo. CIFOR, Jakarta Indonesia
Millenium Ecosystem Assessment (2005) Current state and trends, vol 1. Island Press, Washington, DC
Mora C, Metzger R, Rollo A et al (2007) Experimental simulations about the effects of overexploitation and habitat fragmentation on populations facing experimental warming. Proc Biol Sci 274:1023–1028
Muller-Landau HC (2007) Predicting the long-term effects of hunting on plant species composition and diversity in tropical forests. Biotropica 39:372–384
Myers N (1979) The sinking ark: a new look at the problem of disappearing species. Pergamon Press, Oxford
Myers N (1988) Threatened biotas: 'hotspots' in tropical forests. Environmentalist 8:1–20
Myers N (1993) Questions of mass extinction. Biodivers Conserv 2:2–17
Myers N (2003) Conservation of biodiversity: how are we doing? Environmentalist 23:9–15
Myers N, Mittermeier RA, Mittermeier CG et al (2000) Biodiversity hotspots for conservation priorities. Nature 403:853–858

Newmark WD (1991) Tropical forest fragmentation and the local extinction of understory birds in the eastern Usambara mountains, Tanzania. Conserv Biol 5:67–68
Novotny V, Basset Y, Miller SE et al (2002) Low host specificity of herbivorous insects in a tropical forest. Nature 416:841–844
Orme CDL, Davies RG, Burgess M et al (2005) Global hotspots of species richness are not congruent with endemism or threat. Nature 436:1016–1019
Pimm SL, Brooks TM (2000) The sixth extinction: how large, where and when? In: Raven PH (ed) Nature and human society: the quest for a sustainable world. National Academy Press, Washington, DC, p 46
Pimm SL, Harvey J (2001) No need to worry about the future. Nature 414:149–150
Pimm SL, Raven PH (2000) Extinction by numbers. Nature 403:843–845
Pimm SK, Russell GJ, Gittleman JL, Brooks TM (1995) The future of biodiversity. Science 269:347–349
Pimm SL, Raven R, Peterson A et al (2006) Human impacts on the rates of recent, present, and future bird extinctions. PNAS 103:10941–10946
Possingham HP, Andelman SJ, Burgman MA et al (2002) Limits to the use of threatened species lists. TREE 17:503–507
Purvis A, Agapow P-M, Gittleman JL et al (2008) Nonrandom extinction and the loss of evolutionary history. Science 288:328–330
Raup DM (1986) Biological extinction in Earth history. Science 231:1528–1533
Raven PH (1987) The scope of the plant conservation problem world-wide. In: Bramwell D et al (eds) Botanic gardens and the world conservation strategy. Academic Press, London, pp 19–29
Raven PH (1988) Our diminishing tropical forests. In: Wilson EO, Peter FM (eds) Biodiversity. National Academy of Sciences, Washington, DC, pp 119–122
Raven PH, McNeely JA (1998) Biological extinctions: its scope and meaning for us. In: Guruswamy LD, McNeely JA (eds) Protection of global biodiversity: converging strategies. Duke University Press, Durham, pp 13–32
Reid WV (1992) How many species will there be? In: Whitmore TC, Sayer JA (eds) Tropical deforestation and species extinction. Chapman and Hall, London, pp 55–73
Sharrock JTR (1974) The changing status of breeding birds in Britain and Ireland. In: Hawksworth DL (ed) The changing flora and fauna of Britain. Academic Press for the Systematics Association, London, pp 203–220
Shirt DB (ed) (1987) British Red Data Books. 2. Insects. Nature Conservation Council, Peterborough
Simberloff D (1986) Are we on the verge of a mass extinction in tropical rainforests? In: Elliott DK (ed) Dynamics of extinction. Wiley, New York, pp 165–180
Simberloff D (1992) Do species–area curves predict extinction in fragmented forest. In: Whitmore TC, Sayer JA (eds) Tropical deforestation and species extinction. Chapman and Hall, London, pp 75–89
Simon JL (1995) Why do we hear prophecies of doom from every side? The Futurist 29:19–24
Sloan S (2007) Fewer people may not mean more forest for Latin American forest frontiers. Biotropica 39:443–446
Smith FDM, May RM, Pellew R et al (1993) Estimating extinction rates. Nature 364:494–496
Sodhi NS, Koh PL, Peh KS-H, Tan HTW et al (2008) Correlates of extinction proneness in tropical angiosperms. Divers Distrib 14:1–10
Steadman DW (1995) Prehistoric extinctions of Pacific Island Birds: biodiversity meets Zooarchaeology. Science 267:1123–1131
Stork NE (1988) Insect diversity: facts, fiction and speculation. Biol J Linn Soc 35:321–337
Stork NE (1997) Measuring global biodiversity and its decline. In: Reaka-Kudla ML, Wilson DE, Wilson EO (eds) Biodiversity II: understanding and protecting our biological resources. Joseph Henry Press, Washington, DC, pp 41–48
Stork NE (1999) The magnitude of biodiversity and its decline. In: Cracraft J, Grifo JT (eds) The living planet in crisis: biodiversity, science and policy. Columbia University Press, New York, pp 3–32
Stork NE, Lyal CHC (1993) Extinction or 'co-extinction' rates? Nature 366:307
Stork NE, Coddington JA, Colwell RK et al (2009) Vulnerability and resilience of tropical forest species to land use change. Conserv Biol 23:1438–1447
Thomas CD, Cameron A, Green RE et al (2004) Extinction risk from climate change. Nature 427:145–148
Tilman DR, May RM, Lehman CL et al (1994) Habitat destructions and the extinction debt. Nature 371:65–66
van Roosmalen MGM, van Roosmalen T, Mittermeier RA (2002) A taxonomic review of the titi monkeys, Genus Callicebus Thomas, 1903, with the description of two new species, *Callicebus bernhardi* and *Callicebus stephennashi*, from Brazilian Amazonia. Neotrop Primates 10(Suppl):1–53
Wang BC, Sork VL, Leong MT et al (2007) Hunting of mammals reduces seed removal and dispersal of the Afrotropical tree *Antrocaryon klaineanum* (Anacardiaceae). Biotropica 39:340–347

Williams SE, Bolitho EE, Fox S (2003) Climate change in Australian tropical rainforests: an impending environmental catastrophe. Proc R Soc Lond B 270:1887–1892
Willis KJ, Bennett KD, Walker D (2004) The evolutionary legacy of the ice ages. Philos Trans R Soc B 359:15–158
Wilson EO (1988) The current state of biological diversity. In: Wilson EO, Peters FM (eds) Biodiversity. National Academy Press, Washington, DC, pp 3–18
Wilson EO (1989) Threats to biodiversity. Scientific American, September, pp 108–116
Wright SJ, Muller-Landau HC (2006a) The future of tropical forest species. Biotropica 38:287–301
Wright SJ, Muller-Landau HC (2006b) The uncertain future of tropical forest species. Biotropica 38:443–445
Wright SJ, Stoner KE, Beckman N et al (2007) The plight of large animals in tropical forests and the consequences for plant regeneration. Biotropica 39:289–291
Wright SJ, Muller-Landau HC, Schipper J (2009) The future of tropical species on a warmer planet. Conserv Biol 23:1418–1426

Biodivers Conserv (2010) 19:373–391
DOI 10.1007/s10531-009-9686-3

ORIGINAL PAPER

Late quaternary turnover of mammals in Borneo: the zooarchaeological record

Earl of Cranbrook

Received: 28 March 2008 / Accepted: 18 June 2009 / Published online: 16 July 2009

Abstract The Quaternary has been a period of repeated, oscillating patterns of climate change. Global fluctuations in sea level affected the island status of Borneo, which was probably joined to continental Asia for more than half of the last 250,000 years. Alternating connection and isolation, coupled with the ecological barrier of a savanna corridor running from the Malay Peninsula to Java during periods of marine recession, are reflected in the present mammal fauna of Borneo. 38% of mammal species (excluding bats) are endemic, and some distinctive species or subspecies are confined to the north of the island. No known sites in Borneo match the Early and Middle Pleistocene regional sources in eastern Java. However, caves at Niah, Sireh and Jambusan, Sarawak, and Madai, Sabah, provide a zooarchaeological record covering the past 50,000 years. The Late Pleistocene mammals of Borneo included ten species also present among a Javan Middle Pleistocene savanna-adapted assemblage. Of these, four are categorised as 'megafuana': a giant pangolin, Javan rhinoceros, Malay tapir and tiger; the Sumatran rhinoceros can be added. In addition, there are less secure Pleistocene records of Asian elephant from Sarawak and Brunei. Holocene canid remains from Madai could either be the dhole or an early domestic dog. Palynological data combined with the mammal fauna confirm that around 45,000 years ago the vicinity of Niah was vegetated by closed forest. The continuous presence of a suite of arboreal specialists, including large primates, indicates that forest cover persisted through the terminal Pleistocene. Among local extinctions, the giant pangolin apparently disappeared early in this period, but tiger, Javan rhinoceros and tapir probably survived into the last millennium. Human predation of juveniles may account for the loss of the large ungulates, but the disappearance of tiger needs another explanation. Despite hunting pressure throughout the terminal Pleistocene and Holocene, a population of orangutan survived at Niah until perhaps the last millennium. Size diminution observed

Earl of Cranbrook
Sarawak Museum, Kuching, Sarawak, Malaysia

Earl of Cranbrook (✉)
Great Glemham House, Saxmundham IP171LP, UK
e-mail: lordcranbrook@greatglemhamfarms.co.uk

among large, medium and small mammal species is interpreted as the selective impact of environmental change. Once more is known about their ecology, changes in the bat fauna of Niah cave may provide indicators of environmental impacts affecting the wider mammal community during the later Holocene. In conclusion, it is recommended that the three nations, Brunei Darussalam, Malaysia and Indonesia, should support the WWF sponsored 'Heart of Borneo' as the most hopeful project to provide sustainable management of the rare and threatened forest-adapted wild mammals of the island.

Keywords Pleistocene · Holocene · Borneo · Mammals · Extinctions · Zooarchaeology

Introduction

During the Quaternary period large climatic fluctuations and consequential environmental effects impacted heavily on islands. On a shallow continental shelf, such as the Southeast Asian Sunda Shelf, island shorelines advanced and retreated with changing sea-levels, so that the area of exposed land shrank or expanded, periodically merging with the adjacent continental mainland (Voris 2000). These geographic changes affected the opportunities for immigration by mammals from continental Asia and dispersal between the Sunda Islands. There is also evidence of the periodical appearance of a dry corridor of savanna-like habitat that would have acted as a terrestrial barrier between humid forest zones (Bird et al. 2005). Analysis of the taxonomy and distribution of the present living mammals of Borneo provides supplementary clues to these past events.

Archaeology has its limitations as an evidential source, being dependent on the discovery of productive sites and on conditions of preservation of buried organic remains. In the Malaysian States of Sarawak and Sabah, however, half a century of discovery, excavation and subsequent analysis of the results has provided a rich zooarchaeological resource in the form of animal remains, mostly stored at the State Museums in Kuching and Kota Kinabalu, with a few specimens deposited in the Natural History Museums of London, UK, and Leiden, The Netherlands. Broadening the focus of this symposium from the present biodiversity crisis on tropical islands, and taking as an example our familiar, enormous island of Borneo ($\sim$751,000 km^2 in present subaerial extent), it is useful to review the evidence of past mammal faunas, especially the closing 50 millennia for which the archaeological record is most comprehensive. The survival of some species and elimination of others potentially provides information on past environmental change during this period. Since much of the zooarchaeological material accumulated in caves as a result of human activity, it also provides evidence of the impact of people on the mammal fauna. This background may also assist predictions of the future conservation state of mammals, under the present regime of continuing global climate change and human impacts.

Systematic names of mammals are given at first mention and, unless otherwise stated, follow Wilson and Reeder (2005); thereafter accepted English vernacular names are used alone. SI units are used, unless original measurements were imperial units (e.g., inches). Abbreviations include: Ma = millions of years (ago); ka = thousands of years (ago); BP = before present, defined as before 1950. ^{14}C dates are quoted from the source references. Where given, and important in context, 'uncal.' denotes an uncalibrated age and 'cal.' indicates a calibrated age BP; one ^{14}C date was stated by the originators to be "outside calibration". LGM = Last Glacial Maximum; BPL = below present (sea) level.

Quaternary environments

The start of the Quaternary was the Plio-Pleistocene transition, dated by geomagnetic stratigraphy around 1.8 million years ago (Ma; Gibbard and van Kolfschoten 2004). The subsequent Pleistocene period is conventionally divided into Early (or Lower), Middle and Late (Upper) stages, again based on geomagnetic stratigraphy but also partly on representative faunas. The Pleistocene was characterised by repeated, oscillating patterns of climate change during which cold phases (or glacials) lasting around 100,000 years alternated with briefer warm periods (interglacials) of 10–15,000 years. During glacials the expansion of polar and high altitude ice sequestered a proportion of the world's finite supply of free surface water, resulting in global lowering of the sea level. Rainfall was generally reduced, and wind velocities increased, resulting in enlarged arid zones in many parts of the world (Wilson et al. 2000).

The start of the last interglacial, known as the Eemian (~128 ka until ~115 ka), conventionally dates the transition from Middle to Late Pleistocene. During this period, global mean temperatures fluctuated around 2°C higher than nowadays. The climate subsequently cooled progressively, although continuing to show pronounced instability, with periods of moderate temperature about 105–75 ka and again between 60 and 40 ka separated by a cool period peaking at 70 ka (Johnson 2006). The coldest stage of the cycle began at about 35 ka and reached extreme conditions ~23–18 ka (the Last Glacial Maximum or LGM) when global mean temperatures were around 5°C below present (Wilson et al. 2000).

In the northern hemisphere, and probably worldwide, the return to a warm, interglacial climate was interrupted during the terminal Pleistocene by a cool stadial known as the Younger Dryas around 13–11.5 ka (Wilson et al. 2000). The environmental amelioration following the Younger Dryas marks the onset of the Holocene, the present period (Roberts 1998). During the mid-Holocene, the climate warmed to its greatest extent. Locally, South China Sea levels peaked at almost 5 m above present at 4,850–4,450 BP (Horton et al. 2005).

Based on present bathymetry, Voris (2000) mapped the Pleistocene exposure of the floor of the South China Sea at different stages of marine recession below present level (BPL). At 40 m BPL western Borneo was connected to the continent via land-bridges across the Karimata Straits to Sumatra and thence to the southern Malay peninsula; at 50 m BPL, the Malay peninsula, Sumatra, Borneo and Java were joined, and at 75 m BPL the Indochinese peninsula was also incorporated in a single Sunda landmass. From that level down to 120 m BPL at the LGM, a concave northern shore connected the east point of Vietnam with northwest Borneo. Voris (2000) also estimated that over the past 250 ka marine recessions were at maximum only for short periods of time, but that sea levels stood at or below 40 m BPL (the critical point for the westward connection of Borneo to Sumatra) for more than half of the time period. Using Voris's data, Cranbrook, Earl of (2000) calculated that after the LGM, as the sea rose again, a Karimata land bridge persisted until ~10 ka. Eastward, at all times the Macassar Straits formed a permanent deep sea-barrier between Borneo and Sulawesi.

Drawing on many sources, Bird et al. (2005) concluded that, during the phases of cold climate and lowered sea level, a continuous zone of open 'savanna' vegetation followed a land-bridge running northwest to southeast from the Malay Peninsula to Java. A maximal interpretation of the evidence is compatible with the existence of a broad corridor exceeding 150 km wide, connecting similar open vegetation types north and south of the equator. This corridor therefore served both as a route for dispersal of savanna-adapted mammals from continental Asia and also as a barrier to the movement of forest-dependent species between west and east Sundaland. The authors left open the question whether this savanna corridor was at any time wide enough to include parts of Borneo.

Sundaland regional Pleistocene faunas

Key regional Pleistocene assemblages of mammals occur at fossil sites in Java. The transitional Early/Middle Pleistocene faunas of Trinil (0.9 Ma) and Kedung Brubus (0.8–0.7 Ma) were dominated by large browsing or grazing ungulates of continental Asian affinity; those marked † are extinct: proboscideans *Stegodon trigonocephalus*† and *Elephas hysudrindicus*†, Malay tapir *Tapirus indicus*, Javan rhinoceros *Rhinoceros sondaicus*, hippopotamus *Hexaprotodon sivalensis*†, cattle *Epileptobos groenveldtii*† and *Bibos palaeosondaicus*†, buffalo *Bubalus palaeokerabau*†, a large hog deer or chital *Axis lydekkeri*† , barking deer or muntjac *Muntiacus muntjak* and a regionally endemic antelope *Duboisia santeng*†. Also present were giant pangolin *Manis palaeojavanica*†, tiger *Panthera tigris*, leopard cat *Prionailurus bengalensis*, a large hyaena *Hyaena brevirostris*† and a porcupine *Hystrix* sp. Long-tailed macaque *Macaca fascicularis* and a langur *Presbytis* sp. occurred only at Trinil, in small numbers.

This mammal assemblage denotes an environment of grassy savanna habitat with scattered woodland, watered by lakes or large rivers required by hippopotamus (Storm 2001; van den Bergh et al. 2001). That this assemblage and, presumably, the associated habitat extended across Sundaland to the northwest is shown by a small collection from a cave at Tambun, near Ipoh, Peninsular Malaysia, which included hippopotamus and the antelope *Duboisia santeng* (Hooijer 1962a). Malay tapir and Javan rhinoceros, along with Sumatran rhinoceros *Dicerorhinus sumatrensis*, were present in southern China from Early through Late Pleistocene, pointing to continental Asian origins of these representatives of the regional Middle Pleistocene ungulate 'megafauna' (Jablonski and Whitfort 1999, Louys et al. 2007).

Among arboreal primates, orangutans *Pongo* sp. were present in continental Southeast Asia from Early until Late Pleistocene but so far lack Holocene records, whereas macaques *Macaca* and gibbons *Hylobates* have been present throughout the Quaternary (Jablonski and Whitfort 1999). Although source populations were therefore available much earlier, the strictly arboreal and forest-dependent orangutan and gibbons first appear in Java at Late Pleistocene sites collectively labeled Punung, of Eemian interglacial age between 128 and 118 ka (Westaway et al. 2007). The Punung assemblage included moonrat *Echinosorex* sp., porcupine *Hystrix brachyurus*, long-tailed giant rat *Leopoldamys sabanus* and a *Rattus* sp., orangutan, siamang *Hylobates syndactylus* and gibbon *Hylobates* sp., pig-tailed *Macaca nemestrina* and long-tailed macaques, sun bear *Helarctos malayanus*, tiger, Asian elephant *Elephas maximus*, Malay tapir, Javan rhinoceros, bearded pig *Sus barbatus* and Eurasian wild pig *S. scrofa vittatus*, muntjac, Javan rusa *Cervus timorensis*, serow *Naemorhedus sumatraensis*, banteng *Bos javanicus* and wild buffalo *Bubalus bubalis*. This assemblage has been recognized as a modern, closed tropical rainforest community (Storm et al. 2005). During the warmer, wetter climate around the start of the interglacial, an expansion of forest cover evidently permitted the dispersal of these mammals into eastern Java where, presumably, they supplanted a former archaic, savanna-adapted community (Westaway et al. 2007).

This assemblage is followed in time by Song Gupuh cave, Punung, containing over 16 m of deposits. The deepest faunal remains, with an inferred age of around 70 ka, lacked orangutan, siamang and other gibbons—strictly arboreal primates—but included (listed in order of frequency) long-tailed macaque, a deer *Cervus* sp., muntjac, lesser mousedeer *Tragulus javanicus*, wild cattle cf. banteng *Bos* sp., buffalo *Bubalus* sp., Eurasian wild pig and Suidae indet., Javan rhinoceros, Asian elephant, sun bear and Malay tapir. From about 12 ka human use of the cave intensified and the number and variety of animals represented increased. Deer, pig, bovids, bear, elephant, rhinoceros and tapir remained evident, while macaque frequency increased hugely. The change in relative representation is attributed to alteration in human behaviour, rather than environmental factors (Morwood et al. 2009). Finally, also in eastern Java, the Wajak site, dated at the cusp of the Holocene 10,560 ± 75 BP, shared many mammals with the earlier sites, including a langur *Presbytis* sp., tiger, Asian elephant, Javan rhinoceros, Malay tapir, Javan rusa, banteng, wild buffalo and wild boar, but again lacked the more strictly forest-adapted orangutan, siamang, sun bear and serow. This assemblage has been categorized by Storm (2001) as a 'modern sub-recent open woodland fauna'.

The origins of Borneo mammals

No remains of Early or Middle Pleistocene mammals have been found in situ in Borneo (Cranbrook, Earl of et al. 2006), but two primitive proboscidean teeth of these periods are known. One, the type and unique example of *Stegolophodon lydekkeri* was presented to the Zoological Society of London in 1885 by Everett who claimed that it was found in forest in Brunei (Medway 1972). The genus *Stegolophodon* emerged in the Miocene and diversified over a wide range in continental South and East Asia in the Early Pleistocene, but did not reach Java (Saegusa 1997). The related *Stegolophodon latidens* was an Early Pleistocene species typical of the Upper Irrawaddy deposits of Burma (Hooijer 1952). The second tooth, which was found northeast of Samarinda, East Kalimantan, "about an hour's walk from the shore", was identified as *Elephas* (*Palaeoloxodon*) *namadicus*. Remains of this Middle Pleistocene elephant are widespread in continental Southeast Asia, including southern China, the Indochinese region and the Mogok caves, Burma, (Hooijer 1952) and Perak, Peninsular Malaysia (Andrews 1905), but again absent from the fossil sites of Java (de Vos, personal communication). The provenance of both specimens is questionable but, if truly originating in Borneo, their affinities point to immigration by *Stegolophodon*, in the Early Pleistocene, and *Elephas namadicus*, in the Middle Pleistocene, by a land connection with continental Asia along the northern foreshore of the exposed bed of the South China Sea to Borneo, and not through central Sundaland.

Among extant mammal faunas of the main landmasses of present Sundaland, i.e., the Malay peninsula and the Greater Sunda islands, Sumatra, Borneo and Java, the large number of shared species with an evident common derivation from continental Southeast Asian precursors has long been recognized as the consequence of land connections during the Pleistocene (Darlington 1957). The great majority of naturally occurring mammals of Borneo, including bats, are associated with woodland habitats; many have not been found outside closed tropical rainforest (Payne and Francis 2005). As noted above, orangutan and gibbons, strictly arboreal in their behaviour, were present in continental Asia throughout the Pleistocene. It is therefore not possible to determine when these primates (and other associated forest mammals) originally reached Borneo. The opportunity to do so could have existed at any cool phase when connections existed, provided that the intervening land was vegetated by forest.

More than one route may have been available at different times. A track along the northern foreshore of the exposed South China sea bed is again suggested by the presence in northern Borneo (but not in the Malay peninsula, Sumatra or Java) of two mammal species whose closest relatives occur in the Indochinese region: smooth-tailed treeshrew *Dendrogale melanura* and ferret-badger *Melogale everetti* (Wilson and Reeder 2005). Although there is, as yet, no palaeontological or archaeological confirmation, Harrison et al. (2006) proposed that orangutans have been present in Sundaland throughout the Pleistocene, and that the two extant species, the Sumatran *Pongo abelii* and Bornean *P. pygmaeus* (Groves 2001) arose by vicariant differentiation. An alternative scenario could be two independent invasions from the continental source region. Within Sundaland, Cranbrook, Earl of and Piper (2008a) considered that the distributions of ruddy treeshrew *Tupaia splendidula*, banded leaf-monkey *Presbytis (melalophos) chrysomelas*, agile gibbon *Hylobates (agilis) albibarbis* and tricolored subspecies of Prevost's squirrel *Callosciurus prevostii* are best explained by immigration during the terminal Pleistocene from Sumatra to western Borneo via the Karimata land bridge.

During glacial phases, although subaerial exposure of the Sunda Shelf would have been at its greatest, the central savanna belt could have been a barrier to genetic exchange between populations of forest-adapted mammals. The isolation of Borneo, by a terrestrial ecological barrier during marine recessions and by the sea itself at other times, is reflected by the higher proportion of endemic mammals than found on other Sunda Islands. Current opinion recognises 50 endemics, all forest-adapted, among 132 species of land mammals, excluding bats, i.e., 38% (Table 1). Allopatric sibling species have been confirmed by molecular studies of clouded leopards *Neofelis* (Buckley-Beason et al. 2006), and are recognised by conventional taxonomy among orangutans *Pongo* spp. (Groves 2001), langurs *Presbytis* and treeshrews *Tupaia glis/longipes*. Local selection in (ecological) isolation in northern Borneo is also indicated by distinctive subspecies of arboreal squirrels with ranges corresponding fairly closely to the territorial bounds of Sabah: the giant squirrel *Ratufa affinis sandakanensis*, the black-backed form of Prevost's squirrel *C. prevostii pluto* and the white-bellied form of horse-tailed squirrel *Sundasciurus hippurus pryeri* (distributions from Payne and Francis 2005).

The archaeological record

Archaeological investigations initiated by a former Curator of the Sarawak Museum, the late Tom Harrisson, and subsequently extended by others, have sampled the mammal fauna of northern and northwestern Borneo at dates ranging from Late Pleistocene about 45,000 years ago (45 ka) through the Holocene at three principal sites: Madai caves in eastern Sabah, 118°08′E, 4°40′N (Bellwood 1988; Harrison 1998), and Niah caves 113°48′E, 3°45′N (Harrisson 1958, 1959, 1967; Barker et al. 2002a, b, 2007) and Sireh cave 110°27′E, 1°10′N (Ipoi Datan 1993) in Sarawak. Additional records have been found among material excavated in 1878–1879 by Everett in caves at Jambusan, Sarawak, 110°10′E, 1°23′N, now held in the Natural History Museum, London (Everett et al. 1880; Cranbrook, Earl of 2004).

At Niah, recent attention has focussed on the deep levels of the excavation in the West mouth, Harrisson's 'Hell' trenches (Rabett et al. 2006). Here mammal remain occur in contexts ^{14}C-dated 42,850 $\pm$ 500–33,940 $\pm$ 230, i.e., 46,417 $\pm$ 417–39,044 $\pm$ 543 cal. BP, and artefacts and other signs of human activity continued to ^{14}C 45,900 $\pm$ 800 BP outside calibration (Barker et al. 2007, Table 1). The palynology of a vertical transect extending 215 cm downwards from these levels indicates that the local environment at Niah went through a series of very rapid and intense changes (Hunt and Rushworth 2005;

Table 1 Endemic land mammal species of Borneo. Data from Wilson and Reeder (2005)

Scandentia	Rodentia
Dendrogale melanura	*Rheithrosciurus macrotis*
Tupaia gracilis.	*Petaurillus hosei*
Tupaia dorsalis	*Petaurillus emiliae*
Tupaia longipes	*Aeromys thomasi*
Tupaia montana	*Callosciurus adamsi*
Tupaia picta	*Callosciurus baluensis*
Primates	*Callosciurus orestes*
Presbytis chrysomelas	*Sundasciurus brookei*
Presbytis frontata	*Sundasciurus jentinki*
Presbytis hosei.	*Glyphotes simus*
Presbytis rubicunda	*Lariscus hosei*
Nasalis larvatus	*Dremomys everetti*
Hylobates muelleri	*Exilisciurus exilis*
Hylobates albibarbis	*Exilisciurus whitehedi*
Pongo pygmaeus	*Chiropodomys major*
Soricomorpha	*Chiropodomys muroides*
Suncus ater	*Chiropodomys pusillus*
Suncus hosei	*Haeromys margarettae*
Crocidura baluensis	*Haeromys pusillus*
Crocidura foetida	*Maxomys alticola*
Chimarrogale phaeura	*Maxomys baeodon*
Carnivora	*Maxomys ochraceiventer*
Catopuma badia	*Niviventer rapit*
Diplogale hosei	*Pithecheirops otion*
Melogale everetti	*Rattus baluensis*
Artiodactyla	*Sundamys infraluteus*
Muntiacus atherodes	*Hystrix crassispinis*

Hunt et al. 2007). No dates have been determined but, by reference to global events, the authors estimate an age ~52 ka cal. BP for the basal pollen sample (=Zone H-1 of Hunt et al. 2007). This zone was rich in conifers, dominated by *Podocarpus* (52%), with *Picea*, *Pinus* and *Dacrydium* also present, sedges Cyperaceae (12%) and some grasses Poaceae (5%), as well as broad-leaved trees including oaks *Quercus*; this vegetation is consistent with a cooler and drier climate and lowered tree-line. Zones H-2 to H-4 are dominated by open-ground taxa, mostly sedges and grasses, with herbaceous plants such as goosefoots Chenopodiaceae and composites Asteraceae, Lactuceae and *Artemisia*, suggesting major drought stress and the retreat of forest vegetation. Most forest taxa expand in H-5 to H-7 (~49–48 ka), including taxa typical of lowland, montane, coastal and swamp forest; *Podocarpus* contracted. This probably reflects a rise in rainfall and temperature relative to the previous biozones.

At the base of H-9 a very abrupt change in palynological assemblage composition and a sharp decline in numbers of pollen recovered per sample suggest a period of non-deposition. H-9 and H-10 (at ~47 ka) have very high mangrove pollen and high lowland forest percentages. In H-11, coastal taxa expand, mangroves remain high, while swamp forest taxa are rare. The most important mangrove species in H-9 to H-11 is *Avicennia*, reflecting airfall deposition from nearby mangrove vegetation. Summing up, Hunt et al. (2007) concluded that at ~52–49 ka cal. BP the margin of the savanna corridor proposed by Bird et al. (2005) extended to Niah, and that any forest refugium indicated by other evidence lay further north and/or in the interior uplands where orographic rainfall would have maintained sufficient humidity to support closed forest. Their data also show that, by ~46–45 ka cal. BP, sea levels had risen, coastal mangrove was close to Niah cave where the local environment was forested with trees familiar in lowland habitat today.

Zones H9–H11 are coeval with basal 'Hell' levels at Niah, where mammal remains included the following species, or close relatives, of the archaic savanna-adapted assemblage at Middle Pleistocene sites in eastern Java: giant pangolin *Manis* cf. *palaeojavanica* (Piper et al. 2007b), porcupine, long-tailed macaque, langur *Presbytis* sp, leopard cat, Malay tapir (Piper and Cranbrook, Earl of 2008; Cranbrook, Earl of and Piper 2008a), cattle *Bos* sp. and muntjac (Barker et al. 2007 Table 2; Cranbrook, Earl of and Piper 2008a Appendix 1). Although not represented at these levels, specimens from Madai, Sabah, and Niah have confirmed the presence in the Late Pleistocene and early Holocene of tiger (Harrison 1998; Piper et al. 2007a) and Javan rhinoceros (Cranbrook, Earl of and Piper 2007). Significantly, also present were tree squirrels, orangutan, sun bear and several species of arboreal civets,

Mammal identifications from the West mouth are listed in Table 2. Finds elsewhere in north and northwestern Borneo add two mammals to the Late Pleistocene or early Holocene fauna. Firstly, evidence of Asian elephant is provided by the find in a cave at Bau, Sarawak, of 'a semi-fossilized fragment of an elephant's molar' (Shelford 1899), unfortunately never pictured and now lost (Cranbrook, Earl of et al. 2007a, b), and a whole molar said to have been found in a cave in Belait district, Brunei, considered by Hooijer (1972) to be of Pleistocene age on the basis of its high uranium content (9 ± 1 ppm eU_3O_8). Secondly, in an early Holocene context, among mammal remains from Madai, Cranbrook, Earl of (1988) identified a calcaneum and a canine tooth of a canid attributable either to dhole *Cuon alpinus* or an early, dingo-like domestic dog *Canis familiaris.* Both the Holocene age and the specific identity of these canid specimens remain unsettled (Piper et al. 2009).

People of modern type were present throughout this entire period and much of the zooarchaeological assemblage is the residue of human activity (Barker et al. 2007). In addition, an even larger number of remains represent naturally occurring cavernicolous mammals, i.e., bats and small terrestrial mammals that inhabited the cave, or spontaneously entered it, and died there. In total, 18 taxa in 12 genera of bats have been identified in Late Pleistocene and Holocene contexts (Table 2).

Late Pleistocene and Holocene selective adaptation and faunal change

The palynological evidence at Niah at ~45 ka cal. BP shows maritime and terrestrial forest cover, conformant with a high sea level. The zooarchaeological evidence indicates the presence of ten survivors of the regional Middle Pleistocene fauna of continental Southeast Asia and/or Java: giant pangolin, long-tailed macaque, a langur *Presbytis*, leopard cat, tiger, Malay tapir, Javan rhinoceros, Sumatran rhinoceros, muntjac and porcupine. All other species present at Niah were characteristic of modern tropical rainforest, including the minute pigmy shrew *Suncus hosei* (Piper and Cranbrook, Earl of 2007) and

Table 2 Mammals identified in the archaeological deposit at Niah caves, Sarawak, adapted from Cranbrook, Earl of and Piper (2008a, Appendix 1)

Scandentia	*Rhinolophus trifoliatus**
Tupaia cf. *minor**	*Rhinolophus borneensis**
Tupaia sp.	*Miniopterus* cf. *schreibersi**
Dermoptera	*Miniopterus* cf. *tristis**
Cynocephalus variegatus	*Miniopterus* cf. *phlebotis**
Primates	*Miniopterus* cf. *australis**
Nycticebus coucang	*Myotis* spp.*
Macaca cf. *fascicularis**	Pholidota
Macaca cf. *nemestrina**	*Manis javanica**
Presbytis sp.*	*Manis* sp. cf *palaeojavanica** +
Trachypithecus cristatus	Carnivora
Hylobates cf. *muelleri**	*Canis familiaris*
*Pongo pygmaeus**	*Helarctos malayanus**
Rodentia	*Melogale everetti**
Rattus (Rattus) sp.*	*Martes flavigula**
*Leopoldamys sabanus**	*Lutra sumatrana**
Maxomys sp.*	*Aonyx cinerea**
*Sundamys muelleri**	*Mustela nupides*
Callosciurus prevostii	*Arctictis binturong**
*Sundasciurus lowii**	*Arctogalidia trivirgata**
*Rheithrosciurus macrotis**	*Paguma larvata**
*Ratufa affinis**	*Paradoxurus hermaphroditus**
Petauristinae*	*Viverra tangalunga**
Hystrix cf. *brachyurus**	*Hemigalus* cf. *derbyanus**
*Trichys fasciculata**	*Herpestes* spp.*
Erinaceomorpha	*Felis* cf. *planiceps**
*Echinosorex gymnurus**	*Prionailurus bengalenis**
Soricomorpha	*Neofelis diardi**
*Crocidura foetida**	*Panthera tigris** o
*Suncus hosei**	Perissodactyla
Chiroptera	*Tapirus indicus** o
Pteropus cf. *vampyrus*	*Dicerorhinus sumatrensis**
*Eonycteris spelaea**	*Rhinoceros sondaicus** o
Rousettus sp. cf. *amplexicaudatus**	Artiodactyla
Penthetor lucasi	*Sus barbatus**
Cynopterus cf. *horsfieldi**	*Sus* cf. *scrofa* (domestic)
Emballonura sp.*	*Tragulus napu**
*Hipposideros diadema**	*Tragulus javanicus*
*Hipposideros larvatus**	*Capra hircus*
*Hipposiderus galeritus**	*Bos javanicus**
*Cheiromeles torquatus**	*Cervus unicolor**
*Chaerephon plicata**	*Muntiacus* cf. *muntjak**

Order and species names follow Wilson and Reeder (2005), except clouded leopard *Neofelis diardi* (Buckley-Beason et al. 2006). An asterisk (*) denotes confirmed presence in the Late Pleistocene; cross (+) denotes species globally extinct and (o) species now extinct in Borneo

Bornean shrew *Crocidura foetida* (Cranbrook, Earl of and Piper 2008b), lesser treeshrew *Tupaia minor*, pig-tailed macaque, orangutan, gibbon (Hooijer 1962b) sun bear, several viverrids including the highly arboreal bearcat *Arctictis binturong*, arboreal civets *Paguma larvata* and *Paradoxurus hermaphroditus*, and the more terrestrial Malay civet *Viverra tangalunga*, bearded pig, large mousedeer *Tragulus napu*, banteng, sambar *Cervus unicolor*, and a variety of forest rats and squirrels (Table 2).

Palynology also shows that preceding millennia had been marked at Niah by phases when conifers, sedges and grasses dominated, indicating a drier, cooler climate, and a more open savanna-like landscape. Although zooarchaeological evidence is lacking, analogy with the Punung sites and Song Gupuh implies that, if archaic members of a savanna-adapted Middle Pleistocene fauna had formerly been present, the ecological repercussions of climate change in the last interglacial would have eliminated them. It is therefore surprising that one striking example, a giant pangolin, did survive until the Late Pleistocene at Niah where (unlike Javan sites) it existed alongside its modern congener the Malay pangolin *Manis javanica* (Piper et al. 2007b).

Around 35 ka the global climate began cooling towards the LGM. However, it is clear from the presence of arboreal primates throughout the archaeological profile at the West mouth site (Table 2) that, through the Late Pleistocene and early Holocene there was no reversion to savanna habitat at Niah. Although the hunting habits of the human cave visitors have skewed the relative representation of mammal species in the West mouth and in other occupation sites at Niah (Piper and Rabett 2009), the persistence of tree cover is testified by the presence, throughout the profile, of forest specialists including lesser treeshrew, *Presbytis* langurs, macaques, gibbon and orangutan, tree squirrels, sun bear, binturong, other civets and clouded leopard *Neofelis diardi*. Evidence of floristic changes at other regional sites indicates that at the LGM altitudinal boundaries between lowland and montane forest formations were lowered as much as 900 m (van der Kaars and Dam 1995; Flenley 1998, 2005). Through these coolest millennia (~23–18 ka), the forest of the lowland plain in the neighbourhood of Niah cave would probably have included many tree species that are nowadays characteristic of Upper Dipterocarp or Lower Montane facies. In their present distribution, the macaques, langurs, gibbons and orangutan all frequent terrain up to 1,000 m or higher in interior Borneo (Medway 1977). The submontane forest habitat evidently provides the necessary resources for these arboreal primates, allowing them to thrive at Niah through all climatic vicissitudes of the Late Pleistocene.

Botanical evidence of lowered ambient temperature is supported by the zooarchaeological record at Niah. Thus, an archaeological context consistent with Late Pleistocene age yielded a jaw of ferret badger *M. everetti* (Medway 1958), an endemic species nowadays confined to montane environment on Mt. Kinabalu at 1,070–3,000 m (Payne and Francis 2005). Among Pleistocene long-tailed macaques (Harrison 1996) and colobines of the genera *Presbytis* and *Trachypithecus* at Niah. (Harrison 2000) a slight but significant enlargement in tooth size, measured as occlusal area, has been compared with Bergmann's rule, implying a cooler prevailing climate in the Late Pleistocene. Hooijer (1961) provisionally identified a parallel size diminution among orangutans at Niah, but this has been contradicted by Harrison (2000), although a small (presumed early Holocene) sample from Jambusan and Paku, Sarawak, does show trends to enlarged size.

Among small mammals, specimens of Bornean shrew from the 'Hell' levels, judged by comparative measurements of the teeth of the lower jaw, were larger than modern specimens from the locality which are allocated to *Crocidura f. foetida* (Ruedi 1995). The difference is sufficient to allocate the archaic population to a separate subspecies *Crocidura f. doriae*, now submontane in distribution and therefore adapted to cooler ambient temperatures. Among shrews, larger size is linked to improved thermo-control (Cranbrook, Earl of and Piper 2008b). Similar trends are shown by three rats represented from Late Pleistocene to modern contexts at Niah: Müller's rat *Sundamys muelleri*, long-tailed giant rat *Leopoldamys sabanus* and a *Rattus* species (Piper and Cranbrook, Earl of 2009). The terminal Pleistocene and Holocene evolutionary process towards diminution in size in all these cases is likely to have been driven by selective pressures of a changing environment. Warming of the climate and consequent vegetational change was potentially a critical factor.

Among the large ungulates of the Early/Middle Pleistocene regional megafauna persisting through the Late Pleistocene of Borneo, the Sumatran rhinoceros may also have undergone selective diminution in size. In this case, among a small zooarchaeological sample of early Holocene specimens, the limb bones (including foot bones) were longer than homologues in an equally small sample of modern specimens (Cranbrook, Earl of 1986). One intriguing limb bone, a right radius that formed the 'pillow' in burial #27 in the West mouth, dated by association as terminal Pleistocene or early Holocene, is 13% longer than the median value of modern homologues. Apart from this adult bone, individual ages are indicated by a left metatarsal II, with distal epiphysis unfused (therefore a juvenile), and eight fragmentary cheek teeth, four of which were unerupted, recovered from levels 12–24 in. (~30–60 cm) down to 60–72 (~150–180 cm) in. at the West mouth, Niah, the deepest being dated by association ~29 ka. Other specimens were found in Lobang Hangus, Niah, at Tapadong and Madai caves, Sabah, and in the Bau limestone area, Sarawak (Cranbrook, Earl of 1986, 2004).

While the Sumatran rhinoceros has survived to the present day, there has been no reliable report of the Javan rhinoceros from Borneo in historic times (Cranbrook, Earl of et al. 2007a, b). Yet this rhinoceros is represented by a deciduous 4th molar, two ulna fragments and three foot bones, from terminal Pleistocene and Holocene contexts at Niah and at Madai, Sabah. Potentially the most recent specimen is a central terminal (ungual) phalanx from one of Tom Harrisson's large trenches cutting down 12 in. (~30 cm) from the original ground surface. The position of the bone within this trench was not recorded. If it was at a superficial level, it may have lived in historic times (Cranbrook, Earl of and Piper 2007).

The third large ungulate survivor from the Middle Pleistocene megafauna, the Malay tapir, was also the subject of confused reports in the 19th and early 20th Centuries from west, northwest and northern Borneo but, again, its presence in historic times is unconfirmed by an authenticated specimen. Zooarchaeological material has been recovered from cave sites at Jambusan, Sireh and Niah, Sarawak, and Madai, Sabah, in contexts dated from Late Pleistocene at 40–35 ka uncal., through the terminal Pleistocene and early Holocene. During the excavation at Sireh cave in 1959, one juvenile canine was found on the surface, and is therefore potentially dateable to early historic times. Of nine teeth in total, all but one are from the deciduous dentition or scarcely erupted adult canines, i.e., from young animals (Cranbrook, Earl of and Piper 2009).

The status of the Asian elephant remains controversial. The two specimens (one lost) are equivocal evidence of a Pleistocene presence and, in contrast to the Javan cave sites at Punung, Song Gupuh and Wajak, no finds have been made at any controlled excavation in Borneo. On the basis of their distinctive DNA, Fernando et al. (2003) concluded that the extant elephants in eastern Sabah are derived from endemic (Pleistocene) progenitors. Cranbrook, Earl of et al. (2007a, b), however, have argued that this distinctiveness could also be exhibited if, as local tradition maintains, the present population consists of descendants of the (now extinct) Javan elephant, introduced to eastern Sabah from feral stock that formerly existed on Sulu island, Philippines.

Tiger is the remaining Middle Pleistocene megafauna mammal now extinct in Borneo, yet shown to have been present from the terminal Pleistocene (13,745 ± 55 uncal. BP) to potentially recent at Niah, and early Holocene at Madai, Sabah (Piper et al. 2007a). Finally, the record of a canid from a Holocene context at Madai, Sabah, if confirmed as dhole (Cranbrook, Earl of 1988), would add a middle-sized predatory carnivore to the list of extinct mammals of Borneo.

Among cavernicolous bats, there has been a dramatic change in the frequency of Geoffroy's rousette *Rousettus amplexicaudatus* and/or bare-backed rousette *R. spinalatus* (not confidently separable by dental characters, see Francis 1989). At Niah, remains of rousettes occur in the zooarchaeological record from the deepest Late Pleistocene levels to near superficial, and are numerous, comprising 94% (237 of 253) of a sample of fruit bat jaws (Cranbrook, Earl of and Piper 2008a). At Jambusan, similarly, rousette jaws are by far the most numerous among remains of fruit bats (Pteropodidae; Cranbrook, Earl of 2004). In present times, both rousettes are rare casuals in or near Niah cave (Hall et al. 2002) and recorded at only a few other sites elsewhere in Borneo, not including the Jambusan area (Payne and Francis 2005).

Among large insectivorous bats of the family Molossidae, hairless bats *Cheiromeles torquatus* today roost high in the roof of Niah cave. Their remains are found in archaeological contexts dating from Late Pleistocene, albeit in small number, showing that they have been in occupation for at least 40,000 years. By contrast, the smaller molossid, wrinkle-lipped bat *Chaerephon plicata*, famous in Sarawak for the large colony at Mulu, is unknown in Niah cave today yet was the most abundant microchiropteran throughout Late Pleistocene-Holocene archaeological sequences. At the most superficial, large numbers of wrinkle-lipped bats were represented in deposits towards the rear of the archaeological reserve in the West mouth dated to 1,000 years ago or less (Cranbrook, Earl of and Piper 2008a).

Discussion: ecological factors or human extermination?

The ecology of Early/Middle Pleistocene proboscideans *Stegolophodon lydekkeri* and *E. namadicus* can be inferred from their comparatively low-crowned molars, suggesting that both were browsers on soft, bushy vegetation, rather than silicious sedges and grasses. By implication, both may have been associated with scrub and woodland, or open forest, but not grassy savanna. If the provenance of the specimens proves to be genuine, both may have been susceptible to environmental changes that eliminated favourable habitat, at different times. Given the tenuous Pleistocene status of Asian elephant in Borneo, it is possible that this proboscidean never reached the island or, if it did, was eliminated early in the Late Pleistocene at a time previous to the basal levels of the 'Hell' trenches at Niah. Lacking any archaeological specimen, further conjecture is unprofitable.

In Java, the third large Sunda island, a forest-adapted mammal fauna first appeared in the zooarchaeological record with the climatic amelioration at the onset of the Eemian interglacial, but did not survive in its full composition as progressive cooling and desiccation took hold in the mid-Late Pleistocene. The disappearance of arboreal species such as orangutan and siamang is attributed by van den Bergh et al. (2001) and Storm et al. (2005) solely to environmental change. In Borneo, by contrast, the zooarchaeological record at Niah and other key cave sites tells a different story.

In the Late Pleistocene at Niah at ~52 ka cal. BP, palynology of the basal zone H-1 of Hunt et al. (2007) does indicate a vegetation dominated by conifers, sedges and grasses, implying a savanna-like landscape perhaps comparable to that of transitional Middle

Pleistocene Java. Unfortunately, zooarchaeological investigation did not reach so far, and any mammals associated with this zone remain undiscovered. However, stratified excavations do show that five members of the regional megafauna were present in north and northwestern Borneo from about 45 ka: giant pangolin, Javan and Sumatran rhinoceroses, Malay tapir and tiger. Also present were other mammals occurring in Early/Middle Pleistocene sites in east Java: long-tailed macaque, a langur *Presbytis*, leopard cat, muntjac and a porcupine. The significant difference at Niah was that these species were associated with an assemblage of obligatory arboreal or forest-adapted mammals, comparable with the Punung fauna of Java, and indicative of a tropical rainforest environment.

Three times larger in linear dimensions than its congener the Malay pangolin, the Niah giant pangolin was probably ill-adapted to the closed tropical rainforest environment. All eight skeletal elements discovered at Niah are attributable to a single individual, and no later examples have been found (Piper et al. 2007a, b). Although its presence in the site suggests that it was the prey of human hunters, as was its congener the Malay pangolin, a single specimen is not evidence of overkill. It is more likely that the closed forest environment failed to provide the resources needed by the archaic large-bodied pangolin. This is the only confirmed mammal extinction in Borneo during the Late Pleistocene.

Of other Middle Pleistocene megafauna, the Sumatran rhinoceros showed post-Pleistocene size reduction, in this case in the relative length of the limbs and feet. This may have been a response to increasing forest cover, lowering the selective value of a fast gait. None the less, this rhinoceros, along with other members of the Pleistocene ungulate megafauna (Javan rhinoceros and Malay tapir) survived through the LGM and the Holocene climatic amelioration. Although the changing environment was evidently not lethal to these large browsing mammals, the return of a warm, wet interglacial climate may have stressed them by suppressing low stature vegetation under the shade of the closed canopy of tall lowland tropical rain forest, and thereby restricting their food resources to marginal habitats (Piper and Cranbrook, Earl of 2008).

Under such pressures, a long life-span is advantageous to large mammals. Records (in captivity) are 31 years for Malay tapir, 23 years for Javan and over 32 years for Sumatran rhinoceros. But they are disadvantaged by carrying a single young, and long gestation periods: tapir 13 months, Javan rhinoceros $\sim$17 months, and Sumatran rhinoceros at least 7 months. The duration of dependency of the young on parental care is also long (biodata from Medway 1983). Johnson (2006, chapter 6) has modeled different levels of off-take applied to any large mammal with these characteristics, and shown that a small increase in the mortality of juveniles can hold recruitment rate below levels needed to replace breeding adults. It is notable that those remains indicating individual age, at Niah and at other sites, show a high proportion of juveniles among all three species. Therefore, by adopting hunting strategies that deliberately or unintentionally targeted juveniles, as seems to have been the case, people could have tipped the balance. After the rising sea had severed the Karimata land bridge around 10 ka, new immigrants were unobtainable. Now isolated, and already stressed by environmental change, Borneo populations of these large ungulates would have been progressively depleted by the loss of potential recruits.

This gradual impact would have heightened when trading contacts with China were established. The earliest recorded imperial embassy sent to the Indonesian region was by the Han Emperor Wang Mang (1–23 AD) to acquire a rhinoceros for the Imperial Zoological Garden (Vlekke 1943). Thereafter, hunting for the prized rhinoceros horn and other body parts must have intensified as Iron Age weaponry improved and, ultimately, shotguns were introduced. During the 1930s, the residual population of rhinoceroses in Sarawak was almost exterminated by trophy hunters and with them, perhaps, the last tapirs (Medway 1977).

This scenario fits the case of the ungulate megafauna. It is not easily transferred to the tiger, a large carnivore with short gestation period of 15–16 weeks, and litter size up to six (Medway 1983). The three zooarchaeological specimens are insufficient to demonstrate any pattern of hunting, although one is indeed juvenile in age. The abundance of bearded pig throughout the archaeological deposits, at Niah, Sireh and Madai alike, show that there was never any shortage of the predominant ungulate prey of this versatile large predatory carnivore in the rainforest environment. Based on camera-trapping in tropical rainforest in Peninsular Malaysia, Lyman et al. (2007) computed the (naïve) mean natural density at 0.67 ± 0.25 tigers per 100 km^2. Applied to the present land area of Borneo, this indicates a potential island-wide population around 5,000 individuals. Before recruitment from outside sources was prevented by the final submersion of the Karimata land bridge at the onset of the Holocene, the exposed land area of Borneo was greater and tiger numbers therefore potentially higher. The few zooarchaeological specimens confirm that tigers persisted from the terminal Pleistocene until the near present (Piper et al. 2007a). There is no apparent reason to attribute the elimination of tigers in Borneo either to over-hunting by people or to some unidentified ecological change.

Among larger arboreal mammals, macaques and gibbons remain present in the forest of Niah National Park, but there have been no historical records of a resident orangutan population within 200 km (Harrison et al. 2006). Hooijer (1961) enumerated 89 orangutan teeth and jaw fragments (with or without teeth) from the west mouth of Niah cave recovered at depths down to 84 in.. By the crude measure of depth, these remains are evenly distributed in time with eleven (12%) from levels within 12 in. (~30 cm) of the original surface (Table 3). Although the depth of the excavator's cuts up to 24 in. (~60 cm) obscures the age/depth correlation, the presence of the orangutan at Niah from Late Pleistocene through the mid-Holocene and probably into the past millennium is confirmed. The condition of the remains shows that orangutans were hunted by people throughout this period (Harrison 2000; Piper and Rabett 2009; Piper et al. 2009). Human exploitation has nowadays reached critical levels (Corlett 2007). The representation of juveniles in Hooijer's sample is high, at 28% (Table 3). Like the ungulate megafauna, this large, slow-breeding primate is vulnerable to recurrent loss of young

Table 3 Distribution of teeth or fragments of jaw of orangutan, by dental age, in the West mouth, Niah cave, from surface to 84 in. depth (~213 cm), aggregated into 12 or 24 in. intervals

Depth (inches)	0–12[a]	12–24[a]	24–36[a]	36–48[a]	48–60	60–84	Totals
Adult dentition	11	15	10	13	5[b]	12	64
Juvenile		8	4	4[a]	3	6	25

Data from Hooijer (1961). 'Adult dentition' is recognized as teeth of the permanent set, showing definite signs of wear; 'Juvenile' is recognized as teeth of the deciduous dentition plus unerupted or slightly worn teeth of the permanent dentition

[a] Of two teeth from the trench X/4 0–24 in., one has arbitrarily been assigned to 0–12 in. and the other to 12–24 in. and, similarly, of two teeth from E/C1 and E/C2 24–48 in., one has been assigned to 24–36 in. and the other to 36–48 in. A female lower C dext. slightly worn from E/B2 24–48 in., has been placed as a juvenile at 24–36 in; a P^3 sin from X/3 (A) at 0–24 in has been assigned to 0–12 in.; a P^3 sin in situ in a female maxillary from E/C1 24–48 in. has been assigned to 24–36 in; a P^4 sin in female maxillary from E/C1 24–48 in. has been assigned to 36–48 in; M_2 dext. From X/1 24–48 in. has been assigned to 24–36 in

[b] A female lower C sin., tip broken, at 60 in. (under rock) has been assigned to 48–60 ins

animals. Again, like the Malay tapir and Javan rhinoceros, the population evidently survived at Niah until a recent time when it seems that cumulative hunting pressure, perhaps coupled with a deterioration in environmental resources, led to their local extermination.

As indicators of environmental change, small mammals may give more reliable signals than large species. The diminished size of the modern Borneo shrew at Niah, compared with the Late Pleistocene population, conforms with the pattern of climate change, but may show no more than the selective advantage of larger size in a cooler environment among small insectivores. Parallel size changes in time among rats are likely to represent more complex ecological responses, yet to be identified.

Bats also offer independent evidence of environmental change. Whether representing food debris of people or natural casualties, rousettes were common in Late Pleistocene and Holocene at Niah, and also at Jambusan in presumed early Holocene. Their near-complete disappearance could reflect the loss of some critical resource on which these fruit- and pollen-eaters depend. The change in bat fauna could therefore signal a widespread change in the floristic composition of forests across northwestern Borneo during the later Holocene. In due course, future pollen studies may illuminate this puzzle.

The complete evacuation of Niah cave by a large population of wrinkle-lipped bats, probably within the past millennium, may also reflect the loss of an important insect food resource. On the other hand, the feeding range of these bats is large, and their diet eclectic. Again, human disturbance cannot be discounted, perhaps by fire, or possibly the start of regular harvesting of swiftlet nests at Niah at the beginning of the nineteenth century (Lim and Earl of Cranbrook 2002). The collapse of a large population of wrinkle-lipped bats should convey a message from the past, urging caution, to National Park authorities and site managers at Mulu and Gomantong, where the emergence of these bats is a famous attraction.

In conclusion, the evidence of zooarchaeology points to significant differences in the development and turnover of the Late Pleistocene mammal faunas in northwest and north Borneo compared with Java. At these Borneo sites, covering present Sarawak and Sabah, after the disappearance of the giant pangolin, the remaining survivors of an earlier, savanna-adapted Middle Pleistocene megafuana continued to exist among an assemblage of forest-adapted mammals. Despite human exploitation, and environmental changes that favoured selective size-reduction among large, middle-sized and small mammals, this assemblage persisted through the LGM and well into the Holocene. It appears that only within the last millennium (i.e., 1000–2000 AD) did human activity and climatic factors, acting independently or in concert, lead to local disappearances or, for three large mammals—tiger, tapir and Javan rhinoceros—island-wide extinctions.

The diversity of scenarios for species loss provides relevant background information, in the context of potential climatic shifts, continuing rural development and consequent ecological impacts, for present day Borneo conservationists to consider carefully. Bennett et al. (2000) considered that the prospects for hunted mammals were poor unless severe restrictive measures are introduced. Junau et al. (2008) and Diong et al. (2008) were more optimistic that sustainable harvesting could be managed in the case of bearded pig, still (as in prehistoric times) the principal quarry. The future of the remaining megafauna is now perilous. However, if further elucidation of their genome confirms that the elephants in north-eastern Borneo originated from Java via Sulu, the importance of this population is enhanced as an example (albeit unwitting) of successful translocation. Piper and Cranbrook, Earl of (2008) have proposed that deliberate measures should be taken to reintroduce Malay tapir to suitable locations. With determination, backed by adequate resources, it is possible that this vanished member of the Middle Pleistocene mammal megafauna could be restored to the wild in Borneo.

The social and environmental transformation resulting from rural development since 1963 has been documented for Sarawak by Schatz (2007), who called for more prudent management of forest resources and a more cautious and environmentally sensitive approach to continued expansion of oil palm. Having survived the vicissitudes of the terminal Pleistocene and early Holocene, it is not only the remaining megafuana that are threatened. Many other forest-adapted animals (and plants) are also now vulnerable to extinction. The progressive diminution in the area of natural forests throughout Borneo over the past 50 years is shown pictorially by WWF (2009). This respected organization is currently promoting an international conservation programme, 'Heart of Borneo'. By coordinating the efforts of state, provincial and national organizations of Brunei Darussalam, Malaysia and Indonesia, along with commercial and voluntary sectors, successful implementation of this programme now appears to offer the best hope for the survival of a reasonable proportion of the forest-adapted biota of this great island of Borneo.

Acknowledgments I am grateful to successive Directors of Museums, Sarawak, for encouraging me to remain Hon. Curator of Mammals, and to En. Ipoi Datan, Deputy Director and Curator of Archaeology, Sarawak Museum, for providing facilities to study zoological materials from the Museum's excavations. Prof. Graeme Barker and members of the Niah Caves Project (Barker et al. 2002a, b) have consistently given supportive encouragement. Particular thanks go to Chris Hunt, and to zooarchaeologist colleagues, Philip Piper and Ryan Rabett, who not only shared much of the research reported in these pages but also gave constructive advice on a first draft of this paper. David Lane's careful editing and the comments of two anonymous referees have greatly improved the final version.

References

Andrews CW (1905) Fossil tooth of *Elephas namadicus* from Perak. J Fed Malay States Mus 1:81–82

Barker G, Barton H, Beavitt P, Bird M, Daly P, Doherty C, Gilbertson D, Hunt C, Krigbaum J, Lewis H, Manser J, McClaren S, Paz V, Piper P, Pyatt B, Rabett R, Reynolds T, Rose J, Rushworth G, Stephens M (2002a) Prehistoric foragers and farmers in South-east Asia: renewed investigations at Niah Cave, Sarawak. Proc Prehist Soc 68:147–164

Barker G, Barton H, Bird M, Cole F, Daly P, Gilbertson D, Hunt C, Krigbaum J, Lampert C, Lewis H, Lloyd-Smith L, Manser J, McLaren S, Menotti F, Paz V, Piper P, Pyatt B, Rabett R, Reynolds T, Stephens M, Thompson J, Trickett M, Whittaker P (2002b) The Niah cave project: the third (2002) season of fieldwork. Sarawak Mus J 57(78 ns):87–178

Barker G, Barton H, Bird M, Daly P, Datan I, Dykes A, Farr L, Gilbertson DD, Harrisson B, Hunt CO, Higham T, Krigbaum J, Lewis H, McLaren S, Paz V, Pike PA, Piper P, Pyatt P, Rabett R, Reynolds TEG, Rose J, Rushworth G, Stephens M, Stringer C, Thompson G (2007) The 'human revolution' in lowland tropical Southeast Asia: the antiquity and behavior of anatomically modern humans at Niah Cave (Sarawak, Borneo). J Hum Evol 52:243–261

Bellwood P (eds) (1988) Archaeological research in south-eastern Sabah. Sabah Museum Monograph 2

Bennett EL, Nyaoi AJ, Sompud J (2000) Saving Borneo's bacon: the sustainability of hunting in Sarawak and Sabah. In: Robinson JG, Bennet EL (eds) Hunting for sustainability in tropical forests. Ch. 15. Columbia University Press, New York, pp 305–324

Bird MI, Taylor D, Hunt C (2005) Palaeoenvironments of insular Southeast Asia during the Last Glacial period: a savanna corridor in Sundaland? Quat Sci Rev 24:2228–2242

Buckley-Beason VA, Johnson WE, Nash WG, Stanyon R, Menninger JC, Driscoll CA, Howard J, Bush M, Page JE, Roelke ME et al (2006) Molecular evidence for species level distinction in clouded leopards. Curr Biol 16(23):2371–2376

Corlett RT (2007) The impact of hunting on the mammalian fauna of tropical Asian forests. Biotropica 39:292–303

Cranbrook, Earl of (1986) A review of fossil and prehistoric remains of rhinoceroses of Borneo. Sabah Mus Arch J 1(1):50–110

Cranbrook, Earl of (1988) The contribution of archaeology to the zoogeography of Borneo, with the first record of a wild canid of early Holocene age. Fieldiana Zool NS 42: i–iv, 1–7

Cranbrook, Earl of (2000) Northern Borneo environments of the past 40,000 years: archaeozoological evidence. Sarawak Mus J 55 (NS 76): 61–110

Cranbrook, Earl of (2004) A history of animal diversity of the Bau limestone area. Sarawak Mus J 59 (NS 80), special issue No. 6: 193–220

Cranbrook, Earl of, Piper PJ (2007) The Javan Rhinoceros *Rhinoceros sondaicus* in Borneo. Raffles Bull Zool 55(1):217–220

Cranbrook, Earl of, Piper PJ (2008a) Sarawak through the Ice Ages to present time: environmental change and human impacts on the past and present distribution of mammals. Regional Conference on biodiversity conservation in tropical planted forests in Southeast Asia. Natural History Publications (Borneo), Kota Kinabalu, pp 81–104

Cranbrook, Earl of, Piper PJ (2008b) Post-Pleistocene evolution of Bornean shrews *Crocidura foetida* (Mammalia, Soricidae). Biol J Linn Soc Lond 94:413–419

Cranbrook, Earl of, Piper PJ (2009) Borneo records of Malay tapir, *Tapirus indicus* Desmarest: a zooarchaeological and historical review. Int J Osteoarchaeol 19(4). doi:10.1002/oa.1015

Cranbrook, Earl of, Currant AP, Davison GWH (2006) Quaternary mammal fossils in Borneo: retraction and review. Sarawak Mus J 62 (83 ns): 107–119

Cranbrook, Earl of, Payne J, Leh CMU (2007a) Origin of the elephants *Elephas maximus* L. of Borneo. Sarawak Mus J 63, no. 84(NS): 95–125

Cranbrook, Earl of, Smeenk C, Lenglet G (2007b) The Henrici collection of Bornean mammals at the Royal Belgian Institute for Natural Sciences. Bull de l'Institut R des Sci Naturelles de Belgique 77:5–22

Darlington PJ Jr (1957) Zoogeography: the geographical distribution of animals. John Wiley, New Jersey

Datan I (1993) Archaeological excavations at Gua Sireh (Serian) and Lubang Angin (Gunung Mulu National Park), Sarawak, Malaysia. Sarawak Mus J Spec Monogr No 6

Diong CH, James DA, Stuebing RS (2008) Harvest management strategy for the bearded pig *Sus barbatus* in the Planted Forest Zone, Sarawak, Malaysia. 7th International Symposium on Wild Boar (*Sus scrofa*) and on Sub-Order Suiformes. Sopron, Hungary, pp 30–32

Everett AH, Evans J, Busk G (1880) Report on the exploration of the caves of Borneo. Proc R Soc Lond 203:310–321

Fernando P, Vidya TNC, Payne J, Stuewe M, Davison G, Alfred RJ, Andau P, Bosi E, Kilbourn A, Melnick DJ (2003) DNA analysis indicates that Asian Elephants are native to Borneo and are therefore a high priority for conservation. PLoS Biol 1:1–6

Flenley JR (1998) Tropical forests under the climates of the past 30, 000 years. Clim Chang 39:177–197

Flenley JR (2005) Palynological richness and the tropical rainforest. In: Bermingham E, Dick CW, Moritz C (eds) Tropical rainforest: past, present and future. Ch. 5. Chicago University Press, Chicago, pp 72–77

Francis C (1989) Notes on fruit bats (Chiroptera, Pteropodidae) from Malaysia and Brunei with a description of a new subspecies of *Megaerops wetmorei* Taylor 1934. Can J Zool 67:2878–2882

Gibbard P, van Kolfschoten T (2004) The pleistocene and Holocene epochs. In: Gradstein FM, Ogg JG, Smith AG (eds) A Geologic Time Scale. Ch. 22. Cambridge University Press, Cambridge, pp 441–452

Groves CP (2001) Primate taxonomy. Smithsonian Institute Press, Washington

Hall LS, Richards G, Abdullah MT (2002) The bats of Niah national park, Sarawak. Sarawak Mus J 52 (NS 78):255–282

Harrison T (1996) The palaeoecological context at Niah Cave, Sarawak: evidence from the primate fauna. Indo-Pac Prehist Assoc Bull 14 (Chang Mai Papers) 1:90–100

Harrison T (1998) Vertebrate faunal remains from Madai Caves (MAD 1/28), Sabah, East Malaysia. Bull Indo-Pac Prehist Assoc 17:85–92

Harrison T (2000) Archaeological and ecological implications of the primate fana from prehistoric sites in Borneo. Indo-Pac Prehist Assoc Bull 20 (Melaka Papers) 4:133–146

Harrison T, Krigbaum J, Manser J (2006) Primate biogeography and ecology on the Sunda Shelf islands: a palaeontological and zooarchaeological perspective. Ch. 12. In: Lehman SW, Fleagle JG (eds) Primate biogeography. Springer, New York

Harrisson T (1958) The caves of Niah: a history of prehistory. Sarawak Mus J (NS 12) 8:549–595

Harrisson T (1959) New archaeological and ethnological results from Niah caves. Man 59:1–8

Harrisson T (1967) Niah caves progress report to 1967. Sarawak Mus J 15 (NS 30–31): 95–96

Hooijer DA (1952) *Palaeoloxodon* cf. namadicus (Falconer et Cautley) from Borneo. Verhandelingen Koninklijke Nederlandsche Akademie Wetenschappen afd. Natuurkunde (B) 55:395–398

Hooijer DA (1961) The orang-utan in Niah cave prehistory. Sarawak Mus J 9(1960):408–421

Hooijer DA (1962a) Report upon a collection of Pleistocene mammals from tin-bearing deposits in a limestone cave near Ipoh, Kinta valley, Perak. Fed Mus J 7 (NS):1–5

Hooijer DA (1962b) Prehistoric bones: the gibbons and monkeys of Niah Great Cave. Sarawak Mus J 10(19–20):428–449

Hooijer DA (1972) Prehistoric evidence for *Elephas maximus* L. in Borneo. Nature 239:238
Horton BP, Gibbard PL, Milne GM, Morley RJ, Purintavaragul C, Stargardt JM (2005) Holocene sea levels and palaeoenvironments, Malay-Thai Peninsula, Southeast Asia. Holocene 15(8):1199–1213
Hunt CO, Rushworth G (2005) Airfall sedimentation and pollen taphonomy in the West mouth of the Great Cave, Niah. J Archaeol Sci 32:465–473
Hunt CO, Gilbertson DD, Rushworth G (2007) Modern humans in Sarawak, Malaysian Borneo, during oxygen isotope stage 3: palaeoenvironmental evidence from the Great Cave of Niah. J Archaeol Sci 34:1953–1969
Jablonski N, Whitfort M (1999) Environmental change during the Quaternary in East Asia and its consequences for mammals. Rec West Aust Mus 57:307–315
Johnson C (2006) Australia's mammal extinctions: a 50,000 year history. Cambridge University Press, Cambridge
Junau DJ, Stuebing R, Diong CH (2008) A study of skull parameters related to the demography and sustainable use of the bearded pig (Sus barbatus) in the Planted Forest Zone. Regional Conference on biodiversity conservation in tropical planted forests in Southeast Asia. Natural History Publications (Borneo), Kota Kinabalu, pp 204–209
Lim CK, Earl of Cranbrook (2002) Swiftlets of Borneo: builders of edible nests. Natural History Publications (Borneo), Kota Kinablau, pp 1–12
Louys J, Curnoe D, Tong HW (2007) Characteristics of Pleistocene megafauna extinctions in Southeast Asia. Palaeogeogr Palaeoclimatol Palaeoecol 243:152–173
Lyman AJ, Wan LaidlawR, Shahanuddin WN, Elagupillay S, Bennett EL (2007) Assessing the conservation status of the tiger *Panthera tigris* at priority sites in Peninsular Malaysia. Oryx 41:454–462
Medway L (1958) Food bone in the Niah cave excavations (1958). A preliminary report. Sarawak Mus J 8 (NS 12):627–636
Medway L (1972) *Stegolphodon lydekkeri* Osborn: a reassessment. Sarawak Mus J 40–41: 339–350, pl. xxv–xxviii
Medway L (1977) Mammals of Borneo. Monographs of the Malaysian Branch Royal Asiatic Society 7. Kuala Lumpur
Medway L (1983) The wild mammals of Malaya (Peninsular Malaysia) and Singapore, 2nd edn. Oxford University Press, Petaling Jaya Reprinted with corrections
Morwood MJ, Sutikna T, Sapotmo EW, Westaway KE, Jatmiko Awe Due R, Moore MW, Dwi Yani Yuniawati Hadi P, Zhao J-X, Turney CSM, Fified K, Allen H, Soejono RP (2009) Climate, people and faunal succession on Java, Indonesia: evidence from Song Gupuh. J Archaeol Sci. doi: 10.1016/jas.2007.11.025
Payne J, Francis CM (2005) A field guide to the mammals of Borneo. 4th reprint. Sabah Society, Kota Kinabalu
Piper PJ, Cranbrook, Earl of (2007) Niah cave bone IX: First Pleistocene record of Hose's pigmy shrew *Suncus hosei.* Sarawak Mus J 63, no. 84 (NS): 227–232
Piper PJ, Cranbrook, Earl of (2008) The potential of large protected plantation areas for the secure reintroduction of Borneo's lost 'megafauna' : a case for the Malay tapir *Tapirus indicus.* Regional Conference on biodiversity conservation in tropical planted forests in Southeast Asia. Natural History Publications (Borneo), Kota Kinabalu, pp 182–189
Piper PJ, Cranbrook, Earl of (2009) Pleistocene and Holocene rats of Niah caves, Sarawak. Int J Osteoarchaeol 19(4)
Piper PJ, Rabett RJ (2009) Hunting in a tropical rainforest: evidence from the terminal Pleistocene at Lobang Hangus, Niah cave, Borneo, Int J Osteoarchaeol (in preparation)
Piper PJ, Cranbrook, Earl of, Rabett RJ (2007) Confirmation of the Presence of the Tiger *Panthera tigris* (L.) in Late Pleistocene and Holocene Borneo. Malayan Nature J 59: 259–265
Piper PJ, Rabett RJ, Cranbrook, Earl of (2007) New discoveries of an extinct giant pangolin (*Manis* cf. *palaeojavanica* Dubois) at Niah Cave, Sarawak, Borneo: biogeography, palaeoecology and taxonomic relationships. Sarawak Mus J 63, 84(NS): 205–226
Piper PJ, Rabett RJ, Kurui E bin (2009) Using community, composition and structural variation in terminal Pleistocene vertebrate assemblages to identify human hunting behaviour at Niah Caves, Borneo. Int J Osteoarchaeol 19(4)
Rabett RJ, Piper PJ, Barker G (2006) Bones from Hell: Preliminary results of new work on the Harrisson faunal assemblage from the deepest part of Niah Cave, Sarawak. In: Bacus EA, Glover IC, Pigott VC (eds) Uncovering Southeast Asia's Past: Selected Papers from the 10th International Conference of European Association of Southeast Asian Archaeologists. NUS Press, Singapore, pp 46–59
Roberts N (1998) The Holocene: an environmental history, 2nd edn. Blackwell, Oxford

Ruedi M (1995) Taxonomic revision of shrews of the genus Crocidura from the Sunda Shelf and Sulawesi with description of two new species (Mammalia: Soricidae). Zool J Linn Soc 115:211–265
Saegusa H (1997) Stegodontidae: evolutionary relationships. In: Shoshani J, Tassy P (eds) The Proboscidea: evolution and palaeoecology of elephants and their relatives. Oxford University Press, England, pp 178–190
Schatz RE (2007) Structural transformation and rural development in Sarawak, Malaysia, 1963–2006. Sarawak Mus J 63(84 NS): 1–40
Shelford RWC (1899) Notes from Sarawak Museum: on a fossil tooth found at Bau, Upper Sarawak. J Malay Branch R Asiat Soc 32:218–219
Storm P (2001) The evolution of humans in Australasia from an environmental perspective. Palaeogeogr Palaeoclimatol Palaeoecol 171:363–383
Storm P, Aziz F, de Vos J, Kosasih DD, Baskoro S, Ngaliman HOLW (2005) Late Pleistocene Homo sapiens in a tropical rainforest fauna in East Java. J Hum Evol 49:536–545
van den Bergh GD, de Vos J, Sondaar PY (2001) The Late Quaternary palaeogeography of mammal evolution in the Indonesian Archipelago. Palaeogeogr Palaeoclimatol Palaeoecol 171:385–408
van der Kaars WA, Dam MAC (1995) A 135,000-year record of vegetational and climatic change from the Bandung area, West Java, Indonesia. Palaeogeogr Palaeoclimatol Palaeoecol 117:55–72
Vlekke BHM (1943) Nusantara: a history of the East Indian Archipelago. Harvard University Press, Cambridge
Voris HK (2000) Maps of Pleistocene sea levels in Southeast Asia: shorelines, river systems and time durations. J Biogeogr 27:153–167
Westaway KE, Morwood MJ, Roberts RG, Rokus AD, Storm ZhaoJ-X, Aziz P, van den Bergh F, Hadi G, Jatmiko P, de Vos J (2007) Age and biostratigraphic significance of the Punung rainforest fauna, East Java, Indonesia, and implications for *Pongo* and *Homo*. J Hum Evol 53:709–717
Wilson DR, Reeder DM (eds) (2005) Mammal species of the world: a taxonomic and geographic reference. Vols. 1 & 2, 3rd edn. Johns Hopkins University Press, USA, pp 1–12
Wilson RCL, Drury SA, Chapman DL (2000) The great Ice age: climate change and life. Routledge, London and New York
WWF (2009) www.panda.org/what_we_do/where_we_work/borneo_forests

Biodivers Conserv (2010) 19:393–409
DOI 10.1007/s10531-008-9510-5

ORIGINAL PAPER

Loss of biodiversity and ecosystem functioning in Indo-Malayan peat swamp forests

Catherine M. Yule

Received: 4 May 2008 / Accepted: 20 October 2008 / Published online: 5 November 2008

Abstract The tropical peat swamp forests of Indonesia and Malaysia are unusual ecosystems that are rich in endemic species of flora, fauna and microbes despite their extreme acidic, anaerobic, nutrient poor conditions. They are an important refuge for many endangered species including orang utans. Ecosystem functioning is unusual: microbial decomposition is inhibited because the leaves are sclerophyllous and toxic to deter herbivory in the nutrient poor environment, yet bacteria are abundant and active in the surface layers of the peat, where they respire DOC leached from newly fallen leaves. The bacteria are subsequently consumed by aquatic invertebrates that are eaten by fish, and bacterially respired CO_2 is assimilated by algae, so bacteria are thus vital to carbon and nutrient cycling. Peat swamp forests are highly sensitive to the impacts of logging, drainage and fire, due to the interdependence of the vegetation with the peat substrate, which relies on the maintenance of adequate water, canopy cover and leaf litter inputs. Even minor disturbances can increase the likelihood of fire, which is the major cause of CO_2 emissions from regional peat swamp forests and which impact ecosystems worldwide by contributing to climate change. Indo-Malayan peat swamps affect the hydrology of surrounding ecosystems due to their large water storage capacity which slows the passage of floodwaters in wet seasons and maintains stream base flows during dry seasons. These forests are of global importance yet they are inadequately protected and vanishing rapidly, particularly due to agricultural conversion to oil palm, logging, drainage and annual fires.

Keywords Bacteria · Biodiversity · Carbon sequestration · Hydrology · Indonesia · Malaysia · Microbial communities · Peat fires · Tropical forests

Abbreviations

DOC	Dissolved organic carbon
UNDP	United Nations Development Programme
IUCN	International Union for Conservation of Nature

C. M. Yule (✉)
School of Science, Monash University, Sunway Campus, Jalan Lagoon Selatan, 46150 Bandar Sunway, Selangor, Malaysia
e-mail: catherine.yule@sci.monash.edu.my

Introduction

The global centre of biodiversity of tropical peat swamp forests lies in the Indo-Malayan region where these unique habitats impact the functioning of ecosystems far beyond their borders through their influence on climate and hydrology. These forests support a biological resource of global importance, yet on a daily basis they are being rapidly degraded and destroyed. Their ecology has received little attention: most research has focused on biodiversity (e.g. Anderson 1963, who also looked at vegetation structure; Latiff et al. 2005; Ng et al. 1994; Ng and Ibrahim 2001; Beamish et al. 2003; Prentice and Aikanathan 1989; Wells and Yule 2008), carbon storage (e.g. Maltby and Immirzi 1993; Sorensen 1993; Page et al. 2004; Chimner and Ewel 2005; Jauhiainen et al. 2005; Rydin and Jeglum 2006; Hooijer et al. 2006; Takashi et al. 2007) and agricultural conversion (e.g. Coulter 1957; Kanapathy 1975; Andriesse 1988; Vijarnsorn 1996). Relatively few studies have been published in peer-reviewed journals—most are reported in symposia and workshops (e.g. Aminuddin 1992; Maltby et al. 1996; Chin and Havmoller 1999; Rieley and Page 2002; Latiff et al. 2005). This paper reviews our knowledge of the present state of Indo-Malayan peat swamp forests and examines what is known of the impacts of logging, fire and agricultural conversion on biodiversity and ecosystem functioning.

The tropical peat swamp forests of the Indo-Malayan region are unusual ecosystems and they differ dramatically from north temperate and boreal peat lands which are dominated by *Sphagnum* spp., grasses, sedges and shrubs. Trees in the peat swamp forests can grow to over 70 m high (Peter Ashton, personal communication), despite the extreme conditions of low pH (2.9–4), low nutrients and anaerobic, unstable, spongy substrate of peat that may be 20 m deep or more. The forest floor is flooded during wet seasons, but even during dry seasons the peat remains waterlogged, with pools in amongst the trees (Fig. 1). Streams and rivers only flow out of these forests, so the sole nutrient inputs are from rainfall, dust and marine aerosols, otherwise river inflow would create nutrient rich, freshwater swamps. Peat swamp forests support a diverse flora and fauna with many endemic and endangered species. Unfortunately our knowledge of their flora and fauna is far from complete and these forests are vanishing before we know what we are losing.

Regional distribution and conservation status

Most of the world's tropical peatlands (about 62%) are found in the Indo-Malayan region (80% in Indonesia, 11% in Malaysia, 6% in Papua New Guinea, with small pockets and remnants in Brunei, Viet Nam, the Philippines and Thailand: Rieley et al. 1996; Page et al. 2006). These peat swamp formations are geologically recent, mostly <5,000 years old. They largely overlie marine muds and sands in the coastal lowlands of Peninsula Malaysia, and the islands of Borneo, Sumatra, Papua New Guinea, Sulawesi and the Moluccas. Some older peat swamps (up to 11,000 years old) overlie riverine sediments at slightly higher elevations in central Kalimantan in Borneo (Rieley et al. 1992; Page et al. 2006).

All regional peat swamp forests are threatened either legally or illegally with logging (e.g. for high quality timber such as ramin, *Gonystylus bancanus*), drainage, agricultural conversion (mostly to oil palm, as well as rice, rubber, coconut and pineapple), fire, fragmentation of habitats, hunting and collecting, tin mining (in Malaysia) and finally reclamation for residential centres and industries (e.g. Putrajaya and Cyberjaya—new townships south of Kuala Lumpur in Malaysia). Vast areas of peatland in Indonesian Borneo and Sumatra have been cleared for transmigration programmes. Conversion to agriculture

Fig. 1 North Selangor Peat Swamp Forest, Malaysia. Showing stilt roots and typical forest pool

involves drainage, burning and addition of lime and fertilizers which dramatically changes the characteristics of the peat substrate. Once drained, peat is highly flammable, with the most degraded peat areas being at greatest risk. Once alight, peat fires can burn for months or even years and they are very difficult to extinguish unless the natural high water table is restored. Of the original 27 million ha of peat forests in southeast Asia, by 2006, 12 million ha (45%) had been logged and almost as much had been drained (Hooijer et al. 2006). Since these estimates were made, the push for biofuels from oil palm has driven further logging and drainage, particularly in Indonesia's easternmost province of Papua, where indigenous land ownership is tenuous.

Almost all of the peat swamp forests of Peninsular Malaysia, as in all the other locations, have been destroyed or degraded over the past 50 years and none remain in a pristine condition. Precise estimates are difficult, but they have declined from about 1 million ha (Coulter 1957) to 0.67 million ha in 1981, to 0.34 million ha in 1991 (Rieley et al. 1996), to 0.30 million ha in 2005 (UNDP 2006). Since then there has been further decline and more than half of the remainder is degraded. Only about 100 ha is fully 'protected' in Virgin Jungle Reserve, but this is an area too small to ensure preservation of ecosystem functioning, since the water table cannot be maintained in its original state. About 67% is now in Permanent Forest Reserves but these are targeted for sustainable logging. The rest is available for conversion to other land uses. The only large tract of pristine peat swamp forest remaining on mainland Asia is the Pekan peat swamp forest, in the state of Pahang on the east coast of the Malaysian peninsula. Despite its high biodiversity and ecosystem value, it has been designated as Production Forest Reserve (i.e. logging is allowed). Small areas of peat swamp forest remain in Sabah, and although significant areas remain in Sarawak, they have mostly been degraded by logging.

In Indonesia, less than 3% of the remaining peat swamp forest is in protected forest (but still subject to illegal logging, drainage and agricultural conversion), the rest is available for logging and conversion to other land uses (MacDicken 2001). More than half of the original peat swamp forest in Sumatra and also Indonesian Borneo has been destroyed (Indonesia WWF 2008). Logging concessions cover 80% of the remaining peat forest in Sumatra. Only Brunei has managed to maintain much of its original peat swamp forest relatively intact, although the largest sector, in Belait district, is currently designated as production forest. Destruction of Papua's peat swamp forests has recently commenced with the drive for conversion to oil palm.

Agricultural conversion of peat forests in Kalimantan (Borneo) typically consists of drainage and forest clearing, followed by burning to prepare the seedbed (Anshari and Armiyarsih 2005). This results in a loss of 2–5 cm of peat year^{-1}, with a further 1 cm year^{-1} lost to subsidence yet the benefits are small as the ash from peat burning is low in nutrients, and the farmers abandon the land after one growing cycle. Thereafter the land becomes colonized by ferns and the ecosystem and hydrology are dramatically altered. Water holding capacity is decreased and during the dry season these open peatlands are highly flammable (Anshari and Armiyarsih 2005).

Microbial communities

It has long been assumed that waterlogging and acidity impede microbial decomposition, leading to the build up of organic matter as peat (e.g. Qualls and Haines 1990; Gorham 1991; Zakaria 1992; Whitten et al. 2000; Pahang Forestry Department 2005; UNDP 2006) but our recent studies have revealed a diverse microbial flora in Malaysian peat swamps, some of which are capable of decomposing leaf litter (e.g. Voglmayr and Yule 2006; Gomez 2007; Wantzen et al. 2008; Yule et al. 2008b; Yule and Gomez 2008; Jackson et al. 2008; Yule and Grey 2008). Studies of leaf litter decomposition in North Selangor Peat Swamp Forest (Yule and Gomez 2008) revealed that although the leaves of four endemic species barely changed after 1 year, either when submerged in the water or exposed on the peat forest floor (decay rates of only 0.0006–0.0016 k day^{-1}), leaves from a secondary forest species (*Macaranga tanarius*) broke down relatively rapidly (decay rates of 0.0047–0.005 k day^{-1}) due to microbial decomposition. Fungi grew abundantly on the endemic leaves and bacteria were also present, but there were no signs of decomposition. This shows that it is intrinsic qualities of the leaves that impede microbial decay such as the tough, leathery nature of the leaves and high levels of toxic secondary compounds (Jantzen 1974; Whitten et al. 2000; Lim et al. 2008). These adaptations to deter herbivory also prevent consumption of the leaves by shredding aquatic invertebrates, such as typically occurs in freshwaters with large leaf litter inputs including those in Malaysia (Wantzen et al. 2008; Yule et al. 2008a).

Jackson et al. (2008) used molecular techniques to study microbial communities of the North Selangor Peat Swamp Forest in Peninsular Malaysia, at depths from 0 to 50 cm. We found unusual microbial communities dominated by Acidobacteria (27–54%) and Crenarchaeota, with Archaea limited to, and dominating the deeper samples (20–50 cm). Archaea have also been shown to increase in abundance with depth in Siberian sphagnum peat bogs (Kotsyurbenko et al. 2004; Dedysh et al. 2006). Interestingly, Crenarchaeota have not previously been reported to be abundant in soils (typically only 0.5–3%:

Ochsenreiter et al. 2003). The bacterial communities showed a decrease in diversity with increasing depth, which could be related to increasingly anaerobic conditions and decreasing nutrients.

Another interesting feature of the microbial communities of the Selangor peat swamp was a lack of methanogenic bacteria (Jackson et al. 2008) such as have been reported for boreal peat swamps (e.g. Basiliko et al. 2003; Kotsyurbenko et al. 2004). However, Inubushi et al. (1998) studying peat swamp forest soils in Sarawak, found methanogenic bacteria to be active at about 15 cm deep, and methane to be accumulated in deeper peat layers (28–48 cm), although only low levels of methane were released to the atmosphere unless the soil was disturbed. A lack of methanogenic bacteria in the Selangor peat swamp was supported by stable C isotope analyses of peat down to 1 m depth (Gomez 2007; Yule and Grey 2008). These revealed depleted values of ^{13}C ($\delta^{13}C$ -32.4 to -30.4‰), but they were not the very low values typical of methane oxidising bacteria ($\delta^{13}C < 60$‰, e.g. Grey et al. 2004). Algae, invertebrates and fish from the same peat swamp also recorded lower ^{13}C values ($\delta^{13}C$ -40.4 to -32.7‰) than those typically recorded for freshwaters, indicating that bacterially respired carbon was being assimilated throughout the aquatic food web.

The importance of bacteria in the diets of aquatic invertebrates in the Selangor peat swamp was supported by gut analyses that showed the most abundant invertebrates (mayfly and chironomid larvae) mainly ingested fine particulate organic matter that was largely composed of bacteria. These in turn were eaten by carnivorous invertebrates and fish. Algae were not an important component of fish or invertebrate diets (never more than 5% of total gut contents) and neither did the fauna consume peat, leaves or fruits (Gomez 2007; Yule and Grey 2008). Thus bacteria form the base of the peat swamp food web.

Since there is a lack of methanogenic bacteria and a lack of decomposition, and since the ^{13}C signatures indicated a bacterially respired carbon source, the question remained: from where are the bacteria acquiring the carbon? The black waters of the peat swamps have high levels of dissolved organic carbon (DOC values for Selangor 71.6–84.3 mg l^{-1}, compared with <10 mg l^{-1} for most freshwaters: Gomez 2007). Laboratory studies showed that DOC rapidly leaches from newly fallen leaves but the levels then fall quite quickly, and there is a corresponding increase and decrease in bacterial abundance (Liew 2008; Yule and Gomez 2008). Thus it seems likely that bacteria utilize DOC leached from the leaves within days of their falling to the forest floor. France (1999) also found that DOC from newly shed leaves provided a substrate for bacterial growth in boreal regions. This scenario is also supported by enzymatic studies which revealed that microbial extracellular enzymes were active in carbon, phosphorus and nitrogen cycling in the upper few centimetre (the site of fresh leaf litter deposition), but declined rapidly with depth (Jackson et al. 2008). Furthermore, Weiss et al. (2002) found that levels of P, K, Ca, Mg, Si and Na were highest in the uppermost layers of a Kalimantan peat swamp also suggesting that leaching from freshly fallen leaves is the key to nutrient cycling. Following leaching of the cellular contents and other soluble materials, the recalcitrant physical matrices of the leaves remain intact, resistant to microbial decomposition and they slowly form peat which then probably contributes little to nutrient cycling in the peat swamp. Thus contrary to past suppositions, bacteria are actually vital to the functioning of the tropical peat swamp ecosystem. These peat swamps are unusual in that bacteria support the aquatic foodweb via the respiration of DOC leached from the leaves, not via a decomposition pathway.

Clearance of the peat swamp forest removes the source of freshly fallen leaves providing DOC, while burning, drainage and the application of lime and fertilizers will change the physico-chemical conditions further. Our present studies comparing the microbial communities of disturbed forests and oil palm on peat are revealing changes in the microbial communities (Liew 2008) which are likely to reflect changes in nutrient cycling. This has implications with respect to the rehabilitation of peat swamp forests such as is being attempted in various parts of the region [e.g. the Central Kalimantan Peatlands Project in Borneo run by the World Wide Fund for Nature (WWF), Borneo Orang Utan Survival Foundation, CARE, Wetlands International and the University of Palangka Raya].

Aquatic flora

A survey of the algae of the North Selangor and Pekan Peat Swamp Forests in Peninsular Malaysia recorded 58 species: mostly Chlorophyta, as well as Cyanobacteria, Chrysophyta, Bacillariophyta and Rhodophyta (Supramaniam 2005). Filamentous green algae (e.g. *Mougeotia* sp.) and red algae *Batrachospermum* sp. nov. (North Selangor) and *B. cylindrocellulare* (Pekan) are often abundant in sunny locations, such as pools in forest gaps, rivers (e.g. Bebar River) and forest edges. Algae are restricted by the tannin darkening and only occur in the top 15 cm of water. Bladderworts (*Utricularia* spp.) occur in sunny locations—reflective of the low nutrients. They were seen to ingest chironomid (midge) larvae and cladocerans. The submerged herb *Cryptocoryne cordata* (Araceae) is common in Pekan in forest pools and along the sides of the Bebar River. In regions of fast flow, the leaves often become encrusted with diatoms. Phytoplankton levels are low (levels of chlorophyll a $<0.1\ \mu g\ l^{-1}$), again reflecting low nutrients and light.

Supramaniam (2005) compared algal communities from pristine and disturbed sites, and discovered that the greatest species richness was found in an irrigation drain flowing out of the Selangor peat forest and in pools in a Pekan peat swamp forest that had been burnt 2 years previously and was still seriously degraded. Here increased light and nutrients supported more algae. Conversely, sedimentation of the Bebar River due to logging in the catchment caused a decrease in algal diversity due to decreased light penetration. Algae did not contribute significantly to the diets of fish and aquatic invertebrates in the Selangor peat swamp (Gomez 2007; Yule and Grey 2008), but they provide a habitat and refuge from predation for smaller invertebrates and may contribute to oxygenation of the water during the day.

Terrestrial flora

Due to the extreme nature of the environment, plants require many adaptations to survive such as buttress roots and stilt roots to provide stability in the soft peat substrate (Fig. 1); a thick, shallow root system (Fig. 2); knee roots and pneumatophores to enhance gas exchange in the waterlogged substrate; the root mat also enhances uptake of nutrients since they are more abundant in the surface layers; plus adaptations to reduce herbivory in the nutrient limited environment such as thorns and spikes, thick, leathery leaves, and high levels of toxic secondary compounds (e.g. resins, aromatic oils, latex, tannins and phenols: Anderson 1963). Plants growing in waterlogged sites will exhibit more stilt roots, pneumatophores and knee roots, and have thicker roots with more pore spaces, than the same species in drier sites (Pahang Forestry Department 2005). Peat swamp forests thus support a high proportion of specialized, endemic taxa, yet diversity is surprisingly high despite the inhospitable conditions. Anderson (1963) recorded 927 species of flowering plants and ferns in the peat swamp forests of Borneo. In Peninsular Malaysia 260 species of plants were recorded from the Pekan Peat Swamp Forest (Latiff 2005). Most of the tree families

Fig. 2 Bebar River, Pekan, Malaysia showing thick root mat

present in lowland dipterocarp forests are also represented in peat swamp forests, but many of the species in peat swamp forests are restricted to this habitat (Ng and Ibrahim 2001).

Several authors have reported marked variation in the floristic composition and forest structure of peat swamp forests that are related to hydrology, nutrient status and thickness of the peat (e.g. Anderson 1963; Whitmore 1984; Page et al. 1999). Page et al. (1999) found the tallest forest (canopy up to 45 m) growing on the thickest, oldest peat in a forest in central Kalimantan, with younger, thinner peats supporting low pole forest (up to 20 m high) and mixed swamp forest (upper canopy 35 m). In general, the taller forests with higher floral diversity, provide a wider variety of food sources and microhabitats and support a higher diversity of fauna. These forests typically have four main layers, emergent, canopy, middle and understorey, with seedlings, sedges and ferns on the forest floor. *Pandanus* spp. are often common in the understorey and along riverbanks, where stands can grow so thickly that they block water flow. Disturbed forests can be dominated by *Macaranga pruinosa*, and heavily logged forests may regrow with uniform stands of this species (Fig. 3).

Regional peat swamp forests are primarily managed for timber production (Ng and Ibrahim 2001). In the past, prior to the introduction of mechanization, trees were selectively logged, and only trees larger than 45 cm dbh of high commercial value were taken (Ng and Ibrahim 2001). Selective logging thus removes emergent trees, and when it is extensive it can reduce the canopy height, reduce diversity (pioneer species colonize gaps) and decrease the stratification of forest layers. Mechanization was introduced in 1964. Railway lines were constructed to remove logs by train, and then in the 1980s canal systems, 3 m wide and 4 m deep, were developed for log removal (Ng and Ibrahim 2001). These leave open scars that last for decades and are usually used in peat forests that are to be converted to agriculture. Canals cause drainage of the peat swamps which alters hydrology and this adversely impacts the peat substrate and vegetation, and makes them more susceptible to fires (Hooijer 2006). Forest burning eliminates seeds and seedlings. Regeneration is further impeded because fire destroys the peat as well as the vegetation, reducing water holding capacity, so that the floor becomes flooded (Fig. 4) and seedlings, particularly of trees, cannot re-establish. Lack of trees to provide leaf litter means that the peat will not readily build up again. Logged forests are many times more likely to burn than unlogged forests due to the open canopy, drier microclimate and logging debris to fuel fires (Brown 1998).

Fig. 3 North Kuala Langat Peat Swamp Forest, Malaysia. Regrowth of *M. pruinosa* following logging

Fig. 4 Pekan Peat Swamp Forest 2 years after a fire

Another impact of disturbance on the terrestrial flora is the introduction of weeds. A study in Peninsular Malaysia by Asyraf and Mansor (2002) recorded 130 species of weeds including noxious species such as *Salvinia molesta* and *Mimosa pigra* in oil palm plantations on peat land, 61 weed species from disturbed peat swamps and only 12 species from pristine peat swamps. Thus pristine sites are resistant to invasion by introduced species, but there is a decrease in resistance with increasing disturbance.

Aquatic invertebrates

Aquatic invertebrates are poorly known, but Malaysian peat swamps harbour a diverse fauna, although lacking molluscs, leeches and most taxa restricted to rocky substrates and flowing waters in the tropics such as stoneflies. There are many new and undescribed species, including a chironomid with gills—an adaptation to the anaerobic conditions, and the first known in the world (Gomez 2007; Wells and Yule 2008). Gomez (2007) and Lim (2003) recorded 83 taxa from pools and drains in the North Selangor Peat Swamp Forest. Chironomids, baetid mayflies and cladocerans were the most abundant taxa while predatory Anisoptera, Zygoptera and Hemiptera were also common. The Hemiptera are generalists that are found in a wide variety of aquatic habitats in the region, and since they are usually present in the swamps as winged adults, they can escape disturbances. However, many of the species appear to be restricted to peat swamp forests and they will become extinct if the peat swamps are destroyed.

Fish

A total of 200–300 fish species are recorded from the peat swamps of Peninsular Malaysia, Borneo and Sumatra (Dennis and Aldhous 2004) with 65 species recorded from Pekan, and 47 from North Selangor Peat Swamp Forest (Ng et al. 1992, 1994). The Asian arowana (*Scleropages formosus*) a popular aquarium fish occurs in blackwater rivers in the regional peat swamps. The world's smallest known fish (*Paedocypris progenetica*), <8 mm long, was recently discovered in the peat swamp forests of Sumatra (Kottelat et al. 2006). These forests have been degraded by logging, fires and drainage and several populations of *Paedocypris* have already been exterminated.

Past logging of the Selangor forest has not adversely affected the fish fauna significantly (Ng et al. 1994), but recent extension and deepening of drains may have significant impacts. A large number of the fish species seem to be generally fairly resistant to perturbations, being able to survive in irrigation canals and roadside ditches containing water draining from the peat swamps, but when the peat swamps are completely destroyed, the fish vanish (Ng et al. 1994). During high water periods, fish can leave the main rivers and breed when the forest floor is flooded and nutrient levels are higher, so a decrease in water levels can negatively impact fish life histories (Pahang Forestry Department 2005). The fish are largely carnivorous eating aquatic invertebrates, mostly chironomids and mayfly larvae, and other fish (Chung 1996). Fish in the Bebar River in Pekan are important in the diets of the indigenous Orang Asli.

Vertebrate fauna

Regional peat swamp forests are increasingly becoming refuges for globally significant species from other lowland forests, which are under even greater pressures from logging, hunting and development. Elephants, tapirs, leopards, rhinoceros, orang utans, proboscis monkeys and tigers all live in peat swamp forests. Most of the vertebrate fauna are not restricted to this habitat, apart from the tri-coloured langur, now found only in a small patch of degraded forest in Maludam National Park in Sarawak, and the Borneo roundleaf bat, hairy-nosed otter, hook-billed bulbul and cinnamon-headed pigeon, which are largely confined to peat swamp forests. Sebastian (2002) recorded 57 mammal species and 237 bird species from Malaysian peatswamp forests (excluding Chiroptera and Muridae). Of these, 51% of the mammals and 27% of the birds are on the IUCN red list of globally threatened species. The bird fauna comprise 38% of the overall Malaysian avifauna excluding seabirds. Not surprisingly given the wet season flooding, ground dwelling birds such as pheasants and understorey insectivores are poorly represented in peat swamp forests (Sebastian 2002).

Peat swamp forests are the most important remaining habitats for orang utans, classified as Critically Endangered on the 2002 IUCN red list. The largest known surviving population of orang utans inhabits the catchment of the Sebangau river in central Kalimantan, Indonesia, an area predominantly covered by peat swamp forest (Morrogh-Bernard et al. 2002). In 2001 the orang utan population here was estimated to be about 6,000, but the area is designated as production forest, has no protective status and is subject to extreme levels of illegal logging. Their survival is in extreme jeopardy.

The reptile and amphibian fauna of regional peat swamp forests are not well known. A survey of amphibians at Pekan revealed 19 species (<30% of the species known from Malaysia) of which two are on the IUCN red list as 'near threatened' (Norhayati et al. 2005). Sharma et al. (2005) studied reptiles at Pekan and encountered 22 species, including four species of freshwater turtles. It is likely that there are many species of amphibians and reptiles unrecorded owing to the limited surveys. Sebastian (2002) states that specialist species are the most threatened vertebrates, with generalist species surviving in the increasingly degraded habitats. The loss of specialist species is likely to have major impacts on ecosystem function, particularly those species which have roles in seed dispersal and as top predators. Preservation of the peat swamp forests is of high priority in maintaining a genetic resource pool for endangered vertebrate species.

Role of regional peat swamp forests in the global carbon cycle

The major role of Indo-Malayan peat swamp forests in carbon sequestration is now well established (e.g. Page et al. 2002, 2006; Aldhous 2004; UNDP 2006). Carbon is stored in peat that ranges in depth from less than a meter up to 25 m deep (recorded in Borneo), depending on location and age (Page et al. 1999; Whitten et al. 2000; Hooijer 2006). It has been estimated that 5,800 tonnes of carbon ha^{-1} can be stored in a 10 m deep peat swamp forest compared to 300–500 tonnes ha^{-1} for other types of tropical forest (UNDP 2006). Page et al. (2006) estimate that regional peat swamp forests store 16–21% of the total peat carbon even though they form only 12% of the total peat area, and furthermore they contain 2–3% of total soil carbon (Maltby and Immirzi 1993; Page et al. 2002). Hooijer et al. (2006) state that they store at least 42,000 Megatonnes of soil carbon. Tropical peat typically forms at a rate of 2–5 mm $year^{-1}$ (Sorensen 1993; Pahang Forestry Department 2005), which is much faster than in temperate and boreal peatlands (0.2–1 mm $year^{-1}$ reported in Page et al. 2006) due to the higher productivity and greater amount of litter produced by tropical forests compared with temperate moss and grasses.

Peat fires are common throughout the region, and are a particular threat to ecosystem functioning. They are most frequent during dry periods, typically coinciding with El Niño events. In 1997 more than 2.7 million ha in the region burnt and were the cause of the largest recorded increase in global CO_2 levels since records began in the 1950s (13–40% of the global yearly emissions due to fossil fuel combustion) (Page et al. 2002; Aldhous 2004). The El Niño event of 1982–1983 resulted in the largest forest fire ever recorded, where 4 million ha of forest burnt in Kalimantan and Sabah (Brown 1998). Since the peat is up to

20 m deep, it will be subject to burning for years to come unless drainage canals are blocked and the land is reforested.

Over 1 million ha of peat swamp forests in central Kalimantan (Indonesian Borneo) were cleared for the Mega Rice Project which was a complete failure (Aldhous 2004). These abandoned lands burn annually due to clearing and accidents. Aerobic decomposition and subsidence of peat following agricultural conversion is inevitable, with subsidence rates as high as 1 m in 2 years recorded in Sarawak and Peninsular Malaysia (Pahang Forestry Department 2005). The peat soils eventually completely oxidise, exposing acid sulphate soils beneath which do not support agriculture (Brown 1998).

Hooijer et al. (2006) estimated that between 1997 and 2006, regional peatland fires caused average yearly CO_2 emissions of 1,400 Mt (with 90% of this originating in Indonesia, mostly from Riau in Sumatra). This is equivalent to almost 8% of global emissions from fossil fuel burning. A further 600 Mt year^{-1} is emitted by Indonesia due to the decomposition of drained peat. Following conversion to oil palm plantations, the water is typically drained to a depth of 1 m which causes degradation of peat of about 10 cm year^{-1} which causes emissions of 130–180 tonnes CO_2 year^{-1}. Thus including the loss due to deforestation, about 3,750–5,400 tonnes of CO_2 will be emitted for each hectare of peatland converted to oil palm over the next 25 years (Pearce 2007). If the palm oil is used for biofuel, this will result in 25–36 times more CO_2 emitted from the peatland as would be saved from vehicle emissions. Consequently the total peatland emissions place Indonesia 3rd after China and USA in global CO_2 emissions. Without inclusion of peatland emissions they rank 21st. The emissions show no signs of abating. Instead with the present push from Europe for biofuels, conversion of peat forests to oil palm is increasing.

Air pollution

Another major impact due to peat fires with far reaching effects on other ecosystems is air pollution. Adverse effects on human health in the region have been well documented (Brown 1998; Heil and Goldammer 2001; Sastry 2002). At least 20 million people were exposed to dangerously high levels of air pollution during the 1997 fires, with dramatic increases in asthma, bronchitis and other respiratory illnesses (Brown 1998), and no doubt fauna were similarly affected. In Borneo and Sumatra levels of suspended particulate matter up to 4,000 $\mu g/m^3$ were recorded—15 times higher than the Indonesian air quality standards (Heil and Goldammer 2001). Fine particulate matter is highly carcinogenic. Regional peat fires produce aerosols that have considerably greater light-scattering properties than other forest fires (Gras et al. 1999). This is largely due to the high SO_2 content (almost 10 times above the natural levels), apparently because of long term Indonesian volcanic emissions enriching the peat with sulphur (Langmann and Graf 2003). Langmann and Graf (2003) suggest that Indonesian peat swamps receive wet deposition of sulphur at a rate of 200–500 tonnes S/km^2 1,000 year^{-1}. Resulting air pollution in Malaysia during peat fires have caused values of 48 ppb SO_2 compared to background levels of <5 ppb (Davies and Unam 1999).

Despite elevated CO_2 levels resulting from the fires, this does not enhance tree growth. Davies and Unam (1999) found photosynthesis in three tree species in Borneo was reduced by the smoke-haze due to direct effects of air pollution and indirect effects of reduced photosynthetically active radiation (PAR) levels due to the smoke darkened skies. Hazy skies caused a 92% reduction in PAR levels compared to clear days, and this would greatly decrease fixation of forest-level C throughout the region.

Hydrology

Peat can hold 5–10 times its weight in water (Sabah Forestry Department 2005), so pristine peat swamp forests can store enormous amounts of water. But the popular perception, that peat swamps are like sponges that store flood waters in the wet season and release it during the dry, is not strictly correct. Their high water table and waterlogged conditions mean that, even after the dry season, when water levels may be 40–90 cm below ground level, their capacity to absorb water is limited to about 20–30 mm of rainfall (Pahang Forestry Department 2005; UNDP 2006). However, due to their flat topography and poor drainage, during the wet season they store water above the forest floor (typically about 50 cm deep in Pekan peat swamp forest). Thus they act as a buffer reducing the velocity of the water and minimizing downstream flooding impacts. This water is released gradually, maintaining the base flows of rivers during droughts and reducing saline intrusion up the rivers (UNDP 2006).

The acidity and colour of peat swamp water make it unsuitable for potable use, but peat swamp forests are often drained to supply water for irrigation (e.g. water from the North Selangor Peat Swamp Forest is used to irrigate oil palms and rice). When peatlands are drained, the aerobic conditions lead to breakdown and subsidence of the peat, and water holding capacity is reduced to between 2 and 6 times their weight (Boetler 1969), so the amount of water that can be retained is far less than in the original peat forest and this will decrease the nutrients available to plants (UNDP 2006) and lead to the likelihood of fire. Wösten et al. (2006) studying the relationships between hydrology and fire in a Sumatran peat swamp forest showed that when groundwater levels fell more than 1 m below the soil surface, fire was inevitable. Furthermore, if the vegetation and peat are burnt, high groundwater levels create lakes which prevent the reestablishment of plants, particularly tree species.

Rehabilitation of regional peatlands

Several efforts at rehabilitation of peat swamp forests that have been degraded by drainage, fire and logging or converted to agriculture are under way in central Kalimantan (Australian Orangutan Project 2008; Dohong 2008). Rehabilitation of the Mega Rice Project in Borneo (described by Dohong 2008) is supported by Presidential Instruction No 2/2007, and it aims to restore and conserve 80% of the former Mega Rice area. Rehabilitation involves restoration of hydrology through dam construction, blocking of the canals that were built to drain the peatlands, followed by replanting with indigenous species of commercial value (such as *Dyera lowii*, *Alstonia pneumatophora*, and *Garcinia* spp. which produce latex that is used in chewing gum). Use of commercial species such as these, encourages the local people to conserve the forests in perpetuity. Furthermore, these species have high levels of toxic secondary compounds, which render them resistant to pests, and impede decomposition of their leaf litter and thus help to rebuild the peat layers.

A major feature of rehabilitation is dam construction to restore the peat swamp hydrology. The indigenous inhabitants construct a variety of dams using traditional techniques. The dams slow water flow by up to 55% and prevent leaf litter and plant debris from being washed away, so that gradually the canals become filled in naturally. The dams are constructed of mainly of timber (sometimes with sandbags for added strength) and they can become increasingly permanent as they trap plant debris and become vegetated.

Since peat subsidence occurs following agricultural conversion, logging and fires, flooding of degraded areas can impede seedling regeneration, so various mounting systems are being utilized to keep the seedlings from being submerged. Sites subject to mild degradation (logged over forest where parent trees still remain) can be left to regenerate naturally from wilding and seeds available on site. Where logging has been more intensive, and parent trees are rare and canopy cover reduced, enrichment planting of commercially important species is undertaken. Where the areas have been burnt, degradation is severe and the rehabilitation efforts are more intensive, extensive replanting on mounting systems is needed. Extremely degraded areas result from repeated burning and here the prospects for rehabilitation are low. Monitoring of all rehabilitation efforts is important in order to maximise its potential and learn from mistakes. So far rehabilitation is showing a great deal of promise and for mildly degraded sites ecosystem functioning is expected to return with the restoration of hydrological conditions and full forest canopy.

Research into the restoration of peat lands is also ongoing in Sumatra where 160,000 ha have been cleared for *Acacia* pulp plantations (Gunawan et al. 2007, 2008). Here they have found that regeneration of trees is constrained by the degree of disturbance (particularly with respect to hydrology), depth of peat, seed availability, predation of seeds and seedlings and also competition with other plant species. Where mother trees are still available and the hydrological conditions are not severely altered, then natural regeneration may be possible, but human intervention is required when disturbance is severe and seedlings and seed sources are scarce or absent. Remnant strips of natural forest (up to 2 km long and 0.5 km wide) border the *Acacia* plantations, having been left as windbreaks. The ecosystem functioning of these green belts is severely impacted by lowering of the water table down to 120–150 cm—instead of the natural minimum of 50 cm, raising of pH to 4.5–6.2, high light levels, planting and harvesting activities in the plantations. Gunawan et al. (2007, 2008) found that regeneration results in a dominance of secondary forest species tolerant of these factors, but many of the original canopy species (e.g. *Shorea* spp.) cannot regenerate under these conditions.

Conclusions

The need to conserve pristine peat swamps in the Indo-Malayan region is clearly urgent. They support many unique animals, plants and microbes and they are the last refuge for many endangered species. Their ecosystem functioning is unusual in that bacteria are vital to carbon cycling, but they do not utilize a decomposition pathway. Tropical peat swamp ecosystems are neither resistant nor resilient to perturbations, particularly logging, drainage and fires. Their fragility is due to the dependence of the entire ecosystem on the peat substrate which in turn relies on the maintenance of adequate water, canopy cover and leaf litter inputs. Combustion of peat is the major cause of CO_2 emissions from regional peat swamp forests, but clearly even slight disturbance of peat results in release of stored carbon, and thus impacts ecosystems worldwide by contributing to climate change. The present drive to convert peatlands to oil palm for biofuel is misguided since destruction of peat forests results in CO_2 emissions up to 36 times higher than the amounts saved by using the biofuels. Far greater efforts must be given to peat swamp forest protection and rehabilitation.

Acknowledgments I wish to thank the following collaborators in my studies of Malaysian peat swamps: Dr. Colin Jackson (University of Mississippi, USA: microbial analyses); Dr. Jonathan Grey (Queen Mary University, London, UK: stable isotopes); Dr. Alice Wells (Australian Biological Resources Study, Canberra, Australia: Trichoptera), Dr. Hermann Voglmayr (University of Vienna, Austria: fungi) and students at

Monash University: Lalita Gomez, Chew Eng Chuan, Melissa Chin, John Lim, Meera Supramaniam, Carol Ng, Tan Kian Yong, Khing Su Li, Liew Kong Cheng and Lim Tse Yuen. I am grateful to Dr. Charles Clarke, Dr. Valerie Yule and three reviewers for helpful comments on the manuscript. This work has been supported by grants from Monash University, Sunway Campus, Malaysia.

References

Aldhous P (2004) Land remediation: Borneo is burning. Nature 432:144–146. doi:10.1038/432144a

Aminuddin BY (ed) (1992) Proceedings of the international symposium on tropical peatland, Malaysian Agricultural Research and Development Institute and Department of Agriculture, Sarawak

Anderson JAR (1963) The flora of the peat swamp forests of Sarawak and Brunei. Including a catalogue of all recorded species of flowering plants, ferns and fern allies. Gard Bull Singap 29:131–228

Andriesse JP (1988) Nature and management of tropical peat soils. FAO Soils Bulletin No. 59. Food and Agriculture Organization of the United Nations, Rome

Anshari GZ, Armiyarsih (2005) Carbon decline from peatlands and its implications on livelihood security of local communities. In: Mudiyarso D and Herawati H (eds) Carbon Forestry. Who will benefit? Proceedings of a workshop on carbon sequestration and sustainable livelihoods. Centre for International Forestry Research, Bogor Barat. http://www.cifor.cgiar.org/publications/pdf_files/Books/BMurdiyarso0501.pdf

Asyraf M, Mansor M (2002) The diversity and composition of weeds in disturbed and undisturbed peat swamp forest in Peninsula Malaysia. In: Rieley J, Page S (eds) Peatlands for people: natural resources function and sustainable management. Proceedings of the international symposium on tropical peatland, BPPT and Indonesian Peat Association, Jakarta, pp 7–14, 22–23 Aug 2001

Australian Orangutan Project (2008) Hydrological restoration in the sebangau peat swamp forest, Central Kalimantan. URL: http://www.orangutan.org.au/projects.html

Basiliko N, Yavitt JB, Dees PM, Merkel SM (2003) Methane biogeochemistry and methanogen communities in two northern peatland ecosystems, New York state. Geomicrobiol J 20:563–577. doi:10.1080/713851165

Beamish FWH, Beamish RB, Lim SL-H (2003) Fish assemblages and habitat in a Malaysian Blackwater peat swamp. Environ Biol Fishes 68:1–13. doi:10.1023/A:1026004315978

Boetler DH (1969) Physical properties of peat as related to degree of decomposition. Soil Sci Soc Am Proc 33:606–609

Brown N (1998) Out of control: fires and forestry in Indonesia. Trends Ecol Evol 13:41

Chimner RA, Ewel KC (2005) A tropical freshwater wetland: II. Production, decomposition, and peat formation. Wetlands Ecol Manage 13:671–684

Chin TY, Havmoller P (eds) (1999) Sustainable management of peat swamp forest in Peninsular Malaysia-resource and environment, vol I. Forestry Department of Peninsular Malaysia, Kuala Lumpur

Chung DSL (1996) Feeding ecology of blackwater anabantoid and channoid fishes of north Selangor peat swamp forest, Peninsular Malaysia. Dissertation, National University of Singapore

Coulter JK (1957) Development of the peat soils of Malaya. Malay Agric J 40:188–199

Davies SJ, Unam L (1999) Smoke-haze from the 1997 Indonesian forest fires: effects on pollution levels, local climate, atmospheric CO_2 concentrations, and tree photosynthesis. For Ecol Manag 124:137–144

Dedysh SN, Pankratov TA, Belova SE, Kulichevskaya IS, Liesack W (2006) Phylogenetic analysis and in situ identification of bacteria community composition in an acidic sphagnum peat bog. Appl Environ Microbiol 72:2110–2117. doi:10.1128/AEM.72.3.2110-2117.2006

Dennis C, Aldhous P (2004) A tragedy with many players. Nature 430:396–398. doi:10.1038/430396a

Dohong A (2008) Community-based peat swamp restoration in Block A north of the ex mega rice project in central Kalimantan. Paper presented at the Symposium on peat swamp forests. Towards enhancement of knowledge-based decision making in conservation and management of natural resources. Kuala Lumpur, 11–12 Mar 2008

France RL (1999) Relationships between DOC concentration and epilothon stable isotopes in boreal lakes. Freshw Biol 41:101–105. doi:10.1046/j.1365-2427.1999.00391.x

Gomez LN (2007) Leaf litter decomposition in a tropical peat swamp forest in Peninsular Malaysia. M.Sc. Dissertation, Monash University, Malaysia

Gorham E (1991) Northern peatlands: role in the carbon cycle and probable responses to climate warming. Ecol Monogr 1(2):182–195

Gras JL, Jensen JB, Okada K, Ikegami M, Zaizen Y, Makino Y (1999) Some optical properties of smoke aerosol in Indonesia and tropical Australia. Geophys Res Lett 26:1393–1396. doi:10.1029/1999GL900275

Grey J, Waldron S, Hutchinson R (2004) The utility of carbon and nitrogen isotope analyses to trace contributions from fish farms to the receiving communities of freshwater lakes: a pilot study of Esthwaite Water, UK. Hydrobiologia 524:253–262. doi:10.1023/B:HYDR.0000036144.49795.dd

Gunawan H, Muhammad A, Qomar N, Helentina, Hakim A, Yanti MM, Darmasanti, Lumri, Lisda, Petra (2007) The role of natural peat swamp forest remnants in maintaining native trees diversity in a timber estate, Riau. Paper presented at the international symposium and workshop on tropical peatlands "carbon-climate-human interactions: carbon pools, mitigation, restoration and wise use". Yogyakarta, 27–29 Aug 2007

Gunawan H, Page S, Muhammad A, Qomar N, Helentina, Hakim A, Yanti MM, Darmasanti (2008) Peat swamp forest regeneration using green belts in a timber estate in Riau, Sumatra, Indonesia. Paper presented at the international seminar on the role of biology in sustainable utilization of local natural resources for environmentally friendly industrial purposes. North Sumatra

Heil A, Goldammer JG (2001) Smoke-haze pollution: a review of the 1997 episode in southeast Asia. Reg Environ Chang 2:24–37. doi:10.1007/s101130100021

Hooijer A (2006) Tropical peatlands in southeast Asia. In: Rydin H, Jeglum JK (eds) The biology of peatlands. Oxford University Press, Oxford, pp 233–239

Hooijer A, Silvius M, Wösten H, Page S (2006) PEAT-CO2, Assessment of CO2 emissions from drained peatlands in SE Asia. Delft Hydraulics Report Q3943

Indonesia WWF (2008) Deforestation, forest degradation, biodiversity loss and CO_2 emissions in Riau Sumatra, Indonesia. WWF Indonesia, Technical Report, 27 Feb 2008. wwf.wwf.or.id

Inubushi K, Hadi A, Okazaki M, Yonebayashi K (1998) Effect of converting wetland forest to sago palm plantations on methane gas flux and organic carbon dynamics in tropical peat soil. Hydrol Process 12:2073–2080. doi:10.1002/(SICI)1099-1085(19981030)12:13/14<2073::AID-HYP720>3.0.CO;2-K

Jackson CR, Liew KC, Yule CM (2008) Structural and functional changes with depth in microbial communities in a tropical Malaysian peat swamp forest. Microb Ecol. doi: 10.1007/s00248-008-9409-4

Jantzen DH (1974) Tropical blackwater rivers, animals and mast fruiting by the Dipterocarpaceae. Biotropica 6:69–103. doi:10.2307/2989823

Jauhiainen J, Takahashi H, Heikkinen JEP, Martikainen PJ, Vasanderet H (2005) Carbon fluxes from a tropical peat swamp forest floor. Glob Chang Biol 11:1788–1797. doi:10.1111/j.1365-2486.2005.001031.x

Kanapathy K (1975) Growing crops on peat soils. Ministry of Agriculture and Rural Development, Malaysia

Kotsyurbenko OR, Chin K-J, Glagolev MV, Stubner S, Simankova MV, Nozhevnikova AN, Conrad R (2004) Acetoclastic and hydrogenotrophic methane production and methanogenic populations in acidic West-Siberian peat bogs. Environ Microbiol 6:1159–1173. doi:10.1111/j.1462-2920.2004.00634.x

Kottelat M, Britz R, Tan HH, Witte K-E (2006) *Paedocypris*, a new genus of southeast Asian cyprinid fish with a remarkable sexual dimorphism, comprises the world's smallest vertebrate. Proc R Soc Lond B Biol Sci 273:895–899. doi:10.1098/rspb.2005.3419

Langmann B, Graf HF (2003) Indonesian smoke aerosols from peat fires and the contribution from volcanic sulfur emissions. Geophys Res Lett 30:1547–1553. doi:10.1029/2002GL016646

Latiff A (2005) An overview of the significant findings of the biodiversity expedition to the peat swamp forest of Sungai Bebar, Pahang. In: Latiff A, Hamzah KA, Ahmad N, Said MNM, Toh AN, Gill SK (eds) Biodiversity expedition Sungai Bebar, Pekan, Pahang, summary findings. Peat swamp forest project, UNDP/GEF funded, in collaboration with the Pahang Forestry Department and University Kebangsaan Malaysia

Latiff A, Hamzah KA, Ahmad N, Said MNM, Toh AN, Gill SK (2005) Biodiversity expedition Sungai Bebar, Pekan, Pahang, summary findings. Peat swamp forest project, UNDP/GEF funded, in collaboration with the Pahang Forestry Department and University Kebangsaan Malaysia

Liew KC (2008) Microbial communities of a tropical Malaysian peat swamp forest. Dissertation, Monash University, Malaysia

Lim JEH (2003) A Study of the macroinvertebrate fauna of a tropical peat swamp with consideration of the impact of disturbance. Dissertation, Monash University, Malaysia

Lim TY, Lim YY, Yule CM (2008) Evaluation of antioxidant, antibacterial and anti-tyrosinase activities of four *Macaranga* species. Food Chem. doi:10.1016/j.foodchem.2008.09.093

MacDicken KG (2001) Cash for tropical peat: land use change and forestry projects for climate change mitigation. In: Rieley J, Page S (eds) Peatlands for people: natural resources function and sustainable management. Proceedings of the international symposium on tropical peatland. BPPT and Indonesian Peat Association, Jakarta, pp 1–6, 22–23 Aug 2001

Maltby E, Immirzi CP (1993) Carbon dynamics in peatlands and other wetland soils—regional and global perspectives. Chemosphere 27:999–1023. doi:10.1016/0045-6535(93)90065-D

Maltby E, Immirzi CP, Safford RJ (eds) (1996) Tropical lowland peatlands of southeast Asia. Proceedings of a workshop on integrated planning and management of tropical lowland peatlands held at Cisarua. IUCN, Gland, 3–8 July 1992

Morrogh-Bernard H, Husson S, Page SE, Rieley JO (2002) Population status of the Bornean orang-utan (*Pongo pygmaeus*) in the Sebangau peat swamp forest, central Kalimantan, Indonesia. Biol Conserv 110:141–152. doi:10.1016/S0006-3207(02)00186-6

Ng TP, Ibrahim S (2001) Common trees in peat swamp forests of Peninsular Malaysia. Forest Research Institute Malaysia, Kepong

Ng PKL, Tay JB, Lim KKP, Yang CM (1992) The conservation of fish and other aquatic fauna of the North Selangor Peat Swamp Forest and adjacent areas. Asian Wetland Bureau Publication No 81

Ng PKL, Tay JB, Lim KKP (1994) Diversity and conservation of blackwater fishes in Peninsular Malaysia, particularly in the north Selangor peat swamp forest. Hydrobiologia 285:203–218. doi:10.1007/BF00005667

Norhayati A, Juliana S, Sharma DSK, Ahmad Zafir AW (2005) A survey of amphibians at Sungai Bebar, Pekan Forest Reserve. In: Latiff A, Hamzah KA, Ahmad N, Said MNM, Toh AN, Gill SK (eds) Biodiversity expedition Sungai Bebar, Pekan, Pahang, summary findings. Peat swamp forest project, UNDP/GEF funded, in collaboration with the Pahang Forestry Department and University Kebangsaan Malaysia, pp 149–152

Ochsenreiter T, Selesi D, Quaiser A, Bonch-Osmolovskaya L, Schleper C (2003) Diversity and abundance of Crenarchaeota in terrestrial habitats studied by 16S RNA surveys and real time PCR. Environ Microbiol 5:787–797. doi:10.1046/j.1462-2920.2003.00476.x

Page SE, Rieley JO, Shotyk W, Weiss D (1999) Interdependence of peat and vegetation in a tropical swamp forest. Philos Trans R Soc B Biol Sci 354:1885–1897. doi:10.1098/rstb.1999.0529

Page SE, Siegert F, Rieley JO, Boehm H-DV, Jaya A, Limin SH (2002) The amount of carbon released from peat and forest fires in Indonesia during 1997. Nature 420:61–65. doi:10.1038/nature01131

Page SE, Wüst RAJ, Weiss D, Rieley JO, Shotyk W, Limin SH (2004) A record of late Pleistocene and Holocene carbon accumulation and climate change from an equatorial peat bog (Kalimantan, Indonesia): implications for past, present and future carbon dynamics. J Quat Sci 19:625–635. doi:10.1002/jqs.884

Page SE, Rieley JO, Wüst R (2006) Lowland tropical peatlands of southeast Asia. In: Martini IP, Martínez Cortizas A, Chesworth W (eds) Peatlands: evolution and records of environmental and climate changes. Elsevier, The Netherlands, pp 145–172

Pahang Forestry Department (2005) Pekan peat swamp forest, Pahang, Malaysia. The role of water in conserving peat swamp forests. Pahang Forestry Department, Danida, UNDP/GEF

Pearce F (2007) Bog barons: Indonesia's carbon catastrophe. New Sci 2632:50–53

Prentice C, Aikanathan S (1989) A preliminary faunal survey of the north Selangor peat swamp forest. Asian Wetland Bureau/WWF Malaysia, Report No. 46d. Kuala Lumpur

Qualls RG, Haines BL (1990) The influence of humic substances on the aerobic decomposition of submerged leaf litter. Hydrobiologia 206:133–138. doi:10.1007/BF00018639

Rieley J, Page S (eds) (2002) Peatlands for people: natural resources function and sustainable management. Proceedings of the international symposium on tropical peatland. BPPT and Indonesian Peat Association, Jakarta, 272 pp, 22–23 Aug 2001

Rieley JO, Sieffermann RG, Page SE (1992) The origin, development, present status and importance of the lowland peat swamp forests of Borneo. Suo 43:241–244

Rieley JO, Ahmad-Shah AA, Brady MA (1996) The extent and nature of tropical peat swamps. In: Maltby E, Immirzi CP, Safford RJ (eds) Tropical lowland peatlands of southeast Asia. Proceedings of a workshop on integrated planning and management of tropical lowland peatlands held at Cisarua, IUCN, Gland, 3–8 July 1992

Rydin H, Jeglum JK (eds) (2006) The biology of peatlands. Oxford University Press, Oxford

Sabah Forestry Department (2005) Klias peat swamp forest Sabah, Malaysia. Hydrological processes and strategies for water management. Sabah Forestry Department, UNDP/GEF and Danida, Sabah Malaysia

Sastry N (2002) Forest fires, air pollution, and mortality in southeast Asia. Demography 39:1–23. doi:10.1353/dem.2002.0009

Sebastian A (2002) Globally threatened mammal and bird species in Malaysian peat swamp forests. In: Rieley J, Page S (eds) Peatlands for people: natural resources function and sustainable management. Proceedings of the international symposium on tropical peatland. BPPT and Indonesian Peat Association, Jakarta, pp 23–28, 22–23 Aug 2001

Sharma DSK, Ahmad Zafir AW, Norhayati A, Juliana S (2005) A note on the reptiles at Sungai Bebar, Pekan Forest Reserve, Pahang In: Latiff A, Hamzah KA, Ahmad N, Said MNM, Toh AN, Gill SK (eds) Biodiversity expedition Sungai Bebar, Pekan, Pahang, summary findings. Peat swamp forest project,

UNDP/GEF funded, in collaboration with the Pahang Forestry Department and University Kebangsaan Malaysia, pp 153–155
Sorensen KW (1993) Indonesian peat swamp forests and their role as a carbon sink. Chemosphere 27:1065–1082. doi:10.1016/0045-6535(93)90068-G
Supramaniam M (2005) A survey of the benthic algae of southeast Pahang and north Selangor tropical peat swamp forests, Malaysia. Dissertation, Monash University, Malaysia
Takashi H, Segah H, Harada T, Limin S, June T, Biotrop S, Hirata R, Osaki M (2007) Carbon dioxide balance of a tropical peat swamp forest in Kalimantan, Indonesia. Glob Chang Biol 13(2):412–425. doi:10.1111/j.1365-2486.2006.01301.x
UNDP (2006) Malaysia's peat swamp forests. Conservation and sustainable use. United Nations Development Program (UNDP), Malaysia
Vijarnsorn P (1996) Peatlands in southeast Asia: a regional perspective. In: Maltby E, Immirzi CP, Safford RJ (eds) Tropical lowland peatlands of southeast Asia. Proceedings of a workshop on integrated planning and management of tropical lowland peatlands. Cisarua, IUCN, Gland, 3–8 July 1992
Voglmayr H, Yule CM (2006) *Polyancora globosa* gen.et sp. nov., an aeroaquatic fungus from Malaysian peat swamp forests. Mycol Res 110:1242–1252. doi:10.1016/j.mycres.2006.07.001
Wantzen KM, Yule CM, Mathooko JM, Pringle C (2008) Organic-matter dynamics and processing in tropical streams. In: Dudgeon D (ed) Tropical stream ecology. Academic Press, London, p 43
Weiss D, Shotyk W, Rieley J, Page S, Gloor M, Reese S, Martinez-Cortizas A (2002) The geochemistry of major and selected trace elements in a forested peat bog, Kalimantan, SE Asia, and its implications for past atmospheric dust deposition. Geochim Cosmochim Acta 66:2307–2323. doi:10.1016/S0016-7037(02)00834-7
Wells A, Yule CM (2008) The caddisflies (Trichoptera) from a tropical peat swamp in Selangor, Peninsular Malaysia, including two new species. Aquat Insects 30:69–76. doi:10.1080/01650420701687163
Whitmore TC (1984) Tropical rainforests of the Far East, 2nd edn. Clarendon, Oxford
Whitten T, Damanik SJ, Anwar J, Hisyam N (2000) The ecology of Sumatra. Periplus Editions (HK) Ltd, Singapore
Wösten JHM, van den Berg J, van Eijk P, Gevers GJM, Giesen WBJT, Hooijer A, Aswandi I, Leenman PH, Dipa Satriadi R, Siderius C, Silvius MJ, Suryadiputra N, Tricahyo I (2006) Interrelationships between hydrology and ecology in fire degraded tropical peat swamp forests. Int J Water Resour Dev 22:157–174
Yule CM, Gomez LN (2008) Leaf litter decomposition in a tropical peat swamp forest in Peninsular Malaysia. Wetl Ecol Manag. doi:10.1007/s11273-008-9103-9
Yule CM, Grey J (2008) Microbial respiration of DOC supports the aquatic food web in a tropical Malaysian peat swamp (in press)
Yule CM, Leong MY, Liew KC, Ratnarajah L, Schmidt K, Wong HM, Pearson RG, Boyero L (2008a) Shredders in Malaysia: abundance and richness are higher in cool upland tropical streams. J North Am Benthol Soc (in press)
Yule CM, Lim TY, Liew KC (2008b) Microbial communities are vital to the ecological functioning of Malaysian peat swamp forests. Paper presented at the symposium on peat swamp forests. Towards enhancement of knowledge-based decision making in conservation and management of natural resources. Kuala Lumpur, 11–12 Mar 2008
Zakaria S (1992) Water management in deep peat soils in Malaysia. Vol I. Ph.D. Dissertation, Silsoe College, Cranfield Institute of Technology, UK

Biodivers Conserv (2010) 19:411–423
DOI 10.1007/s10531-009-9624-4

ORIGINAL PAPER

Invasive aliens on tropical East Asian islands

Richard T. Corlett

Received: 2 November 2008 / Accepted: 16 March 2009 / Published online: 31 March 2009

Abstract Tropical East Asia (TEA) has numerous islands, both continental and oceanic. This study uses information on invasive aliens in terrestrial habitats on these islands to test the generality of the continental-oceanic contrast in invasibility, assess the conservation impacts of invasive species, and suggest ways to mitigate these. The continental islands of Hong Kong and Singapore are worst-case scenarios for continental invasibility and alien species often dominate in chronically disturbed sites, but very few have successfully invaded closed forests, with the exception of birds in Hong Kong. On other, less densely populated, continental islands, closed-canopy forests appear to resist invasions by all taxa, with few known exceptions. Forests on oceanic islands isolated by <100 km during the last glacial maximum appear no more susceptible to plant and invertebrate invasions than those on continental islands, but invasions by mammals are widespread. Snake invasions may be under-recognized. The remote oceanic Ogasawara (Bonin) Islands, >1000 km from the nearest continent, have a native biota of largely tropical East Asian origin and are suffering from alien forest invasions across the taxonomic spectrum. These patterns of invasibility are consistent with the idea that alien invasion is facilitated by the absence of native species in the same functional group. Alien invasives are not yet a major conservation problem in TEA, except on remote islands, but their dominance on disturbed sites may slow or prevent recovery of native biodiversity. Strict quarantine is impractical in TEA, although some major introduction routes could be blocked. Management efforts should focus on early recognition and immediate control of potential problem species.

Keywords Exotics · Invasive species · Tropical Asia · Island biogeography · Conservation · Hong Kong · Singapore

R. T. Corlett (✉)
Department of Biological Sciences, National University of Singapore, 14 Science Drive 4, Singapore 117543, Singapore
e-mail: corlett@nus.edu.sg

Introduction

The invasive species literature highlights an apparent contrast between the invasibility of tropical oceanic islands, where alien species are a major conservation problem and can invade ecosystems with little or no human disturbance, and the continental tropics, where aliens often dominate the spontaneous communities of urban and agricultural areas, but are rarely a significant problem in undisturbed sites (e.g. Simberloff 1995; Denslow 2003). Currently, the most plausible general explanation for this contrast is that the higher diversity of species and functional groups on continents and the resulting efficiency of resource capture excludes invasive species, except where resources have been freed by disturbance (Denslow 2003; Teo et al. 2003; Daehler 2006; Funk et al. 2008). Herben (2005) suggests that the important difference between islands and the mainland is not local species richness, but rather the size of the species pools from which these local communities are drawn, since a larger pool increases the likelihood of finding a better adapted or more competitive species for any particular set of conditions. The importance of propagule pressure ('introduction effort') in both plant and animal invasions (Reaser et al. 2008) is an additional complication, since large differences in propagule pressure could potentially mask differences in invasibility.

Tropical East Asia (hereafter TEA) is a biogeographical region that consists of political Southeast Asia, minus eastern Indonesia, plus southern China, the Andaman Islands and the Ryukyu Islands (Corlett 2009) (Fig. 1). It is a region with numerous islands. A key distinction can be made between continental (or 'land-bridge') islands that were connected to the Asian mainland by dry land at the last glacial maximum (LGM), when water levels were 120 m lower than today, and oceanic islands that have never been connected. The biotas of continental islands are controlled largely by extinction after isolation, since an island of a given size can support fewer species than a similar area on the mainland. The biotas of oceanic islands, by contrast, are limited by factors that affect dispersal and establishment. The marine barrier that terrestrial organisms must cross to reach such an island acts as a filter, resulting in floras and faunas that are unbalanced or 'disharmonic' in comparison with those of the mainland. The variety of practical dispersal mechanisms declines as the width of the barrier increases, so the biotas of oceanic island, unlike those of continental islands, vary with distance from the mainland. There are also islands in tectonically active areas of TEA that do not fit clearly in either group, including several that appear to have been connected at some time in the past, but not at the LGM (Corlett 2009).

In addition to the diversity of islands in the region, TEA has two other advantages for the study of invasive aliens. Firstly, the region supports approximately one billion people and is a global centre for both tropical agriculture and international trade, resulting in a massive regional propagule pressure from both deliberate and accidental alien introductions. Secondly, almost the entire land area of tropical East Asia, including probably all the islands, was covered in forest in the early Holocene (Corlett 2009), which considerably simplifies interpretation of subsequent changes in the biota. The aims of the study reported here were to use the diversity of islands in the region to test the generality of the continental-oceanic contrast in invasibility for terrestrial habitats, to assess the current and potential future conservation impacts of invasive species, and to suggest ways in which these impacts could be mitigated.

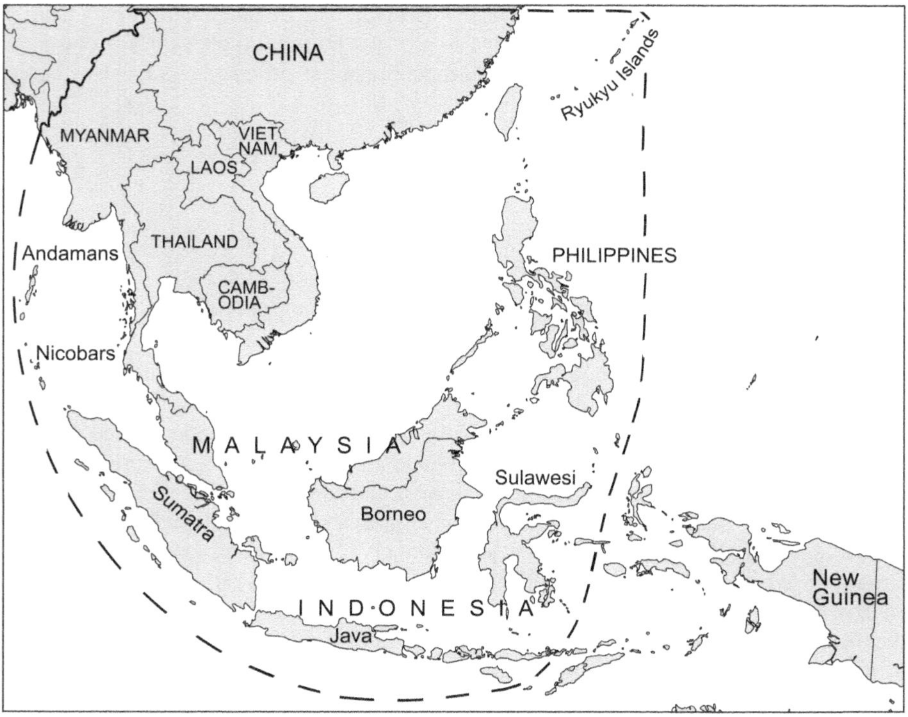

Fig. 1 Map of Tropical East Asia as defined for this study

Methods

I have lived for 5 and 20 years, respectively, on the continental islands of Singapore and Hong Kong Island, so I have been able to make use of both published and unpublished information from these areas. Information for the rest of the region comes largely from a review of the published literature, but my own experience of many of the sites has made it possible to correct for the tendency of some authors to exaggerate the impacts of their particular invasive species. I have used the presence or absence of a land connection at the last glacial maximum (LGM) as the criterion for distinguishing continental and oceanic islands, and excluded some islands (e.g. several off the tectonically active west coast of Sumatra) where this is unclear. Note, however, that the LGM climate was cooler and, in many places, considerably drier than today, so not all species that could occupy an island today could have made use of the land exposed at the LGM (e.g. Meijaard 2003). The oceanic islands have been further divided into near-oceanic (largest water-gap that had to be crossed from the nearest continent at the LGM <100 km), and far-oceanic (>500 km), with the far-oceanic islands (the Ogasawara and Palau groups) all outside tropical East Asia as defined above, but dominated by species that originated in this region.

A general description of the physical geography of the region and the island biogeography of the native biota is available in Corlett (2009). Where they have been assessed, both local species diversities and the sizes of species pools decline from continental islands to near-oceanic islands to far-oceanic islands. The diversities of functional groups have not been directly assessed, but the progressive loss of families and higher taxa as one moves away from the continental shelf suggests that functional group diversity also declines sharply.

Results

Continental islands

Tropical East Asia has numerous continental islands, ranging in size from tiny rocky outcrops to the giant islands of Borneo (750,000 km^2) and Sumatra (440,000 km^2). The densely populated islands of Hong Kong (particularly the 80-km^2 Hong Kong Island) and Singapore (the 600-km^2 main island) can be considered as examples of the 'worst case scenario' for continental invasibility, with pervasive human impacts, numerous accidental and deliberate alien introductions through their ports and botanic gardens, and a massive propagule pressure arising from the close proximity of urban areas and native vegetation. They differ in that Hong Kong (22° N) has a subtropical monsoon climate and was largely deforested centuries ago (Dudgeon and Corlett 2004), while Singapore (1° N) has an aseasonal climate and was largely covered in primary forest until less than 200 years ago (Tan et al. 2007). On both islands, alien plants dominate only in disturbed urban and agricultural areas, although this dominance can apparently persist on some sites after disturbance is halted (e.g. for Hong Kong—Corlett 1992b; Dudgeon and Corlett 2004; Leven and Corlett 2004; Chung and Corlett 2006; for Singapore—Corlett 1988, 1992a; Tan et al. 2007). In contrast, only single alien plant species, *Syzygium jambos* in Hong Kong (Leung et al. 2009) and *Clidemia hirta* in Singapore (Teo et al. 2003), have significantly invaded native, closed-canopy forests. Additional aliens are established in the margins of some forest areas and along forest paths on both islands. Several species have invaded fire-maintained native grassland and shrubland in Hong Kong along streamsides (particularly *Mikania micrantha*) and in areas disturbed by feral cattle, which are not present on Hong Kong Island, but occur on nearby Lantau Island and the adjacent mainland (Leung et al. 2009).

The information available suggests that alien invertebrates in both Hong Kong and Singapore are also largely confined to areas that experience chronic human disturbance (Dudgeon and Corlett 2004; Fellowes 1999; Tan et al. 2007). There have, however, been two dramatic exceptions to this in Hong Kong, where the pinewood nematode (*Bursaphelenchus xylophilus*) and pine needle scale (*Hemiberlesia pitysophila*) have largely eliminated the native *Pinus massoniana* since the 1980s (Dudgeon and Corlett 2004). It is very likely that other, less conspicuous, alien invertebrates have been overlooked in the forests, since the native faunas are also poorly known.

Alien amphibians and reptiles have also been largely excluded from forest and other semi-natural vegetation in both areas, but a number of mammal and bird species have been successful invaders, particularly in Hong Kong. Indeed, more than half the >20 alien bird species established in Hong Kong (including the islands and adjacent mainland) are found in forests and both their diversity and abundance is increasing (Leven and Corlett 2004; Kwok 2007; Leven pers. com.). This pattern of invasion reflects the origin of most alien bird populations in Hong Kong from Buddhist bird releases. Wild-caught birds from similar forests in southwest China are imported into Hong Kong as cage-birds and have been released in large numbers over recent decades, resulting in optimum conditions for establishment (Chan 2007). Buddhist releases have been less common in Singapore, but may be the source of the only widespread avian forest invader there, the white-crested laughing-thrush (*Garrulax leucolophus*) (Wang and Hails 2007). The exotic squirrels in both Hong Kong (*Callosciurus flavimanus*; Chung and Corlett 2006) and Singapore (*C. finlaysonii*; Baker and Lim 2008) were probably also established from deliberate releases. In Hong Kong, where no native squirrel species survived deforestation, *C. flavimanus* has successfully invaded native forests, but in Singapore, which has high densities of native tree squirrels, the exotic species is currently confined to suburban areas, although this may simply reflect its relatively recent introduction. Both areas have free-ranging dogs, but these appear to include truly wild populations only in Hong Kong, where feral cats are also established on Lantau Island, which lacks the native leopard cat, *Prionailurus bengalensis* (Dudgeon and Corlett 2004). Feral cattle (*Bos taurus*) are also established in fire-maintained grasslands on Lantau Island.

Scattered published information and personal observations on other continental islands in the region suggest the same general pattern for plants and invertebrates, with alien species prominent in urban and agricultural settings, but almost entirely excluded from undisturbed forests (e.g. Pfeiffer et al. 2008). Moreover, the forest invasions by vertebrates in Hong Kong appear to be exceptional. The Central American brown anole, *Anolis sagrei,* is abundant in betelnut plantations in southern Taiwan, where it reduces the diversity and abundance of terrestrial arthropods, but it is rare in secondary forests (Huang et al. 2008). However, the capture of two rats from the commensal *Rattus rattus* complex in primary lowland rain forest in Borneo may be a warning sign (Wells et al. 2006). Plantations of alien trees provide a massive propagule pressure on all the larger islands and concerns have been raised about the invasive potential of *Acacia* spp., particularly on nutrient-poor soils, such as those of the extensive heath forests of Borneo (Osunkoya et al. 2005). There is no evidence that these species can invade intact forests, but they may replace native pioneers in disturbed sites and slow or prevent the restoration of native forest cover. Again, the information available for most groups of invertebrates is insufficient to definitely rule out successful forest invasions. An American ambrosia beetle, *Euplatypus parallelus*, has been found in primary forest at Pasoh, on the Malaysian mainland (Maeto and Fukuyama 2003), and a range of exotic forest pests are established in southern China.

Near-oceanic islands

Near-oceanic islands in tropical East Asia range in size from rocky islets to the large islands of Luzon (110,000 km^2) and Mindanao (98,000 km^2), in the Philippines. Sulawesi (180,000 km^2) is also oceanic, in that it remained an island during Pleistocene low sea-level episodes, but it is possible that the ancestors of much of its diverse flora and some of its fauna were stranded there by the opening of the Makassar Straits in the late Eocene and did not cross an oceanic barrier (Inger and Voris 2001; Morley 2003). Nothing in the literature suggests that near-oceanic islands in TEA are more vulnerable to plant invasions than continental islands. This is despite significantly lower local plant diversities on the smaller, more remote islands, such as those in the Andaman and Nicobar groups, Nusa Tenggara, and Maluku (Corlett 2009). On Little Andaman Island (733 km^2), for example, alien plants are confined to forests heavily disturbed by the 2004 tsunami, cattle, or other human impacts (Rasingam and Parthasarathy 2009). There is also no indication of increased vulnerability to invertebrate invasions. On Sulawesi, the yellow crazy ant (*Anoplolepis gracilipes*) is abundant in agroforests, but apparently excluded from natural forests, just as it is in Borneo and Peninsula Malaysia, although this species invades native forests on remote oceanic islands elsewhere (Bos et al. 2008; Pfeiffer et al. 2008)

In contrast to the plants, many alien mammals have successfully invaded native vegetation on near-oceanic islands in the region. In some cases, these invasions have been so successful that the invader would now be treated as native, were it not that archaeological and/or molecular evidence shows otherwise. The likelihood of many more such 'cryptic invasions' makes it very difficult to assess the overall impact of invasive aliens. For example, the archaeological record shows that the only large-bodied animals on Flores (14,100 km^2) at the start of the Holocene were people and Komodo dragons, and that people have subsequently introduced first the Sulawesi warty pig (*Sus celebensis*), ca. 7000 years ago, and then the Eurasian pig (*S. scrofa*), the masked palm civet (*Paguma larvata*), the Javan porcupine (*Hystrix javanicus*), and the long-tailed macaque (*Macaca fascicularis*) (van den Bergh et al. 2008). The archeological record of south Sulawesi shows that deer (*Cervus timorensis*) first appeared there about 4000 years ago (Simons and Bulbeck 2004). Within the last century, three species of deer and the Asian elephant have been introduced to the Andaman Islands, while pigs were probably introduced in prehistoric times. Unlike Flores, which had elephant-like stegodonts until the late Pleistocene, the Andamans had no large native herbivores and the introduced species are having a significant impact on forest structure and species composition (Ali 2004, 2006). The elephant population on Borneo is probably also a recent introduction, probably from the now extinct population on Java, which would explain its genetic distinctiveness (Cranbrook et al. 2008).

Pigs, deer, civets, and macaques occur in native vegetation on many of the near-oceanic islands in the region, where they are often treated as native, but many, possibly all, of these populations are probably introduced (e.g. Flannery 1995). In most cases, the consequences of these introductions for other plants and animals are unknown, although the introduction of mammalian carnivores to islands that previously lacked them must have had a significant impact. More recently, thirty individuals of the mongoose, *Herpestes javanicus*, were released on Amami Island (712 km^2), in the Ryukyus, in 1979, to control a native poisonous snake (*Protobothrops flavoviridis*), but the expanding population now threatens the endangered endemic rabbit, *Pentalagus furnessi*, as well as native birds, reptiles, and amphibians (Watari et al. 2008). Several species of human-commensal rodents and shrews have also been dispersed throughout the region. Their ability to invade native forests is largely unknown, but studies in the Philippines suggest that alien species are unable to penetrate undisturbed or regenerating forests with a rich native small mammal fauna (Balete et al. 2009).

Identifying mammal invasions is made difficult by the fact that all the species mentioned above are native at least somewhere in or near TEA. Translocations of snakes within the region are equally difficult to detect (Brown and Alcala 1970; Lever 2003; de Lang and Vogel 2005) and may have had a significant impact on native ecosystems, particularly on smaller islands with an impoverished native snake fauna. The region's largest snake, *Python reticulatus*, for example, is probably introduced in most of the Philippines (Brown and Alcala 1970). Alien frogs and birds are also established on all of these islands, but information on their ability to invade native forests is currently lacking.

Far-oceanic islands

The subtropical (27° N) Ogasawara (Bonin) Islands are a group of small (<24 km^2) islands of Tertiary volcanic origin, 1,300 km east of the Ryukyus, which until recently were covered in subtropical evergreen broad-leaved forest (Shimizu 2003). The native biota is largely of East Asian origin, has a high level of endemism, and lacks frogs, snakes, non-flying mammals, social bees, and several poorly dispersed plant families. The Palau Islands

(7° N) are islands of varied size (<365 km^2) and origin, 845 km east of Mindanao, but also a similar distance from New Guinea, so their biogeography is more complex. The native biota is similarly disharmonic, although unlike the Ogasawara Islands, the Palau Islands have native frogs and snakes (Crombie and Pregill 1999).

The Ogasawara Islands suffer from many, well-studied, conservation problems as a result of alien invasive species. In contrast to the islands considered so far, alien plants, invertebrates, and vertebrates not only dominate chronically disturbed sites, but have also invaded native forests, although some of these invasions appear to have been promoted by human disturbance. Several introduced tree species threaten the native forests by competitively replacing native trees, suppressing regeneration, and dominating regrowth (Yamashita et al. 2003; Hata et al. 2006; Sugiura et al. 2008a, b). Introduced honeybees (*Apis mellifera*) threaten native pollination systems by displacing endemic bees (Kato et al. 1999; Abe 2006) and introduced wood-feeding beetles (Cerambycidae) have integrated into the native tree-cerambycid association matrix (Sugiura et al. 2008b). Introduced predators, including rats and an invasive flatworm, threaten the diverse endemic land snail fauna (Sugiura et al. 2006; Chiba 2007; Sugiura and Yamaura 2009). An introduced lizard, *Anolis carolinensis* threatens many native insects (Karube and Suda 2004; Abe et al. 2008a). Feral cats eat native birds (Kawakami and Higuchi 2002), introduced rats eat the fruits and seeds of native trees and shrubs (Yamashita et al. 2003; Abe 2007; Abe et al. 2008b), and feral goats threatened native vegetation until they were controlled (Shimizu 2003).

Information on invasive species in the Palau Islands is largely in the form of lists, but it is clear that while many aliens are confined to inhabited areas, others, including the omnivorous *Rattus rattus*, have invaded even the most pristine of the remaining forests (Crombie and Pregill 1999; Space et al. 2003).

Discussion

Although the available information is largely anecdotal, this survey supports the widespread belief that natural ecosystems on remote oceanic islands are more easily invaded by alien plants, invertebrates, and vertebrates than those on continents. This is so despite the far higher human population densities and much greater international trade on the continental islands of Hong Kong and Singapore than on the remote oceanic Ogasawara Islands. While the most remote islands appear to be vulnerable across the taxonomic spectrum, invasibility varies between major taxonomic groups on less isolated oceanic islands. Among the well-studied groups, invasibility is greatest for the mammals, which have the poorest water-crossing ability and thus the most impoverished and unbalanced native faunas on even near-oceanic islands. Mammal diversities have increased on oceanic islands worldwide, with disastrous impacts on endemic avifaunas from introduced predators and the omnivorous rats and pigs (Blackburn et al. 2004).

The simplest general explanation for these patterns is that alien invasion is facilitated by the absence from the species pool of competitive native species in the same functional group. On continental islands, a diverse native forest biota excludes most invaders, except where extinctions resulting from past human impacts have created 'empty niches'. The secondary forests of Hong Kong, for example, may be particularly susceptible to invasion by alien birds and squirrels because of the drastic impoverishment of the native forest fauna as a result of deforestation and the subsequent isolation from potential re-invasion sources (Kwok and Corlett 1999). More generally, the absence of native species adapted to open, non-coastal habitats provides a simple explanation for the prominence—often dominance—of alien species in urban and agricultural habitats. On oceanic islands, the progressive loss of functional groups with distance from continental source areas has resulted in progressively increasing opportunities for alien species. For trees, for example, the failure of shade-tolerant taxa, which are typically relatively large-seeded, to reach remote islands may make forests particularly vulnerable to shade-tolerant invaders (Shimizu 2006).

Correlation is not causation, however, and other potentially significant variables, including local species diversity, show the same pattern. It is also important to note that the presence of native competitors is unlikely to provide an absolute defense if the invasive propagule pressure is sufficiently high. A new species entering a community as rare propagules is unlikely to establish unless it can survive and grow on resources that are left unconsumed by the existing species (Tilman 2004), but if this recruitment limitation is overcome, then invaders with superior traits—including lower susceptibility to local natural enemies—could potentially replace natives. The on-going spread of alien species throughout the region, along roads and into expanding plantations and human settlements, may thus provide a springboard for future invasions into currently resistant continental forest ecosystems. Future studies of invasive species in the region need to go beyond merely listing their presence or absence in different habitats and start to investigate the ecological processes involved in the early stages of invasion.

There is also the problem of the sometimes prolonged lag times between the first introduction of a species and the first reports of invasiveness (Crooks 2005; Corlett unpublished observations). In some cases this probably reflects the buildup of propagule pressure described above, but in other cases evolutionary change and/or the subsequent introduction of more invasive genotypes may be involved. Whatever the cause, the consequence is that the early behaviour of a species may be a poor predictor of subsequent invasiveness.

The actual and potential conservation impacts of invasive species on oceanic islands have been little studied in the region, with the exception of the remote Ogasawara Islands and the few studies in the Ryukyus and Andamans cited above. Rapid extinctions of native species are most likely with introduced predatory or omnivorous mammals and snakes, but many of these introductions occurred in prehistoric times, so their impacts may not be obvious without archaeological information on the pre-introduction native fauna. The impacts of invasive plants and invertebrates will probably be slower, in general, and thus less easy to detect. Sax et al. (2002) show that the establishment of alien plant species on oceanic islands has not, so far, resulted in compensatory extinctions among the native flora, so that total plant diversity has, on average, doubled. Whether this simply reflects a long lag time before competitive exclusion or is a sign that prolonged coexistence is possible is not clear, but there is likely to be a limit to how many aliens can be incorporated into island ecosystems without loss of native species.

The major threat from alien plants on continental and near-oceanic islands appears to be their ability to dominate on disturbed and abandoned land, potentially reducing the biodiversity benefits of native forest recovery. This may also be true for invertebrates, such as ants, where the exotic dominance in disturbed areas appears to restrict the persistence of forest taxa (Bos et al. 2008; Pfeiffer et al. 2008), and probably also of vertebrates. Alien plants and animals also penetrate forest edges to a greater or lesser extent and may, by competition or predation, reduce the effective sizes of forest patches for native species. Unfortunately, there is currently no information available on the crucial question of whether or not the exotic species will be progressively eliminated during forest succession. This information gap means that we cannot yet decide between two extreme scenarios for the continental tropics: invasional meltdown as increasing propagule pressure overwhelms the resistance of native ecosystems, or slow recovery as human impacts decline over the next century. Observations and experiments at abandoned agricultural sites undergoing forest succession are urgently needed.

What can be done to mitigate the actual current impacts and potential future impacts of invasive exotic species on tropical East Asian islands? Alien introductions are not a simple all-or-nothing process, since propagule pressure is a major driver for both establishment and invasive spread (Reaser et al. 2008). TEA is a region of multi-island countries with trade-dependent economies and porous international borders, making the strict border quarantine practiced in countries like Australia and New Zealand impractical. Improvements to quarantine in TEA need to focus therefore on the major pathways by which potential invasives enter the region. International experience leaves little doubt that the most important of these pathways are the horticultural trade—for the ornamental plants themselves, as well as associated weeds, pests, and pathogens—and the pet trade. In both trades, unfortunately, novelty is an attraction, so blacklists of potential invasives are rapidly outdated. The alternative is a 'whitelist', restricting imports to species considered low risk on the basis of previous experience in TEA or in similar environments elsewhere. For novel plants, with no track record, the Australian Weed Risk Assessment system appears to be an effective screening tool in varied geographies (Gordon et al. 2008). An additional problem with plants, however, is the ease with which seeds can be purchased on the Internet and imported by post. For vertebrates, where the ecological arguments are reinforced by concerns for public health and animal welfare, a whitelist, backed up by a blanket ban on the trade in wild-caught animals, would make a lot of sense. Captive-bred animals are, in general, much less successful as invaders (Carrete and Tella 2008) and the great majority of existing vertebrate invasions in TEA were started by wild-caught animals.

Improvements to border controls will need to be backed up by practices aimed at reducing propagule pressure *after* arrival, in the sometimes prolonged lag period when an exotic species exists only in cultivation or captivity, and as a few casual escapes. This requires early detection of potential problems followed by immediate action to eliminate or minimize the threat (Simberloff 2009). Neither is happening at present. New invasions are detected by chance, if at all, and attempts at control are ad hoc and usually insufficient (personal observations in several countries). Species that have already become invasive in one part of TEA are ignored in other parts. What is needed is a regional early warning system, whereby reports of new invasive threats are encouraged and acted upon. Concentrations of exotic species, such as zoos and botanical gardens, need to have strategies in place to minimize the risk of escapes and to detect and control those that occur (Dawson et al. 2008). Deliberate releases of captive animals—a widespread practice in TEA, particularly among the region's Buddhists—must be banned, along with the even more common practice of dumping unwanted pets. As with many other ecologically harmful but socially accepted behaviors, public education is essential, but will usually need to be reinforced by prosecutions of the worst offenders.

The eradication of invasive rodents from small oceanic islands is becoming routine (Howald et al. 2007) and there have been successes with other vertebrates, pest invertebrates, and some plant weeds (Simberloff 2009). Biological control has had a bad press in recent decades, as the risk of side-effects on non-target organisms have become apparent, but the deliberate release of carefully screened natural enemies (predators, pathogens, parasitoids, and herbivores) is still an indispensable tool for the management of invasive species (Messing and Wright 2006; Van Driesche et al. 2008), and one that is probably underutilized in TEA. An invasive tree in the Ogasawara Islands, *Pinus luchuensis*, has been controlled, accidentally, by the introduction of the pinewood nematode, *Bursaphelenchus xylophilus* (Hata et al. 2006).

Eradication of well-established aliens is less likely to be successful on larger islands and risks collateral damage to native species (Denslow 2007). In such cases, the focus should be on reduction of adverse impacts, by managing ecosystems to reduce their invasibility (e.g. encouraging canopy closure in forests). Where invasions have apparently been facilitated by local extinctions—as for example in Hong Kong—the reintroduction of 'missing' native species could be the best defense (Corlett 2006). Funk et al. (2008) suggest that resistance to invasion can be enhanced by restoring vegetation with native species that have traits similar to likely invaders, or by using species with a diversity of functional traits. However, this may not be possible on oceanic islands, where trait combinations present in invasive species may have been filtered out of the native flora during colonization.

A striking feature of the region's current alien biota is that, while numerous extra-regional plant and invertebrate species are established, almost all the invasive vertebrates are native somewhere in or near TEA. This could reflect the differences in water-crossing ability highlighted above, which may mean that native island biotas already include the most competitive plant and invertebrate taxa from the region, but not the most competitive vertebrates. Alternatively, it may simply reflect the composition of the regional trade, with the plant trade dominated by extra-regional ornamentals (probably contaminated by extra-regional invertebrates), while the live vertebrate trade is overwhelmingly dominated by wild-caught animals from the region, or captive-bred, more or less domesticated, species which are less likely to establish in the wild. If the second explanation is true, then the growing regional interest in exotic pets may pose an increasing threat in the future.

Acknowledgments I would like to thank the following people for assistance, information, ideas, and inspiration: Aidia Chan, David Dudgeon, John Fellowes, Nina Ingle, David Lane, Mike Leven, Hugh Tan, Ng Sai-chit, and Laura Wong.

References

Abe T (2006) Threatened pollination systems in native flora of the Ogasawara (Bonin) Islands. Ann Bot (Lond) 98:317–334. doi:10.1093/aob/mcl117

Abe T (2007) Predator or disperser? A test of indigenous fruit preference of alien rats (*Rattus rattus*) on Nishijima (Ogasawara Islands). Pac Conserv Biol 13:213–218

Abe T, Makino S, Okochi I (2008a) Why have endemic pollinators declined on the Ogasawara Islands? Biodivers Conserv 17:1465–1473. doi:10.1007/s10531-008-9355-y

Abe T, Wada K, Nakagoshi N (2008b) Extinction threats of a narrowly endemic shrub, *Stachyurus macrocarpus* (Stachyuraceae) in the Ogasawara Islands. Plant Ecol 198:169–183. doi:10.1007/s11258-007-9393-7

Ali R (2004) The effect of introduced herbivores on vegetation in the Andaman Islands. Curr Sci India 86:1103–1112

Ali R (2006) Issues relating to invasives in the Andaman Islands. J Bombay Nat Hist Soc 103:349–355

Baker N, Lim K (2008) Wild animals of Singapore: a photographic guide to mammals, reptiles, amphibians and freshwater fishes. Draco Publishing, Singapore

Balete DS, Heaney LR, Veluz MJ, Rickart EA (2009) Diversity patterns of small mammals in the Zambales Mts., Luzon, Philippines. Mamm Biol doi:10.1016/j.mambio.2008.05.006

Blackburn TM, Cassey P, Duncan RP, Evans KL, Gaston KJ (2004) Avian extinction and mammalian introductions on oceanic islands. Science 305:1955–1958. doi:10.1126/science.1101617

Bos MM, Tylianakis JM, Steffan-Dewenter I, Tscharntke T (2008) The invasive Yellow Crazy Ant and the decline of forest ant diversity in Indonesian cacao agroforests. Biol Invasions 10:1399–1409. doi:10.1007/s10530-008-9215-4

Brown WC, Alcala AC (1970) The zoogeography of the herpetofauna of the Philippine islands, a fringing archipelago. Proc Calif Acad Sci 38:105–139

Carrete M, Tella J (2008) Wild-bird trade and exotic invasions: a new link of conservation concern? Front Ecol Environ 6:207–211. doi:10.1890/070075

Chan SW (2007) Religious release of birds in Hong Kong. Dissertation, University of Hong Kong

Chiba S (2007) Species richness patterns along environmental gradients in island land molluscan fauna. Ecology 88:1738–1746. doi:10.1890/06-1735.1
Chung KPS, Corlett RT (2006) Rodent diversity in a highly degraded tropical landscape: Hong Kong, South China. Biodivers Conserv 15:4521–4532. doi:10.1007/s10531-005-5102-9
Corlett RT (1988) The naturalized flora of Singapore. J Biogeogr 15:657–663. doi:10.2307/2845443
Corlett RT (1992a) The ecological transformation of Singapore, 1819–1990. J Biogeogr 19:411–420. doi:10.2307/2845569
Corlett RT (1992b) The naturalized flora of Hong Kong—a comparison with Singapore. J Biogeogr 19:421–430. doi:10.2307/2845570
Corlett RT (2006) Conservation of biodiversity in a highly degraded landscape: problems and prospects. In: Jim CY, Corlett RT (eds) Sustainable management of protected areas for future generations. Friends of the Country Parks, Hong Kong, pp 77–92
Corlett RT (2009) The ecology of Tropical East Asia. Oxford University Press, Oxford
Crombie RI, Pregill GK (1999) A checklist of the herpetofauna of the Palau Islands (Republic of Belau), Oceania. Herpetol Monogr 13:29–80. doi:10.2307/1467060
Crooks JA (2005) Lag times and exotic species: the ecology and management of biological invasions. Ecoscience 12:316–329. doi:10.2980/i1195-6860-12-3-316.1
Daehler C (2006) Invasibility of tropical islands by introduced plants: partitioning the influence of isolation and propagule pressure. Preslia 78:389–404
Dawson W, Mndolwa AS, Burslem D, Hulme PE (2008) Assessing the risks of plant invasions arising from collections in tropical botanical gardens. Biodivers Conserv 17:1979–1995. doi:10.1007/s10531-008-9345-0
de Lang R, Vogel G (2005) The snakes of Sulawesi: a field guide to the land snakes of Sulawesi with identification keys. Edition Chimaira, Frankfurt am Main
Denslow JS (2003) Weeds in paradise: thoughts on the invasibility of tropical islands. Ann Mo Bot Gard 90:119–127. doi:10.2307/3298531
Denslow JS (2007) Managing dominance of invasive plants in wildlands. Curr Sci India 93:1579–1586
Dudgeon D, Corlett RT (2004) The ecology and biodiversity of Hong Kong. Joint Publishing, Hong Kong, China
Earl of Cranbrook, Payne J, Leh CMU (2008) Origin of the elephants *Elephas maximus* L. of Borneo. Sarawak Mus J 63(84):1–25
Fellowes JR (1999) Exotic ants in Asia: is the mainland at risk? Aliens 9:5–6
Flannery TF (1995) Mammals of the South West Pacific and Moluccan Islands. Reed, Sydney
Funk JL, Cleland EE, Suding KN, Zavalleta ES (2008) Restoration through reassembly: plant traits and invasion resistance. Trends Ecol Evol 23:695–703. doi:10.1016/j.tree.2008.07.013
Gordon DR, Onderdonk DA, Fox AM, Stocker RK (2008) Consistent accuracy of the Australian weed risk assessment system across varied geographies. Divers Distrib 14:234–242. doi:10.1111/j.1472-4642.2007.00460.x
Hata K, Suzuki JI, Kachi N, Yamamura Y (2006) A 19-year study of the dynamics of an invasive alien tree, *Bischofia javanica*, on a subtropical oceanic island. Pac Sci 60:455–470. doi:10.1353/psc.2006.0029
Herben T (2005) Species pool size and invasibility of island communities: a null model of sampling effects. Ecol Lett 8:909–917. doi:10.1111/j.1461-0248.2005.00790.x
Howald G et al (2007) Invasive rodent eradication on islands. Conserv Biol 21:1258–1268. doi:10.1111/j.1523-1739.2007.00755.x
Huang S-C, Norval G, Wei C-S, Tso I-M (2008) Effects of brown anole invasion and betelnut palm planting on arthropod diversity in southern Taiwan. Zool Sci 25:1121–1129. doi:10.2108/zsj.25.1121
Inger RF, Voris HK (2001) The biogeographical relations of the frogs and snakes of Sundaland. J Biogeogr 28:863–891. doi:10.1046/j.1365-2699.2001.00580.x
Karube H, Suda S (2004) A preliminary report on influence of an introduced lizard *Anolis carolinensis* on the native insect fauna of Ogasawara Islands. Res Rep Kanagawa Prefectural Mus Nat Hist 12:21–30
Kato M, Shibata A, Yasui T, Nagamasu H (1999) Impact of introduced honeybees, *Apis mellifera*, upon native bee communities in the Bonin (Ogasawara) Islands. Res Popul Ecol (Kyoto) 41:217–228
Kawakami K, Higuchi H (2002) Predation by domestic cats on birds of Hahajima Island of the Bonin Islands, southern Japan. Ornithol Sci 1:143–144. doi:10.2326/osj.1.143
Kwok HK (2007) Changes of a forest bird community in Hong Kong of China in 10 years. Acta Ecol Sin 27:3993–4001. doi:10.1016/S1872-2032(07)60085-7
Kwok HK, Corlett RT (1999) Seasonality of a forest bird community in Hong Kong, South China. Ibis 141:70–79. doi:10.1111/j.1474-919X.1999.tb04264.x
Leung GPC, Hau BCH, Corlett RT (2009) Exotic plant invasions in the highly degraded upland landscape of Hong Kong, China. Biodivers Conserv 18:191–202. doi:10.1007/s10531-008-9466-5

Leven MR, Corlett RT (2004) Invasive birds in Hong Kong, China. Ornithol Sci 3:43–55. doi:10.2326/osj.3.43
Lever C (2003) Naturalized reptiles and amphibians of the world. Oxford University Press, Oxford
Maeto K, Fukuyama K (2003) Vertical stratification of ambrosia beetle assemblage in a lowland rain forest at Pasoh, Peninsular Malaysia. In: Okuda T, Monokaran N, Matsumoto Y, Niiyama K, Thomas SC, Ashton PS (eds) Pasoh: ecology of a lowland rain forest in Southeast Asia. Springer-Verlag, Tokyo, pp 325–336
Meijaard E (2003) Mammals of south-east Asian islands and their Late Pleistocene environments. J Biogeogr 30:1245–1257. doi:10.1046/j.1365-2699.2003.00890.x
Messing RH, Wright MG (2006) Biological control of invasive species: solution or pollution? Front Ecol Environ 4:237. doi:10.1890/1540-9295(2006)004[0132:BCOISS]2.0.CO;2 237
Morley RJ (2003) Interplate dispersal paths for megathermal angiosperms. Perspect Plant Ecol 6:5–20. doi:10.1078/1433-8319-00039
Osunkoya OO, Othman FE, Kahar RS (2005) Growth competition between seedlings of an invasive plantation tree, *Acacia mangium*, and those of a native Borneo heath-forest species, *Melastoma beccarianum*. Ecol Res 20:205–214. doi:10.1007/s11284-004-0027-4
Pfeiffer M, Tuck HC, Lay TC (2008) Exploring arboreal ant community composition and co-occurrence patterns in plantations of oil palm *Elaeis guineensis* in Borneo and Peninsular Malaysia. Ecography 31:21–32. doi:10.1111/j.2007.0906-7590.05172.x
Rasingam L, Parthasarathy N (2009) Diversity of understorey plants in undisturbed and disturbed tropical lowland forests of Little Andaman Island, India. Biodivers Conserv. doi:10.1007/s10531-008-9496-z
Reaser JK, Meyerson LA, Von Holle B (2008) Saving camels from straws: how propagule pressure-based prevention policies can reduce the risk of biological invasion. Biol Invasions 10:1085–1098. doi:10.1007/s10530-007-9186-x
Sax DF, Gaines SD, Brown JH (2002) Species invasions exceed extinctions on islands worldwide: a comparative study of plants and birds. Am Nat 160:766–783. doi:10.1086/343877
Shimizu Y (2003) The nature of Ogasawara and its conservation. Glob Environ Res 7:3–14
Shimizu Y (2006) Population dynamics of an invasive plant (*Psidium cattleianum*) after two continuous disturbances in the *Pinus-Schima* secondary forest on Chichijima in the Ogasawara (Bonin) Islands. Reg Views 19:1–15
Simberloff D (1995) Why do introduced species appear to devastate islands more than mainland areas? Pac Sci 49:87–97
Simberloff D (2009) We can eliminate invasions or live with them. Successful management projects. Biol Invasions 11:149–157. doi:10.1007/s10530-008-9317-z
Simons A, Bulbeck D (2004) Late Quaternary faunal successions in South Sulawesi, Indonesia. Mod Quaternary Res Southeast Asia 18:167–190
Space JC, Waterhouse BM, Miles JE, Tiobech J, Rengulbai K (2003) Report to the Republic of Palau on invasive plant species of environmental concern. USDA Forest Service, Honolulu
Sugiura S, Yamaura Y (2009) Potential impacts of the invasive flatworm *Platydemus manokwari* on arboreal snails. Biol Invasions. doi:10.1007/s10530-008-9287-1
Sugiura S, Okochi I, Tamada H (2006) High predation pressure by an introduced flatworm on land snails on the oceanic Ogasawara Islands. Biotropica 38:700–703. doi:10.1111/j.1744-7429.2006.00196.x
Sugiura S, Tsuru T, Yamaura Y, Hasegawa M, Makihara H, Makino S (2008a) Differences in endemic insect assemblages among vegetation types on a small island of the oceanic Ogasawara Islands. Entomol Sci 11:131–141. doi:10.1111/j.1479-8298.2008.00263.x
Sugiura S, Yamaura Y, Makihara H (2008b) Biological invasion into the nested assemblage of tree-beetle associations on the oceanic Ogasawara Islands. Biol Invasions 10:1061–1071. doi:10.1007/s10530-007-9184-z
Tan HTW, Chou LM, Yeo DCJ, Ng PKL (2007) The natural heritage of Singapore. Prentice Hall, Singapore
Teo DHL, Tan HTW, Corlett RT, Wong CM, Lum SKY (2003) Continental rain forest fragments in Singapore resist invasion by exotic plants. J Biogeogr 30:305–310. doi:10.1046/j.1365-2699.2003.00813.x
Tilman D (2004) Niche tradeoffs, neutrality, and community structure: a stochastic theory of resource competition, invasion, and community assembly. Proc Natl Acad Sci USA 101:10854–10861. doi:10.1073/pnas.0403458101
van den Bergh GD, Meijer HJM, Due Awe R, et al. (2008) The Liang Bua faunal remains: a 95 kyr sequence from Flores, East Indonesia. J Hum Evol. doi:10.1016/j.jhevol.2008.08.015
Van Driesche R, Hoddle M, Center T (2008) Control of pests and weeds by natural enemies: an introduction to biological control. Blackwell Publishing, Malden
Wang LK, Hails CJ (2007) An annotated checklist of the birds of Singapore. Raffles Bull Zool 15(Suppl):1–179

Watari Y, Takatsuki S, Miyashita T (2008) Effects of exotic mongoose (*Herpestes javanicus*) on the native fauna of Amami-Oshima Island, southern Japan, estimated by distribution patterns along the historical gradient of mongoose invasion. Biol Invasions 10:7–17. doi:10.1007/s10530-007-9100-6

Wells K, Kock D, Lakim MB, Pfeiffer M (2006) Is *Rattus rattus* invading the primary rainforest on Borneo? Malay Nat J 59:73–79

Yamashita N, Tanaka N, Hoshi Y, Kushima H, Kamo K (2003) Seed and seedling demography of invasive and native trees of subtropical Pacific islands. J Veg Sci 14:15–24

Biodivers Conserv (2010) 19:425–434
DOI 10.1007/s10531-008-9524-z

ORIGINAL PAPER

Restoration of a tropical island: Cousine Island, Seychelles

Michael J. Samways · Peter M. Hitchins · Orty Bourquin · Jock Henwood

Received: 1 April 2008 / Accepted: 14 November 2008 / Published online: 3 December 2008

Abstract Tropical island species and ecosystems are threatened worldwide as a result of increasing human pressure. Yet some of these islands also lend themselves to restoration, as they are physically defined units that can be given focused attention, as long as resources are available and clear conservation targets are set. Cousine Island, Seychelles, is a tropical island that has received such intensive restoration. From a highly degraded island in the 1960s, the island has now been restored to what is believed to be a semblance of the natural state. All alien vertebrates have been eradicated, as have 25 invasive alien plants. Cultivated plants are now confined to one small section of the island. Poaching of nesting marine turtles has been stopped, leading to an increase in turtle breeding numbers. The shearwater population has increased in size with poaching activities under control. The Sooty tern has also returned to the island to breed. The coastal plain has been restored with over 2,500 indigenous shrubs and trees, which have now grown into a forest carpet. There are strict quarantine procedures on the island, keeping it free of rats, mice, various alien

This paper is dedicated to Fred Keeley, patron of Cousine Island, whose insight bequeaths to future generations a jewel of an island.

M. J. Samways (✉)
Department of Conservation Ecology and Entomology and Centre for Invasion Biology, Stellenbosch University, Matieland 7612, South Africa
e-mail: Samways@sun.ac.za

P. M. Hitchins
P.O. Box 101834, Meer en See 3901, South Africa
e-mail: pmhitchins@mweb.co.za

O. Bourquin
172 Stillwater Loop, Columbus, MT 59019, USA
e-mail: ortyb@imt.net

J. Henwood
P.O. Box 977, Cousine Island, Victoria, Mahé, Seychelles
e-mail: cousine@seychelles.net

invertebrates and potentially invasive alien plants. Three threatened Seychelles endemic land birds (Seychelles warbler, Seychelles magpie robin and Seychelles white-eye) have been introduced and are thriving, with these introductions contributing to both the magpie robin and the white-eye being downgraded from CR to EN (the warbler remains at VU). Ecotourism, and nature conservation for the local inhabitants, have been introduced in a way that does not reduce the improved compositional, structural and functional biodiversity of the island. The result of the restoration effort appears to be sustainable in the long term, although challenges still remain, especially with regards to adequate clean water and a non-polluting power supply on the island. Cousine is thus paving the way in the art and science of tropical island restoration as a legacy for future generations.

Keywords Tropical islands · Restoration · Seychelles · Cousine Island

Introduction

Tropical islands worldwide are undergoing an erosion of their indigenous biodiversity from various, often synergistic, impacts (Primack 2006). With their generally simple food webs, honed by local selection pressures, they are often highly susceptible to novel pressures. The impact caused by invasive alien organisms, in combination with other threats, may even lead to a 'meltdown' of the ecological character of the island (O'Dowd et al. 2003).

While there has been much focus on negative impacts on tropical islands, very little attention has been given to their restoration. Indeed, there have been few attempts around the world to totally restore a whole island. Here, we review the restoration of a 27 ha island, Cousine, in the Seychelles. The reason for this focus is to demonstrate that with determined effort, it is indeed possible to reverse the devastation so often seen on small tropical islands.

Any planned restoration project has a particular starting point and a goal. This applies to islands as much as anywhere else. The starting point here is the condition and composition of an island's current ecosystem (Samways 2000). Some islands have been through a catastrophic regime shift (Scheffer and Carpenter 2003) where restoration can never achieve the 'original state', as former inhabitants are now extinct, and the island is on a new ecological trajectory. Easter Island is a case in point (Bahn and Flenley 1992). On the other hand, some islands have the ingredients for recovery, including an absence of certain keystone invasive alien species (Samways 2000).

Prior to the early 1970s, Cousine Island was partially highly denuded of woody vegetation and had been exploited agriculturally. Nevertheless, there were remnants of the original ecosystem that formed a baseline for recovery (Fig. 1). This was largely because it has granite outcrops among which many indigenous plants and animals still existed, albeit at much reduced population levels. In addition, Cousine had the very good fortune of apparently never having had rats or mice, nor the aggressive crazy ant (*Anoplolepis gracilipes*) (Hill et al. 2003). This no doubt was partly due to the fact that the island has always had a low human visitation rate coupled to sparse surface water supplies.

Nevertheless, while Cousine presented a good case for restoration, there were many restoration management decisions that had to be made. These are embedded in Cousine's Mission Statement "To promote and practice nature conservation and the wise use of natural resources of the island and its surroundings" (Hitchins and Bourquin 1995).

It is worth mentioning here the significance of "surroundings". Any small island is in intimate contact with its marine matrix (Polis and Hurd 1996). For Cousine, this is manifested in particular by the influx of nesting sea birds. Their arrival brings nutrients, which

Fig. 1 Cousine Island, seen here in the 1960s, was overgrazed, planted to alien crops and largely denuded of indigenous vegetation, which only remained in pockets

improve the poor soil, and, in turn, support the plants and the animals of this ecosystem, including the endemic skinks (Ball 2001). Furthermore, Cousine is part of the granitic Seychelles archipelago, and its restoration goes hand in hand with conservation of neighbouring islands. With good planning, it is a win–win situation, where advice can be exchanged, species translocated and metapopulation dynamics continued.

In general terms, there have been three phases in the restoration of Cousine: (1) mid-1970s to early 1990s, 'partial clean up' phase, (2) early 1990s to late 1990s, 'poaching controls, major alien plant removal, replanting with indigenous vegetation, initiation of monitoring programmes, and first re-introductions' phase, and (3) 2000s, 'follow-up alien plant removal, continued indigenous plant establishment, second re-introductions phase'. Throughout all these stages, research, information gathering and planning were in effect. While the first two stages involved very minor ecotourism, the third phase introduced ecotourism as a means of the island 'paying its way' (Klocke and Klocke 2006). Hand in hand with this increase in visitors, is the development of more eco-friendly methods of generating power on the island, particularly the use of wind turbines to reduce the consumption of diesel fuel. Cousine, in co-operation with the local NGO Nature Seychelles, and a World Bank-administered GEF fund, built a conservation and education centre on the neighbouring island of Praslin. This enables local school learners, adults and tourists, to see what has been achieved on the islands of Cousine and Cousin. It also allows learners to be educated about Cousine, if it is not possible for them to visit the island.

There is one further introductory comment of global significance. World wide, there is an 'extinction of experience' (Miller 2005). That is, the cultural loss of the experience of nature. This is being remedied in urban areas with the development of ecological landscape planning. However, as biologists know, while it may well be possible to green urban areas and encourage the return of some wildlife, such species are mostly widespread generalists, well-adapted to living in disturbed conditions. The real key with rescuing the extinction of experience is to bring back the rare and threatened species in their natural setting (Samways 2007). This is where Cousine made an enormous contribution. One simply has to step onto the island to see, hear and smell the myriad of seabirds, extremely rare land birds, as well as many other forms of teeming wildlife. In short, the island is now a safe haven, and is a tropical island experiential paradise.

Fig. 2 By the mid-1990s, the plateau on Cousine Island had a vegetation cover but mostly of alien plants, many of which had been harvested in earlier years

The detailed philosophy surrounding the second, planned phase of the restoration began in the early 1990s. Nevertheless, it is appropriate firstly to briefly review the early 'clean up phase', when a broad brush attempt was made to deal with some of the pressing issues facing the island.

The early phase of restoration

In the early twentieth century, Cousine Island was covered with coconut plantations (Diamond 1975). At various times it also had various other crop plants, including tobacco and cotton, all of which were grown wherever suitable, and with little regard to conservation of natural resources on the island (Fig. 2).

Various livestock was also kept on the island, including cattle, pigs and chickens. Feral cats were also present, as were domestic dogs. One of the greatest concerns was the heavy poaching of shearwaters (up to 8,000 chicks per season) (*Puffinus pacificus* and *P. lherminieri*) and turtles (*Eretmochelys imbricata* and *Chelonia mydas*), as well as the legal (at the time) harvesting of Sooty tern (*Sterna fuscata*) eggs of up to 30,000 per season.

The early clean up phase involved putting a stop to poaching of all indigenous wildlife, ad hoc removal of some of the lowland coconut trees, and eradication of the cats (81 in total, using traps and the poison "1080") and domestic livestock. Although some progress in terms of restoration was made by the early 1990s, it largely came to fruition only from the mid 1990s.

During the 1970s, new villas were constructed on the hillside, but the rock blasting required for their construction resulted in loss of an introduced population of the Seychelles warbler (*Acrocephalus sechellensis*) which was re-introduced in June 1990 (Komdeur 1994). The bird has since thrived on the island, the current estimate being 350–400 birds on Cousine. However, owing to it still being threatened elsewhere in the archipelago, it is retained as VU on the IUCN Red List.

Second phase of restoration

This phase began in 1995, with the long-term aim of restoring the coastal plain to what was believed to be its former natural state. Estimation of the original coastal vegetation was done by surveying the natural coastal vegetation remnants on the neighbouring islands of Cousin, Aride and Praslin. Coupled to this baseline study were surveys of seeds that washed up onto the drift line of the beach, as this appeared to be a method by which seeds appear on the various Seychelles islands. Seeds dropped by the visiting Seychelles fruit bats (*Pteropus seychellensis*) from Praslin Island were also monitored.

A scientifically based management plan was then put in place, which combined various activities. These included intense removal programme against some of the most ecologically significant invasive alien plants to the island, as well as all the 243 planted individuals of Casuarina (*Casuarina equisetifolia*), which dominated the plain and dune crest. Areas were also cleared to encourage nesting by the Sooty tern. At the same time, a propagation programme was established to provide seedlings for planting on the plain. All progress was monitored. While much of the activity was carried out by the resident staff on the island, international volunteers were also involved. During this time, and continuing today, is also the ongoing removal of alien birds such as the Barn owl (*Tyto alba*) and the Indian mynah (*Acridotheres tristis*). Poaching was also monitored coupled to anti-poaching initiatives. Strict quarantine of incoming goods and equipment was also introduced, with particular focus on aliens such as rats, mice, crazy ants and any invasive alien plants. Non-disturbance of the nesting seabirds was also enforced.

Throughout this time, plans were also being put in place to re-introduce certain highly threatened, Seychelles endemic land birds, primarily the Seychelles magpie robin (*Copsychus sechellarum*) (Komdeur 1996; LeMaitre 2002) and the Seychelles white-eye (*Zosterops modestus*) (Ward et al. 2007). Certain other endemic birds and indigenous organisms were also being considered for introduction, although these introductions have not been achieved to date. These potential include the Seychelles kestrel (*Falco area*), Seychelles black paradise flycatcher (*Terpsiphone corvine*), as well as the Robber crab (*Birgus latro*), which had largely disappeared from the Seychelles granitic islands probably as a result of human predation (Bowler 1999). During this phase, the Seychelles magpie robin was successfully introduced.

The introduction of the magpie robin went hand-in-hand with habitat restoration through removal of the alien plants and the replanting programme, all of which had the effect of increasing the available habitat, not just for these birds, but also for other endemic land birds, such as the Seychelles blue pigeon (*Alectroenas pulcherrima*), Seychelles sunbird (*Nectarinia dussumieri*) and the Seychelles fody (*Foudia sechellarum*) (the latter of which had reached about 470 birds by 1997). As this bird community is also dependent on invertebrate food resources, the restoration aimed to increase habitat for these resources, as well as providing for invertebrate conservation per se (Kelly and Samways 2003; Lawrence and Samways 2003).

During this phase of restoration was the confinement of agricultural plants (necessary for Cousine's sustenance) to a designated 0.3 ha horticultural area, and removal of domesticated plants outside this area. In some cases, such as Papaya (*Carica papaya*) and Soursop (*Annona muricata*), this has meant the ongoing removal of newly established plants which have grown from seeds dropped by various birds on the island.

From 1995 onwards, chosen plants for propagation were established in a plastic/shade cloth covered tunnel from seeds or cuttings collected from both Cousine and from neighbouring islands. All cuttings were treated with the rooting hormone Seradix and given micro-jet irrigation. As part of the quarantine programme, there was no introduction of any bagged, rooted seedlings.

Planting took place during the north-west Monsoon rainy season (November–March). Temporary irrigation was put in place for some of the seedlings and cuttings that were exposed to the dry effects of the south-east Trade winds. Temporary shade-cloth barriers were also erected to protect some species of the plants from the adverse impact of salt spray. As 21 Aldabra giant tortoises (genetically mostly *Dipsochelys dussumieri*; see Gerlach 2004) had been introduced onto the island as a rescue operation for individuals caged or penned on the larger islands, sometimes under appalling conditions, it meant that nearly 2,000 of the seedlings had to be protected from tortoise grazing and trampling using logs from felled Casuarina and Coconut palm trees, and also using in total about 58,000 fallen coconuts to retain moisture around their bases. These products of the alien plant removal activities were therefore put to good use as protective collars around the newly established shrubs and trees.

A total of 2,543 trees and shrubs were planted mainly on the plain from 1993 to 2001. These consisted of various species characteristic of lowland granitic Seychelles vegetation (Friedmann 1987; Robertson 1989). However, of particular note was the establishment of some Mapou trees (*Pisonia grandis*). Forests with this tree as a dominant species are extremely rare, and are certainly threatened as a forest type in the Seychelles. The point is that these trees are also a favoured nesting tree for the Lesser noddy (*Anous tenuirostris*) (Burger 2005). Besides encouraging the bird for its own sake, the planting of the Mapou trees and other species, once grown, would also produce a considerable leaf litter layer, which encourages colonization of forest invertebrates, which, in turn, provides food for the Seychelles magpie robin in particular (LeMaitre 2002).

During this second phase of restoration, a programme was introduced to monitor the growth of the seabird nesting populations, the increase in abundance of the introduced land birds and the increase in numbers of nesting turtles, especially the Hawksbill turtle. Also, various scientists were encouraged to visit the island to undertake various studies of key species and processes, such as reptiles (Bourquin and Hitchins 1998), Seychelles giant millipede biology (Lawrence and Samways 2003), assessment of the invertebrates of the leaf litter and logs (Kelly and Samways 2003), spider assessments (Saaristo 1999), colonization by dragonflies (Samways 1998, 2003a, b), Seychelles warbler (Komdeur 1994, 1995) and Seychelles magpie robin biology (Komdeur 1996; LeMaitre 2002), turtle biometrics and nesting biology (Hitchins et al. 2003a, b, 2004a, b, 2006), and marine assessments (Samways 2003a, b; Samways and Hatton 2001).

The third phase

The third phase was a natural extension of the second phase, and represents a change in management rather than any change in philosophy or direction. Besides, the trees on the plain had begun to naturally grow to form a closed canopy within a matter of only 10 years or so (Fig. 3). Nevertheless, at least another 755 trees were planted on the plain after 2001. One of the aims of this was to increase the diversity of plants, and hence dependent invertebrates as well. This ongoing planting programme will continue, especially along parts of the dune creast where Casuarinas have been removed, into the foreseeable future, with at 200 saplings of various species available at any one time for planting out.

Starting earlier but continuing during this third phase, shearwater, noddy and land bird numbers increased considerably, with the re-appearance in 2003, of the Sooty tern, after an

Fig. 3 The same view as in Fig. 2, but taken in 2007. The plain of Cousine Island has been largely cleared of alien vegetation and planted to indigenous trees which had already, within 10 years, formed a substantial canopy. This re-establishment of the coastal forest encouraged colonization by many invertebrates and various endemic and indigenous land and nesting sea birds

absence of 31 years. Meanwhile the Seychelles magpie robin had increased to 39 birds and the Seychelles warbler to 350–400 birds, which contributed to its IUCN Red List status being downgraded from CR in 2000 to EN in 2007. In 2007, the Seychelles white-eye was introduced, and it immediately established very well, having already colonized much of the island, and started to breed (Jolliffe et al. 2007; Ward et al. 2007), and also contributing to its downlisting from CR to EN.

Cousine's contribution to the conservation of turtles has been enormous, with increasing numbers nesting on Cousine each year. There were, on average, 24 (19–43) nests per year from 1992 to 1995, increasing to 81 nests (66–108) per year from 1996 to 2006.

During this phase, there was a growing awareness that Cousine must also generate income to sustain the conservation effort and to open up Cousine's successes to the outside world. This led to the establishment of four villas on the island.

Of special note is that to this day, and without doubt partly arising from strict quarantine and control measures, there are no rats, mice nor crazy ants on Cousine.

Cousine: a bird's eye view

Restoration is rarely final, and ongoing maintenance and monitoring activities are still often required. There were originally 78 species of alien plants on Cousine, of which 18 are cultivated. Of the 60 wild aliens (46 invasive and 14 non invasive), 25 have been eradicated, with the remainder in the process of being eradicated (17) or being controlled (26)—a remarkable achievement. When this, coupled with the increase in seabird, land bird, turtle and forest invertebrate numbers, we might well ask what have been the disadvantages of the restoration activity. First of all, not all invasive alien plants have been removed, although they are scheduled to be so. Nor have any invasive alien invertebrate been removed, although it is uncertain as to what is their ecological role, and whether in fact they are damaging. The Big-headed ant (*Pheidole megacephala*) is present in huge numbers, and in view of its adverse impact on other tropical islands, particularly in reducing endemic invertebrates, there is currently a research programme underway to ascertain its impact with the view to possibly eradicating it using the selective bait hydramethylnon. However, not all invasive alien invertebrates are deleterious. The Indian cockroach (*Pycnoscelus indicus*), for example, is a major food item for the Seychelles magpie robin.

Above all, the restoration programme has led to a shift in habitat type, with more forest and less open ground. Interestingly, this has been to the benefit of most of the endemic species. However, this habitat shift is, arguably, back to a more natural one. At least in terms of the invertebrates, most of those of the sunny coastal plain habitats are vagile generalists. Nevertheless, even their habitat is being kept open to some extent by the foraging behaviour of the tortoises, and certainly enough to maintain the African grass blue butterfly (*Zizeeria knysna*). Yet other sun-loving species have been lost to the island, such as the Common citril damselfly (*Ceriagrion glabrum*) and the Blue percher dragonfly (*Diplacodes trivialis*), which are both very common across wide continental areas. However, these two species were not originally native to Cousine, as it had no open water. These, and some other species as well, only appeared on Cousine once artificial ponds for the tortoises had been established.

The original aim was to restore Cousine to its natural state (while accommodating humans to some extent). This has been very successful, with some species being deliberately introduced (tortoises, land bird and some plant species), while others have increased in abundance as a result of cessation of poaching, reduced disturbance and re-establishment of natural lowland forest or clearing of grassland (seabirds and forest invertebrates).

Cousine is also acting as precautionary ark for certain species. For example, it is not clear whether the Seychelles magpie or the Seychelles white-eye ever actually occurred on Cousine in historical times. Certain Seychelles endemic plants [*Erthroxylum sechellarum*, *Mimusops sechellarum* and *Rothmannia annae*; Bowler et al. (2000)] are also being maintained on the island. The philosophy of this could be argued extensively, but we must remember that Cousine is part of an archipelago. Only 17,000 years or so ago the inner grantic islands were connected as the sea level was more than 100 m lower than it is today (Camoin et al. 2004; Johnson and Baarli 2005). Thus, any modern assessment of which species were there is very likely the outcome of ecological relaxation as the islands came into flooded isolation.

There is the additional issue of the introduced tortoises of which the genetic pool collectively is largely the Aldabran tortoise, likely with some genetic stock from the endemic granitic Seychelles tortoises. None of the tortoises on Cousine have yet produced offspring and their presence is mostly one of animal care, although they may breed as they mature further. However, there is another viewpoint, and this posits that restoration is not just about compositional biodiversity but also functional biodiversity. It is likely that tortoises were present on Cousine in former times, especially when we consider the gradual changes that have occurred since the Pleistocene, at a time when Cousine was joined to neighbouring islands, and was therefore more extensive. The important point, however, is that the tortoises bring back a *function* to Cousine. They graze and trample the plateau vegetation, opening up some areas and provide sunlit habitats required by certain flying insects. They are also involved in the distribution of preferred plant species through ingestion and defaecation of seeds.

Cousine's future

A balance has been struck on Cousine between nature conservation and tourism. Cousine provides an *experience* which attracts certain tourists, some of whom come for the wildlife and others for the sun and solitude. This tourism also begets employment for the local people on Praslin and Mahé, who not only maintain the island but also supply the kitchen, with fish as well as fruits and vegetables. In turn, the island has extended a hand to neighbouring islands with the construction of an information and educational centre, based on Praslin, in a joint venture with its sister island Cousin, which is administered by Nature Seychelles.

Purist conservationists may well argue that this combining of people and nature on Cousine is not fully natural. But we have to be realistic, and Cousine must pay its way and integrate with the archipelago community. The restoration of Cousine has shown that people and nature can indeed be combined in a way that has a long-term future.

As with all small islands, there is a shortage of both water and fuel on Cousine. Tourism is demanding on both these commodities, as tourists to the island expect conditions equal or even better than what they experience at home. This is accommodated to a certain extent by using rainwater collected from the roofs, rather than putting demands on boreholes by excessive abstraction and risking salination of the freshwater lens. Roof-harnessed water is also generally more suitable for human consumption.

An additional consideration is that of the power supply for the island. Currently, there are diesel generators supplying the power, the fuel for which is pumped ashore from a supply ship. This is not acceptable as a long-term solution, as generators are a pollutant on the island. Furthermore, there is always risk of spillage as the fuel is pumped ashore. Thus, one of the major challenges, and currently being addressed, is to seek alternative power supplies, based on environmentally more acceptable methods. Should this be achieved, Cousine would have demonstrated to the world not only its value as a restored microcosm for an amazing array of wildlife and employment for people, but also would have earned its niche as a sustainable ecosystem of which future generations can be proud.

Acknowledgments Restoration of an island is the product of a team effort, and in this regard we are very thankful for the enormous input from the local Seychellois people, and from many international volunteers. Also the restoration has benefited enormously from various inputs from Dylan Evans, Selwyn Gendron, Quentin and Debbie Hagens, Stella Hitchins (née Le Maitre), Frankie Hobro, Kevin and San-Marie Jolliffe, James Lawrence, Nirmal Jivan Shah and the late Danielle St Jorre.

References

Bahn P, Flenley J (1992) Easter Island: Earth Island. Thames and Hudson, London

Ball E (2001) Population density, distribution, habitat preference and relationship with nesting seabirds of the skinks *Mabuya sechellensis* and *M. wrightii* on Cousine Island, Seychelles. Unpublished MSc thesis, University College, London

Bourquin O, Hitchins PM (1998) Annotated checklist of the reptiles on Cousine Island, Seychelles. Afr Herp News 28:8–15

Bowler J (1999) The robber crab *Birgus latro* on Aride Island, Seychelles. Phelsuma 7:56–58

Bowler J, Bullock I, Cadbury J, Gerlach J, Hunter J (2000) Conservation status and management of Wright's gardenia *Rothmannia annae*. Phelsuma 8:23–27

Burger AE (2005) Dispersal and germination of seeds of *Pisonia grandis*, an Indo-Pacific tropical tree associated with insular seabird colonies. J Trop Ecol 21:263–271. doi:10.1017/S0266467404002159

Camoin GF, Montaggioni LF, Braithwaite CJR (2004) Late glacial to post glacial sea levels in the western Indian Ocean. Mar Geol 206:119–146. doi:10.1016/j.margeo.2004.02.003

Diamond AW (1975) Cousine Island nature reserve management plan 1975–1979. BirdLife International, Cambridge. Unpublished Report

Friedmann F (1987) Flowers and trees of Seychelles. ORSTOM, Paris

Gerlach J (2004) Giant tortoises of the Indian Ocean. Andreas S. Brahm, Frankfurt am Main

Hill M, Holm K, Vel T, Shah NJ, Matyot P (2003) Impact of the introduced crazy ant *Anoplolepis gracilipes* on Bird Island, Seychelles. Biodivers Conserv 12:1969–1984. doi:10.1023/A:1024151630204

Hitchins PM, Bourquin O (1995) Cousine Island mission statement and management plan. Cousine Island, Internal Report, 7 p

Hitchins PM, Bourquin O, Hitchins S, Piper SE (2003a) Factors influencing emergences and nesting sites of hawksbill turtles (*Eretmochelys imbicata*) on Cousine Island, Seychelles, 1995–1999. Phelsuma 11:59–69

Hitchins PM, Bourquin O, Hitchins S (2003b) Inter-island nesting by hawksbill turtles (*Eretmochelys imbricata*) in Seychelles. Phelsuma 11:70–71

Hitchins PM, Bourquin O, Hitchins S, Piper SE (2004a) Biometric data on hawksbill turtles (*Eretmochelys imbricata*) nesting at Cousine Island, Seychelles. J Zool (Lond) 264:371–381. doi:10.1017/S0952836904005850

Hitchins PM, Bourquin O, Hitchins S (2004b) Nesting success of hawksbill turtles (*Eretmochelys imbricata*) on Cousine Island, Seychelles. J Zool (Lond) 264:383–389. doi:10.1017/S0952836904005904

Hitchins PM, Bourquin O, Hitchins S (2006) Distances covered and times taken for nesting of hawksbill turtles (*Eretmochelys imbricata*), Cousine Island, Seychelles. Phelsuma 13:93–101

Johnson ME, Baarli BG (2005) Erosion and burial of granite rocky shores in the recent and late Pleistocene of the Seychelles Islands: physical and biological perspectives. J Coast Res 21:867–879, 973. doi:10.2112/05-0019.1

Jolliffe K, Ward G, Jolliffe S-M, Henwood J (2007) Monitoring of the translocated Seychelles white-eye on Cousine Island. Internal Report of Cousine Island

Kelly J, Samways MJ (2003) Diversity and conservation of forest-floor arthropods on a small Seychelles island. Biodivers Conserv 12:1793–1813. doi:10.1023/A:1024161722449

Klocke T, Klocke M (2006) Hidaway hotels: the 100 most beautiful hotels and resorts in the world. Klocke Publishing, Bielefeld

Komdeur J (1994) Conserving the Seychelles warbler *Acrocephalus sechellensis* by translocation from Cousin Island to the islands of Aride and Cousine. Biol Conserv 67:143–152. doi:10.1016/0006-3207(94)90360-3

Komdeur J (1995) Influence of helping and breeding experiments on reproductive performance in the Seychelles warbler: a translocation experiment. Behav Ecol 7:326–333

Komdeur J (1996) Breeding of the Seychelles magpie robin *Copsychus seychellarum* and implications for its conservation. Ibis 138:485–498. doi:10.1111/j.1474-919X.1996.tb08069.x

Lawrence JM, Samways MJ (2003) Litter breakdown by the Seychelles giant millipede and the conservation of soil processes on Cousine Island, Seychelles. Biol Conserv 113:125–132. doi:10.1016/S0006-3207(02)00355-5

LeMaitre S (2002) Food and density limitations of the Seychelles magpie robin *Copsychus sechellarum*, on Cousine Island. Ostrich 73:119–126

Miller JR (2005) Biodiversity conservation and the extinction of experience. Trends Ecol Evol 20:430–434. doi:10.1016/j.tree.2005.05.013

O'Dowd DJ, Green PT, Lake PS (2003) Invasional 'meltdown' on an oceanic island. Ecol Lett 6:445–461. doi:10.1046/j.1461-0248.2003.00512.x

Polis GA, Hurd SD (1996) Linking marine and terrestrial food webs: allochthonous input from the ocean supports high secondary productivity on small islands and coastal communities. Am Nat 147:396–423. doi:10.1086/285858

Primack RB (2006) Essentials of conservation biology, 4th edn. Sinauer, Sunderland

Robertson SA (1989) Flowering plants of Seychelles. Royal Botanic Gardens, Kew

Saaristo MI (1999) Arachnid species lists for Silhouette, Cousine and Mahé. Phelsuma 7(Suppl A):1–12

Samways MJ (1998) Establishment of resident Odonata populations on the formerly waterless Cousine Island, Seychelles: an Island Biogeography Theory (IBT) perspective. Odonatologica 27:253–258

Samways MJ (2000) A conceptual model of ecosystem restoration triage based on experiences from three remote oceanic islands. Biodivers Conserv 9:1073–1083. doi:10.1023/A:1008974702634

Samways MJ (2003a) Conservation of an endemic dragonfly fauna in the Seychelles archipelago. Odonatologica 32:177–182

Samways MJ (2003b) Breakdown of butterflyfish (Chaetodontidae) territories associated with the onset of an intense coral bleaching event. Aquat Conserv Mar Freshwat Syst 15:5101–5107

Samways MJ (2007) Rescuing the extinction of experience. Biodivers Conserv 16:1995–1997. doi:10.1007/s10531-006-9144-4

Samways MJ, Hatton MJ (2001) An appraisal of two coral reef rapid monitoring manuals for gathering baseline data. Bull Mar Sci 69:471–485

Scheffer M, Carpenter SR (2003) Catastrophic regime shifts in ecosystems: linking theory to observation. Trends Ecol Evol 18:648–656. doi:10.1016/j.tree.2003.09.002

Ward G, Jolliffe K, Jolliffe S-M, Henwood J (2007) Monitoring of the threatened Seychelles white-eye on Cousine Island. Internal Report of Cousine Island

Biodivers Conserv (2010) 19:435–447
DOI 10.1007/s10531-008-9522-1

ORIGINAL PAPER

Ecological aspects of endemic plant populations on Klang Gates quartz ridge, a habitat island in Peninsular Malaysia

K. M. Wong · M. Sugumaran · D. K. P. Lee · M. S. Zahid

Received: 4 March 2008 / Accepted: 14 November 2008 / Published online: 5 December 2008

Abstract The population characteristics of *Aleisanthia rupestris* (Rubiaceae), *Eulalia milsumii* (Poaceae) and *Ilex praetermissa* (Aquifoliaceae), endemic to the Klang Gates quartz ridge north-east of Kuala Lumpur, Malaysia were studied. Whereas their narrow, natural distribution within species-poor communities in drier, less fertile conditions on the ridge can be explained by the selection of adaptive traits and reduced plant competition compared to the more extensive, surrounding richer forest communities on more fertile soils, recent disturbances altering site conditions have diminished their populations. Populations were compared among various terrain and site conditions (ridge spine or rockface, exposed or sheltered) on less disturbed and very disturbed portions of the ridge. *A. rupestris* was scarce or absent in sites shaded by taller vegetation, probably an inability to establish in conditions associated with deeper soil development. *I. praetermissa* appeared specially adapted to rockfaces where unstable substrates and poor soil development may restrict competition with other plants. All three species were adversely affected to varying degrees by disturbance and altered site conditions; invasive, fast-spreading, thicket-forming, weeds in disturbed sites on gentler terrain on the ridge spine appear to be especially detrimental to *A. rupestris* and *E. milsumii* establishment. Although the larger plant size of *I. praetermissa* compared to the other two species could mean it is less likely to be shaded out by invading weedy species, residual adults as well as new regeneration may not adapt to changed site conditions following disturbance. Distinguishing between adaptation (of both adults and new regeneration) to changed environmental conditions following disturbance, and ability in competing for space and resources against invading weedy species, is relevant.

Keywords Conservation biology · Endemic plants · Habitat islands · Invasives · Malaysia · Rare plants · Quartz ridges · Tropical biodiversity

K. M. Wong (✉) · M. Sugumaran · D. K. P. Lee · M. S. Zahid
Institute of Biological Sciences, University of Malaya, 50603 Kuala Lumpur, Malaysia
e-mail: wong@um.edu.my

Present Address:
M. S. Zahid
Universiti Pendidikan Sultan Idris, Tanjong Malim, Perak Darul Ridzuan, Malaysia

Introduction

The Klang Gates quartz ridge system just outside the north-eastern part of the city of Kuala Lumpur and within the state of Selangor Darul Ehsan in Peninsular Malaysia, is over 14 km long and reaches up to about 380 m high. The narrow wall-like formation of this quartz reef (Figs. 1, 2) is due to crystallisation of the quartz solution that infilled a large fault in the Main Range granite, with the present feature revealed when the surrounding weathered material eroded away. This ridge is the most prominent of a series of such ridges in this part of Peninsular Malaysia.

Current collecting indicates that inventory of the plant life on this ridge remains incomplete since the earlier listings by Ridley (1922) (21 species) and Henderson (1928) (265 species), and a later enumeration by Kiew (1982) (175 species). Many records do not distinguish between an occurrence on the quartz ridge proper and one on the abutting (non-quartz) slopes. Five plant species are known only from the quartz ridge (Reid 1951, 1959; Kiew 1982): *Aleisanthia rupestris* (Ridl.) Ridl. (Rubiaceae) (Fig. 3a, b), *Borreria pilulifera* Ridl. (Rubiaceae), *Eulalia milsumii* Ridl. (Poaceae) (Fig. 3c), *Henckelia primulina* (Ridl.) A. Weber (Gesneriaceae), and *Ilex praetermissa* Kiew (Aquifoliaceae) (Fig. 3d).

Fig. 1 The Klang Gates quartz ridge, with regenerated lowland mixed dipterocarp forest to its north (*left* in the picture), part of the Klang Gates reservoir catchment forest, and cultivated land or young secondary forest to the right

Fig. 2 The precipitous south-facing rockface of the ridge, showing a portion much exposed by previous vegetation clearance for cultivation and housing development

Fig. 3 Endemic plant species of the Klang Gates quartz ridge investigated in the present study: *Aleisanthia rupestris*, habit (**a**) and leafy branch (**b**); *Eulalia milsumii* growing in wefts (**c**); *Ilex praetermissa*, fruiting branch (**d**)

Conservation interest in the ridge (Wong et al. 2005) focusses on a number of aspects. The low, species-poor plant community that resembles *kerangas*-type vegetation on thin, poor sandy soils elsewhere in the peninsula (Wong et al. 1987) is here distinctive by the dominance of the trees *Baeckea frutescens* L. (a small tree often 2–4 m high, with 1-several stems and fine needle-leaves; Myrtaceae) and *Rhodoleia championii* Hook.f. (a small tree, commonly 4–8 m high; Hamamelidaceae). This was formerly surrounded by lowland mixed dipterocarp forest, much of which has been transformed by development (Fig. 2). The occurrence of five narrowly distributed endemics in a relatively small area along the ridge emphasises the special nature of habitat islands (Wong 1998), although the impacts of human activities have affected the integrity of the original vegetation after the 1970s (Kiew 1982). The quartz reef, with a craggy ridge spine, cutting through an otherwise low, undulating landscape presents a spectacular geological feature. Trending roughly southwest to northeast, this ridge also forms a natural boundary between the catchment forest for the Klang Gates reservoir to its north and the area transformed by tree plantations and other crops, and housing and road development to its south (Figs. 1, 2). Earlier, the ridge had been established as the Klang Gates Wildlife Sanctuary (130 ha) in 1936, primarily for conservation of the Serow *Naemorhedus sumatrensis* (Bechstein). Presently, the ridge is proposed for inclusion as a special part of the Taman Warisan Selangor (Selangor Heritage Park), declared in 2005, with the intention of better managing its conservation.

The most spectacular part of the ridge is the middle section, about 4–5 km long, just behind the Taman Melawati residential neighbourhood and near to the National Zoo. Here, access is easiest along the southwest part adjacent to a dip in the ridge where the Klang Gates Dam is situated; the northeast part is precipitous and not accessible and has a largely intact vegetation cover at the top. A preliminary survey in August, 2006 found that the

Fig. 4 Weedy species invading disturbed, open sites on the Klang Gates ridge spine: *Dicranopteris linearis*, the *resam* fern (**a**, *left*) and *Pteridium esculentum*, bracken (**b**)

immediate section of the ridge just southwest of the Dam was conspicuously more disturbed than the section farther southwest from this, which is less visited by people because of its distance. For convenience, these portions are simply referred to as the "very disturbed" and "less disturbed" parts of the ridge. Greater impacts on the vegetation in the very disturbed part are a result of trampling, woodcutting, site clearing and burning in the past. Many disturbed patches are colonized by weedy species, including the ferns *Dicranopteris linearis* J. Underw. (Gleicheniaceae) (Fig. 4a) and *Pteridium esculentum* (G. Forst.) Cockayne (Pteridaceae) (Fig. 4b); the grasses *Eriachne pallescens* R.Br., *Imperata cylindrica* L. and *Pennisetum purpureum* Schumach. (Poaceae); and the low woody dicots *Clidemia hirta* (L.) D.Don and *Melastoma malabathricum* L. (Melastomataceae), and *Eupatorium odoratum* L. (Asteraceae). This work considers various ecological aspects of plant conservation on the Klang Gates ridge by comparing some population characteristics of the endemic plants *A. rupestris*, *E. milsumii* and *I. praetermissa* in the very disturbed and less disturbed parts of the ridge.

From general observation, the distribution of *Aleisanthia* and *Eulalia* appeared to be different between the very disturbed and less disturbed portions of the ridge. It appeared that whereas *A. rupestris* still occurred on the very disturbed portion of the ridge, *E. milsumii* was difficult to sight on this portion and only became more conspicuous in the more remote, less disturbed part. Hence, it appears that *E. milsumii* may be highly sensitive to disturbance. In the preliminary survey, *I. praetermissa* was more frequently found in shaded sites along the northern rock face of the ridge, and less so in exposed sites. Thus, it could be expected that *I. praetermissa* prefers sheltered sites along the rock face.

Methods

Site classification

Ridge spine and rockface were recognised as the two main terrain categories. On the ridge spine, exposed sites are those in which the ground is unshaded or scantily shaded by low bushes or herbs, whereas sheltered sites are those lightly shaded by tree cover mostly formed by *B. frutescens* and *R. championii*. For rockfaces, an especially contrasting feature was the largely sheer and less vegetated southern face (Fig. 2), a result of intense exposure

Table 1 Mean irradiance (klux), temperature (°C) and humidity (%) in different conditions on the Klang Gates quartz ridge

Site category	Mean irradiance (klux)	Mean temperature (°C)	Mean humidity (%)
A. Very disturbed exposed ridge spine	98.9	35.2	56.3
B. Less disturbed exposed ridge spine	90.6	36.0	53.0
C. Very disturbed sheltered ridge spine	17.9	33.8	57.5
D. Less disturbed sheltered ridge spine	10.6	29.8	65.0
E. Disturbed exposed rockface (south-facing)	*	35.8	50.2
F. Undisturbed sheltered rockface (north-facing)	**	30.1	57.6

Means are of three readings taken at different sites for each habitat category, obtained between 12.30 p.m. and 2.30 p.m. on a clear sunny afternoon (5 January 2007)

* not recorded but similar to conditions in habitat categories A and B

** not recorded but similar to conditions in habitat categories C and D

after clearance of the original forest on the slopes, as opposed to the shaded and more vegetated northern face, sheltered by intact forest on the slopes there (and which form part of the protected watershed of the Klang Gates reservoir) (Fig. 1). Site categories accessible for study on the ridge top or ridge spine include: (a) exposed sites on the very disturbed ridge portion; (b) exposed sites on the less disturbed ridge portion; (c) sheltered sites on the very disturbed ridge portion; (d) sheltered sites on the less disturbed ridge portion. Rockfaces were distinguished as (e) the disturbed, exposed south-facing side, and (f) the undisturbed, sheltered north-facing side. Conditions present in these sites are represented in Table 1.

Sampling

Reference transects were used for the terrain categories ridge top (along both very and less disturbed portions of the ridge) and rockface (along the south-facing and north-facing sides). For *A. rupestris* (a small shrub) and *E. milsumii* (a grass), 10 replicate 1 × 1 m quadrats were randomly selected for each site category along the transect. In the quadrats, the density and cover of plant species were recorded. The occurrence of *I. praetermissa* (a small tree mostly 3–6 m high) on the rockfaces rather than flatter sites on the ridge spine implied that quadrats could not be used efficiently, so a transect survey was adopted. A 250-m survey transect was followed along the ridge spine, enumerating the number of *Ilex* individuals seen within 50-m intervals, at the uppermost 5-m zone of both the north-facing and south-facing rock faces.

Results

Table 2 summarises the frequency, mean cover and mean density of *A. rupestris*, as well as mean relative cover of *A. rupestris* and other plant species in the different site categories. *A. rupestris* is reasonably common (60% frequency) on both types of rock faces (site categories E and F) compared to all ridge spine categories. On the ridge spine, there was a higher frequency in exposed (A and B) than in the sheltered sites, in which no *A. rupestris* was

Table 2 Frequency, mean cover per quadrat, mean density per quadrat, and mean relative cover (%) of *Alesanthia rupestris* and other plant species in different site categories studied on the Klang Gates quartz ridge

Site category	Frequency of *A. rupestris* (%)	Mean cover per quadrat[a] (%) of *A. rupestris*	Mean density per quadrat[a] of *A. rupestris*	Mean relative cover (%) of *A. rupestris*	Mean relative cover (%) of other plant species
A. Very disturbed exposed ridge spine	20	1.0	0.2	3.2	96.8
B. Less disturbed exposed ridge spine	30	3.3	0.4	9.8	90.2
C. Very disturbed sheltered ridge spine	0	0	0	0	100
D. Less disturbed sheltered ridge spine	0	0	0	0	100
E. Disturbed exposed rockface	60	9.1	0.7	34.4	65.6
F. Undisturbed sheltered rockface	60	6.6	0.8	20.6	79.4

Ten quadrats were sampled for each site category. Mean relative cover of *A. rupestris* is calculated by $[C_A/(C_A + C_N)] \times 100$ where C_A is mean percent cover of *A. rupestris* for all the quadrats sampled in a category and C_N is mean percent cover of all other plant species. This value subtracted from 100 then gives the mean percent relative cover of all other species

[a] Comparisons by *t*-tests indicated that mean covers and mean densities of *A. rupestris* in each of the categories A, B, E and F were significantly higher ($P = 0.05$) than in C or D

recorded. Both mean cover and mean density values were significantly higher (*t*-test, $P = 0.05$) on rockfaces and exposed ridge spine sites compared to sheltered ridge spine sites. A similar relative establishment success in the various sites is reflected in the mean relative cover of *A. rupestris* compared to that of other plant species (Table 2).

The mean density and cover of weedy species per quadrat, and the total number of plant species found in the different site categories is shown in Table 3. The highest number of plant species (34) was found in category A (very disturbed, exposed ridge spine). Mean density and mean cover of weedy species were not significantly different (*t*-test, $P = 0.05$) between A and C (very disturbed, exposed or sheltered ridge spine), B and D (less disturbed, exposed or sheltered ridge spine), or E and F (exposed and sheltered rockfaces). Ridge spine sites and rockfaces also did not necessarily have significantly different weed densities or cover (compare B, D, E and F). However, mean cover of weeds, indicating intense smothering of ground area, was significantly higher in A than in B and D (the less disturbed ridge spine sites), or E (disturbed, exposed rockface).

The frequency, mean cover and mean density of *E. milsumii*, and the mean relative cover of *E. milsumii* and other plant species in the various sites is shown in Table 4. Frequency is highest on the less disturbed exposed ridge spine (site category B), moderately common in the less disturbed sheltered ridge spine (D) and both rockfaces (E, F), and the species was not encountered in very disturbed sites on the ridge spine, whether exposed or sheltered. Its mean relative cover is also highest in site B, the only site where other plant cover is not greater.

The frequency of *I. praetermissa* along the north-facing and south-facing rockfaces is shown in Table 5. The northern face of the ridge had clearly more individuals (19 stems), compared to the southern face (only 1 stem).

Table 3 Mean values of density and cover per quadrat of weedy species and total number of plant species recorded in different site categories on the Klang Gates quartz ridge

Site category	Mean density per quadrat of weedy species	Mean cover per quadrat (%) of weedy species	Total number of plant species recorded
A. Very disturbed exposed ridge spine	1.6	34.1	34
B. Less disturbed exposed ridge spine	0.3	4.9	23
C. Very disturbed sheltered ridge spine	1.2	23.7	26
D. Less disturbed sheltered ridge spine	0.4	5.3	22
E. Disturbed exposed rockface	0.3	2.3	24
F. Undisturbed sheltered rockface	0.7	13.0	26

'Weedy species' are exotic species or those known to spread quickly following disturbances such as fires

Comparisons by t-tests indicated that the mean density of weedy species in category A was significantly higher ($P = 0.05$) than in B, D, E or F, and higher in C than E. The mean cover of weedy species in A was also significantly higher than in B, D and E

Discussion

Ecological distribution of *Aleisanthia rupestris*

The similar and relatively high cover and density on both disturbed, exposed and undisturbed, sheltered rockfaces (Table 2: E and F) suggest that neither disturbance nor shading per se could account for significant diminishment in *A. rupestris* populations. On the other hand, the differential distribution (or survival) of this species between exposed sites (whether less or more disturbed) compared to sheltered sites suggests that other factors could be important in discouraging its establishment or commonness within sheltered sites on the ridge spine, that were absent or less severe on the sheltered rockface.

It is not easy to confirm if competition with other plants is an important factor although the mean density and cover of weedy species (Table 3) indicate that weedy species are not less likely to establish well on the rockface or in sheltered sites. An observed physical difference between the sheltered ridge spine sites and the exposed ridge spine sites or rockfaces was the accumulation of organic material and development of a thin soil over the rocks in the former, due to the presence of taller plants (that provide shade and organic material) and a more level terrain (that discourages rapid erosion of accumulated substrate). This implies that rooting characteristics, such as intolerance to a deeper, moister substrate, or increased activity of soil organisms, or root competition with other plants, could restrict *A. rupestris* establishment. These were not specifically investigated, but the inability of *A. rupestris* to spread into the adjacent forest on earth slopes and the limited growth and subsequent death of rooted cuttings in moist sandy or clayey substrates in the botanic garden of the University of Malaya (Zahid 1998) may be indications that such factors operate. However, weedy growth was significant in sites on the very disturbed, exposed ridge spine (Table 3: site category A) and fast-spreading, smothering thickets of *resam* fern (*D. linearis*) or bracken (*P. esculentum*) were found to be common only in disturbed, exposed patches on more level terrain on the ridge spine (Fig. 4). This could help explain the much lower mean relative cover of *A. rupestris* there (Table 2).

Table 4 Frequency, mean cover per quadrat, mean density per quadrat, and mean relative cover (%) of *Eulalia milsumii* and other plant species in the different site categories studied on the Klang Gates quartz ridge

Site category	Frequency of *E. milsumii* (%)	Mean cover per quadrat[a] (%) of *E. milsumii*	Mean density per quadrat[a] of *E. milsumii*	Mean relative cover (%) of *E. milsumii*	Mean relative cover (%) of other plant species
A. Very disturbed exposed ridge spine	0	0	0	0	100
B. Less disturbed exposed ridge spine	80	22.7	1.2	52.6	47.4
C. Very disturbed sheltered ridge spine	0	0	0	0	100
D. Less disturbed sheltered ridge spine	30	12.1	0.5	23.1	76.9
E. Disturbed exposed rockface	30	6.9	0.3	18.7	81.3
F. Undisturbed sheltered rockface	40	23.7	0.4	30.9	69.1

Ten quadrats were sampled for each habitat category. Mean percent relative cover of *E. milsumii* is calculated by $[C_E/(C_E + C_N)] \times 100$ where C_E is mean percent cover of *E. milsumii* for all the quadrats sampled in a category and C_N is mean percent cover of all other plant species. This value subtracted from 100 then gives the mean percent relative cover of all other species

[a] Comparisons by *t*-tests indicated that mean covers of *E. milsumii* in each of the categories B, D, E and F were significantly higher ($P = 0.05$) than in A or C, and higher in B than E. For mean densities, values in B, D, E and F were significantly higher than in A or C, and higher in B than in either E or F

Table 5 Frequency of *Ilex praetermissa* along a 250-m transect sampling the uppermost 5-m zone of rockfaces (both north-facing and south-facing) on the Klang Gates quartz ridge

50-m intervals from origin	North-facing rockface	South-facing rockface
0–50 m	0	1
51–100 m	5	0
101–150 m	5	0
151–200 m	8	0
201–250 m	1	0
	Total 19	Total 1

The north-facing side is adjacent to good regenerated lowland forest and so is more sheltered and possibly less disturbed, compared to the south-facing side, which has little shelter from adjacent vegetation and overlooks orchards established in land formerly cleared of forest

Disturbance and *Eulalia milsumii*

Eulalia milsumii was reasonably common in the less disturbed habitats (whether exposed or sheltered ridge spine (Table 4: site categories B and D) and both rockfaces (E, F), and absent in very disturbed sites. This suggests that it may be, like *A. rupestris*, unable to establish or survive on very disturbed, more gentle terrain where very aggressive smothering weeds are common. Indeed, the mean cover per quadrat of weedy species (Table 3) is highest in the very disturbed sites on the ridge spine (A, C). Unlike *A. rupestris*, *E. milsumii* occurs in sheltered sites, but only in the less disturbed or undisturbed areas. This is also reflected in the mean relative cover of *E. milsumii* compared to other plant species (Table 4).

Ilex praetermissa along the disturbed and undisturbed rockfaces

Ilex praetermissa was surveyed along both (undisturbed, sheltered) north-facing and (disturbed, exposed) south-facing rock faces, the only site categories available for study of this species, an understory small tree not found on the main ridge spine. Its higher frequency along the northern face of the ridge (Table 5), where conditions are cooler and slightly more humid (Table 1), suggests that *I. praetermissa* either preferred conditions along the northern part for establishment, or survived better there (i.e. failed to establish or survive along the south-facing rock face).

Distribution of three endemic species compared

As *A. rupestris* is a small woody shrub, *E. milsumii* is a grass, and *I. praetermissa* a small understorey tree, their possible similarities and differences in distribution are of interest. The restriction of these species to the quartz ridge system here is probably most easily understood in terms of the contrasting environment on such ridges (drier conditions, poorer soil development and site fertility on a rocky outcrop) compared to the original surrounding lowland mixed dipterocarp forest (moister conditions, better soil development and fertility), where there would have been a more intense competition with a greater number of plant species. Plant communities on the quartz ridge are indeed simpler, with fewer species than in mixed dipterocarp forest (Reid 1959; Kiew 1982; Whitmore 1984). The relative intolerance of *A. rupestris* and *E. milsumii* for moister, better developed soils and its ability to thrive in hot, dry, exposed rock crevices appear to be well-selected adaptations for life on a quartz ridge. The apparent preference of *I. praetermissa* for rockfaces and rocky slopes may also be related to reduced competition, as relative slope instability and a rocky substrate present problems for plant establishment, as is also probably true of plants (rheophytes) restricted to the banks and flood zone of swift-flowing streams, where a similar substrate instability exists and the specialized plant diversity is reduced to a suite of well-selected species (Wong 1999).

Apart from their narrow endemism, disturbance (notably clearing, fire and trampling) seems to be the major factor that has impacted the distribution of all three species. *A. rupestris* grows well on both exposed and sheltered rock faces, where it had the highest frequencies recorded. On the ridge spine, *A. rupestris* seems to establish well only in the exposed sites. It appears to be sensitive to weedy growth in very disturbed, less steep sites, and unable to establish well on moister, better developed substrates in tree-sheltered sites. Unlike *A. rupestris*, which persisted in at least some very disturbed sites, *E. milsumii* was absent from all such sites and appears to establish and maintain itself best in less disturbed or undisturbed rock face and ridge spine sites, where the growth of weedy species was relatively less aggressive. *Ilex praetermissa* is more common along the less disturbed and more sheltered north-facing rock face and did not establish well along the disturbed, exposed south-facing rock face.

Generally, the population decline of the three endemic species on very disturbed sites could be due to at least two possible factors. The altered conditions after disturbance could be no longer suitable for the establishment, optimal growth or continued survival of these species. Alternatively, or additionally, these changed conditions favoured the colonisation or invasion by other (more weedy) species, which were superior competitors for space and resources. This study has established that fast-growing, smothering weedy growth of other species was important where *A. rupestris* and *E. milsumii* were either reduced in occurrence or absent, notably on the very disturbed, exposed ridge spine (site category A). In the case of *I. praetermissa*, limited comparisons here (restricted to two site categories) suggest that competition from weedy growth was not of overriding importance, as the extent of growth of weedy plant species was in fact greater along the undisturbed north-facing rockface (where there were more individuals of *I. praetermissa*) than the disturbed, south face (Table 3). Thus, it is possible that the more altered environment of the south face was unsuitable for establishment and continued survival of *I. praetermissa*. Unlike *E. milsumii* and *A. rupestris*, existing *I. praetermissa* are bigger plants not easily shaded out by the invading weedy species present; furthermore, its occurrence in the shade of taller trees suggests an extent of shade tolerance. A woody dicot tree, *I. praetermissa* develops a deeper and more extensive rooting system, and thus has potentially better survival against competition from smaller, invading weedy plants.

Plant size and other factors affecting survival of rare or endemic species in changed environments

The survival probability for rare or endemic species in disturbed environments could be influenced by plant size. Plant size can be categorized into small-sized plants (such as *A. rupestris* or *E. milsumii*) and bigger plants (e.g. *I. praetermissa*). Table 6 shows the various postulates for survival probability, with size. This shows the most likely adaptation of residual adults (either small plants or bigger plants) and new regeneration (essentially newly germinating or small plants) and their expected ability to compete with invading weedy species. For small-sized plants, remaining adults may not adapt to the new environmental

Table 6 The possible effect of plant size (considering residual adults versus new regeneration on site) on the ability of rare or endemic plants to adapt to changed site conditions following disturbance and ability to compete against other invading weedy species

Adult size	Generation	Adaptation to new/ changed environmental conditions	Ability to resist competition from invading weedy species
Small-sized plants	Remaining adults	May not adapt	May not resist
	New regeneration in the changed environment	May not adapt	May not resist
Bigger plants	Remaining adults	May not adapt	Likely to resist
	New regeneration in the changed environment	May not adapt	May not resist

See text for Discussion

conditions and also may not compete well with invading weedy species. For bigger plants, such as *Ilex*, the remaining adults are more likely to resist competition from invading weedy species (which are typically smaller ferns, grasses, or shrubs), but may not adapt to the changed new environment and so decline in performance. New regeneration of both small-sized and big-sized plants lack a size advantage in resisting competition from invading weedy species, and also may not adapt to the changed environmental conditions.

It was obvious that a number of species establishing in the disturbed sites on the Klang Gates quartz ridge were introduced species, e.g. *Clidemia hirta* and *Pennisetum purpureum*. Many biological invasions, i.e. the establishment of organisms in habitats where they are not native, with negative effects on the population already there, have been caused by invasive introduced species (Alpert 2006). Invading weedy species may affect the performance of residual plant populations in various ways. Thomson (2005), studying *Oenothera deltoidea* Torrey & Frém. ssp. *howellii* (Munz) W.Klein (Onagraceae) at the Antioch Dunes National Wildlife Refuge, California, found that the strongest impact of invasive weedy plants on this endemic was most likely the effect of invasive species thatch (accumulation of poorly decomposed leaf litter) that reduced soil disturbance (necessary for germination of the endemic plant seed in this case). As such, endemic plant recruitment, rather than superior resource competition on the part of the invasive, could be more important. She noted that many studies have generally presumed resource competition was the main impact of invasive plants on rare or endemic species. For *A. rupestris*, we have discussed the possibility that taller or fast-invading vegetation may alter substrate conditions, besides imposing greater competition for space and resources. There is thus a need to distinguish between adaptation of both remaining adults and new regeneration to changed environmental conditions following disturbance, and the ability of the resident species to compete for resources against invading weedy species.

It should be emphasized that although optimal habitat may be important in sustaining populations of such endemic species on site, still other possible factors need to be considered. For example, a required critical population size, from which some populations can expand into suboptimal habitats through continued subsidy from a core population (Primack and Miao 1992; Pulliam 1998), could also be important. Further work, especially experimental studies, are needed to demonstrate the factors operating in the potential diminishment of populations of *A. rupestris*, *E. milsumii* and *I. praetermissa* on the Klang Gates quartz ridge. In particular, the reproductive and physiological ecology of these species would repay further study. At a basic level, these could investigate the reproductive success of individuals and population samples, modes and efficacy of dispersal, comparative establishment success (both between sites and on different substrates), and the genetic diversity of populations in different habitat conditions.

Conservation importance of the Klang Gates quartz ridge

The Klang Gates ridge is the main feature in a series of quartz ridges around the city of Kuala Lumpur, and includes others such as in the Kanching, Kuang, South Ulu Gombak and Sungai Tua areas (Reid 1959, 1961). Although several of these ridges carried vegetation similar to that found on the Klang Gates quartz ridge, the one at Klang Gates is among the best developed and most important, and some of the others have already been adversely affected by development.

The present study demonstrates that populations of rare, endemic plant species such as *A. rupestris*, *E. milsumii* and *I. praetermissa* are unable to tolerate severe changes in the environment, especially that which encourages weedy, fast-smothering species invading disturbed sites. There are many other rare plant species found at Klang Gates that have not been adequately studied, including *Acrymia ajugiflora* Prain (Lamiaceae), *Anerincleistus pauciflorus* Ridl. (Melastomataceae), *Begonia clivalis* Ridl. (Begoniaceae), *Capparis larutensis* King (Capparaceae), *Carallia suffruticosa* Ridl. (described by Ridley as *C. euryoides* Ridl.) (Rhizophoraceae), *Licuala kiahii* Furtado (Arecaceae) and *Mesua elegans* (King) Kosterm. (Clusiaceae) (Kiew 1982, 1983).

The Klang Gates quartz ridge is a remarkable natural feature with unique plant communities, including a suite of endemic species that are best protected on site (Wise and Zahid 1999; also highlighted in the local press by Tan 1998; Cheang 2004). Its adequate conservation management requires good legislation and enforcement, appropriate regulation of visitation on site, careful restoration of severely transformed sites, and provision of sufficient interpretation and information to encourage public awareness for the special features of the ridge.

Acknowledgments University of Malaya Science Fund project 04-01-03-SF0175 supported facilities, logistics and research materials. V. W. C. Wong, S. P. Ong, Y. W. Low and Zulkapli Ibrahim assisted in the field. Serena Lee (Herbarium of the Singapore Botanic Gardens) assisted with the verification of *Eulalia* material. Comments by two anonymous reviewers and the editors have benefited revision of the original manuscript.

References

Alpert P (2006) The advantages and disadvantages of being introduced. Biol Invas 8:1523–1534. doi:10.1007/s10530-005-5844-z

Cheang M (2004) Klang Gates Quartz Ridge- Precious Monument. The Star (Malaysia), 23 March 2004

Henderson R (1928) The flowering plants of Kuala Lumpur, in the Malay Peninsula. Gard Bull Straits Settlements 4(6–10):211–373

Kiew R (1982) The Klang Gates Ridge. Malay Nat 36(1):22–28

Kiew R (1983) Conservation of Malaysian plant species. Malay Nat 37:1–5

Primack RB, Miao SL (1992) Dispersal can limit local plant distribution. Conserv Biol 6:513–519. doi:10.1046/j.1523-1739.1992.06040513.x

Pulliam HR (1998) Sources, sinks and population regulation. Am Nat 132:652–662. doi:10.1086/284880

Reid JA (1951) Klang Gates and Bukit Takun: Reflections of an Amateur Botanist. Malay Nat J 5:109–123

Reid JA (1959) Plants of the Quartz Ridges. Malay Nat J 14:22–32

Reid JA (1961) Conservation and the Quartz ridges. In: Anon (ed) Nature Conservation in Western Malaysia. Special issue to mark the occasion of the twenty-first anniversary. Malayan Nature Soc, Kuala Lumpur

Ridley HN (1922) The Flora of Klang Gates, Selangor. J Federated Malay S Museums 10:247–251
Tan CL (1998) The Giant that needs protection. The Star (Malaysia), 14 April 1998
Thomson DM (2005) Measuring the effects of invasive species on the demography of a rare endemic plant. Biol Invas 7:615–624. doi:10.1007/s10530-004-5853-3
Whitmore TC (1984) Tropical rain forests of the far east, 2nd edn. Clarendon Press, Oxford
Wise R, Zahid MS (1999) Aleisanthia rupestris: botanical gem of Kuala Lumpur. Plant Talk 18:30
Wong KM (1998) Patterns of plant endemism and rarity in Borneo and the Malay Peninsula. In: Peng CI, Lowry PPII (eds) Rare, threatened and endangered floras of Asia and the Pacific Rim, Monograph, vol 16. Inst of Botany, Academia Sinica, Taipei
Wong KM (1999) The nature of the Brunei flora. In: Wong KM, Kamariah AS (eds) Trees and forests of Brunei Darussalam. University of Brunei Darussalam, Forestry Department Brunei & Brunei Shell Petroleum Co. Sdn. Bhd, Brunei
Wong KM, Saw LG, Kochummen KM (1987) A survey of the forests of the Endau-Rompin area, Peninsular Malaysia: principal forest types and floristic notes. Malay Nat J 41:125–144
Wong KM, Zubaid A, Lee PKY, Sukumaran J (2005) Biodiversity and Ecological Aspects in the proposed Selangor State Park. Special significance, potential and relevant management approaches. Unpublished report. Selangor Town & Country Planning Department, Treat Every Environment Special (TrEES), Selangor
Zahid MS (1998) Ekologi Pemuliharaan Aleisanthia rupestris (Ridl.) Ridl. (Rubiaceae), Species yang Endemik di Permatang Kuarza Genting Kelang (Klang Gates). [Conservation ecology of Aleisanthia rupestris (Rubiaceae), endemic to the Klang Gates quartz ridge.] Dissertation, University of Malaya

Biodivers Conserv (2010) 19:449–469
DOI 10.1007/s10531-008-9482-5

ORIGINAL PAPER

Bats of Borneo: diversity, distributions and representation in protected areas

Matthew J. Struebig · Lenny Christy · Dorothea Pio · Erik Meijaard

Received: 1 April 2008 / Accepted: 17 September 2008 / Published online: 3 October 2008

Abstract Protected areas are valuable in conserving tropical biodiversity, but an insufficient understanding of species diversity and distributions makes it difficult to evaluate their effectiveness. This is especially true on Borneo, a species rich island shared by three countries, and is particularly concerning for bats, a poorly known component of mammal diversity that may be highly susceptible to landscape changes. We reviewed the diversity, distributions and conservation status of 54 bat species to determine the representation of these taxa in Borneo's protected areas, and whether these reserves complement each other in terms of bat diversity. Lower and upper bound estimates of bat species composition were characterised in 23 protected areas and the proposed boundaries of the Heart of Borneo conservation area. We used lower and upper bound estimates of species composition. By using actual inventories, species representation was highly irregular, and even if some reserves were included in the Heart of Borneo, the protected area network would still exhibit low complementarity. By inferring species presence from distributions, composition between most reserves was similar, and complementarity was much higher. Predicting species richness using abundance information suggested that bat species representation in reserves may lie between these two extremes. We recommend that researchers better sample biodiversity over the island and address the conservation threats faced in Borneo both within and outside protected areas. While the Heart of Borneo Initiative is commendable, it should not divert attention from other conservation areas.

Keywords Heart of Borneo · Chiroptera · Gap analysis · Species richness prediction · Southeast Asia · Forest · Tropical conservation · Indonesia · Malaysia · Brunei

M. J. Struebig (✉)
School of Biological and Chemical Sciences, Queen Mary University of London,
London E1 4NS, UK
e-mail: m.struebig@qmul.ac.uk

L. Christy · E. Meijaard
The Nature Conservancy, East Kalimantan Program, Jalan Polantas no. 5, Markoni,
Balikpapan, East Kalimantan, Indonesia

D. Pio
Department of Ecology and Evolution, Biophore, University of Lausanne, 1015 Lausanne, Switzerland

Introduction

The survival of tropical biodiversity is threatened in most regions, so its conservation requires that it receives adequate protection and that habitats are sustainably managed (du-Toit et al. 2004). The majority of protected areas have proved valuable in this regard (Bruner et al. 2001). However, many protected areas were designed for purposes other than conservation, and hence diversity, particularly of low-profile species, is under-represented (Rodrigues et al. 2004). Identifying the species that are represented in reserves and those that are missing is therefore crucial for conservation planning, and the distributions of species and spatial patterns of species richness have become widely used in evaluations of conservation effectiveness (Jennings 2000). These 'gap' analyses range from overlays of species distributions and protected area boundaries to generate inventories, to sophisticated assessments using niche modelling. In addition to identifying gaps in protected area networks these analyses can also determine the extent to which reserves complement each other in terms of biodiversity protection, and whether unique diversity remains outside the system. Unfortunately, analyses are often restricted to well-known taxa because species distribution information for most taxa is limited or unavailable.

Borneo is well known for its high species richness (Mackinnon et al. 1996; Sodhi et al. 2004), but high deforestation rates are of serious conservation concern and place enormous pressures on the island's protected area system (Curran et al. 2004; Fuller et al. 2004). Nine percent of Borneo is within protected areas of IUCN categories I–IV (IUCN 1994), administered by Indonesia (in Kalimantan, four provinces), Malaysia (in Sabah and Sarawak) and Brunei Darussalam. However, these reserves have become increasingly isolated (Curran et al. 2004), leading the three countries to recently commit to the Heart of Borneo Initiative (HoB), which seeks to connect 22 parks and reserves into a continuous 24 million ha transboundary conservation area (WWF 2005). While this initiative is of clear conservation merit, there is little biodiversity information available for much of the area, and no formal assessment has yet determined whether island-wide levels of faunal diversity are adequately represented—particularly for lesser known, low-profile animal groups.

Bats exhibit many of the required features of indicators for biodiversity appraisals: they are diverse (at least 93 species in Borneo, Simmons 2005; ca. 40% of the island's land mammals, Payne et al. 2000); perform ecosystem services (Fujita and Tuttle 1991; Hutson et al. 2001); and comprise assemblages with clear and predictable guild/ensemble structure (Kingston et al. 2003; McKenzie et al. 1995). Many of these animals also depend on forests to some extent and are sensitive to disturbance and landscape changes (Danielsen et al. 2008; Struebig et al. 2008; Zubaid 1993); as a result losses of bat species in Southeast Asia are predicted to exceed 40% by 2100 (Lane et al. 2006). Most of Borneo's old world fruit bats (Pteropodidae) exhibit dispersal capabilities and generalist feeding habits that support their persistence in disturbed landscapes (Meijaard et al. 2005). Likewise, insectivorous species that forage in forest edges and tree fall gaps (e.g. Vespertilionid sub-families Miniopterinae and Vespertilioninae, *Emballonura* spp.) or the open spaces above or outside forests (e.g. Molossidae, *Taphozous* spp.), may be readily adapted to exploit disturbed habitats (Kingston et al. 2003; Meijaard et al. 2005). However, insectivorous species that forage in the forest interior (Hipposideridae, Megadermatidae, Nycteridae, Rhinolophidae, vespertilionid sub-families Kerivoulinae and Murininae) are likely to be particularly sensitive to landscape change because of ecomorphological traits that restrict their use of more open habitats (reviewed in Kingston et al. 2003).

Knowledge of bat distributions in Borneo has been historically biased towards easily captured species, with many appearing confined to the north of the island where most research and specimen collection has been undertaken (reviewed in Payne et al. 2000; Suyanto and Struebig 2007). However, research effort has increased in recent years, and improved communication amongst local and regional experts has led to distributions being substantially updated via the Southeast Asian Mammal Databank (SAMD) (Boitani et al. 2006). As part of the IUCN Global Mammal Assessment, the SAMD team worked with a network of institutions and taxonomic experts to compile and disseminate all information on the distribution, basic ecology and conservation status of Southeast Asian mammals. The main output is an impressive database on mammal diversity that includes distribution maps of the extent of occurrence of each species in a freely available GIS format. However, the database is still limited by a paucity of information for a substantial number of taxa, and will quickly become outdated as more empirical mammal research is undertaken in the region. This is particularly evident for bats, of which much research undertaken in recent years was not considered by the SAMD (e.g. in Borneo, Anwarali et al. 2007; Jayaraj et al. 2006; Struebig et al. 2006a, b; Suyanto and Struebig 2007).

We sought to determine the known and predicted representation of bat species in Borneo's protected areas (including the proposed Heart of Borneo conservation area) using reviewed and updated species distributions. We used lower and upper bound estimates of species composition derived from simple gap analyses to explore the extent to which each reserve protected a unique bat fauna, and whether this would be represented in the Heart of Borneo. We then tested the quality of these estimates and their value in describing true species diversity by predicting the likely species richness at a site based on existing assemblage data. Finally, we identified conservation issues relevant to bats inside and outside protected areas on Borneo to guide future conservation research efforts.

Methods

Data

Analyses were based on a database of bat locality records from all seven administrative units of Borneo, compiled after consultation with international and regional bat specialists. We included records from published and unpublished sources, as well as specimens collected since 1995 in the zoological collections of Bogor, Indonesia; Bandar Seri Begawan, Brunei Darussalam; and Kuching, Malaysia. For species that are notoriously hard to identify, or for which taxonomy is not well defined, we consulted the authors/collectors for additional information to confirm their identity. Because of a general paucity of mammal records in Indonesian territory (Meijaard and Nijman 2003), additional data were obtained from surveys of 11 sites undertaken between 2002 and 2008 in Central and East Kalimantan (Struebig et al. 2006a, b; Suyanto and Struebig 2007; D. Pio et al., Unpublished data). We considered the only conspicuous species likely to be correctly identified by non-specialists to be the large flying-foxes, *Pteropus vampyrus* and *P. hypomelanus*. Hence, we included opportunistic records provided by local people and conservation practitioners for these two species, but only when information was given by more than one source in an area (89 combined records). To our knowledge, and to the agreement of the specialists we consulted, our database represented the most up-to-date knowledge of bat distribution records in Borneo. However, because the majority of records arose from captures in mist-nets or harp-traps set at ground level, some species records remained particularly infrequent. Thus, subsequent analyses focused on a subset of species that comprised the majority of records. These included the 17 Bornean fruit bats within the Pteropodidae, and 37 forest interior insectivorous species (sensu Kingston et al. 2003). These species are readily captured in mist-nets (Pteropodidae) or harp-traps (bats of the forest interior) used for most bat diversity surveys on the island. Eleven of these species are currently IUCN red listed (Boitani et al. 2006).

Bat species composition in protected areas

Bat species composition was determined in 23 protected areas distributed throughout the island (Fig. 1), and also within the proposed boundaries of the Heart of Borneo (According to WWF in 2007). We did not consider all protected areas on Borneo since the boundaries of many reserves have changed substantially over time, and not all are recognised by IUCN/WCMC (http://www.wdpa.org/). The reserves we selected comprised a substantial proportion of those in the existing Borneo network, and included the largest parks both

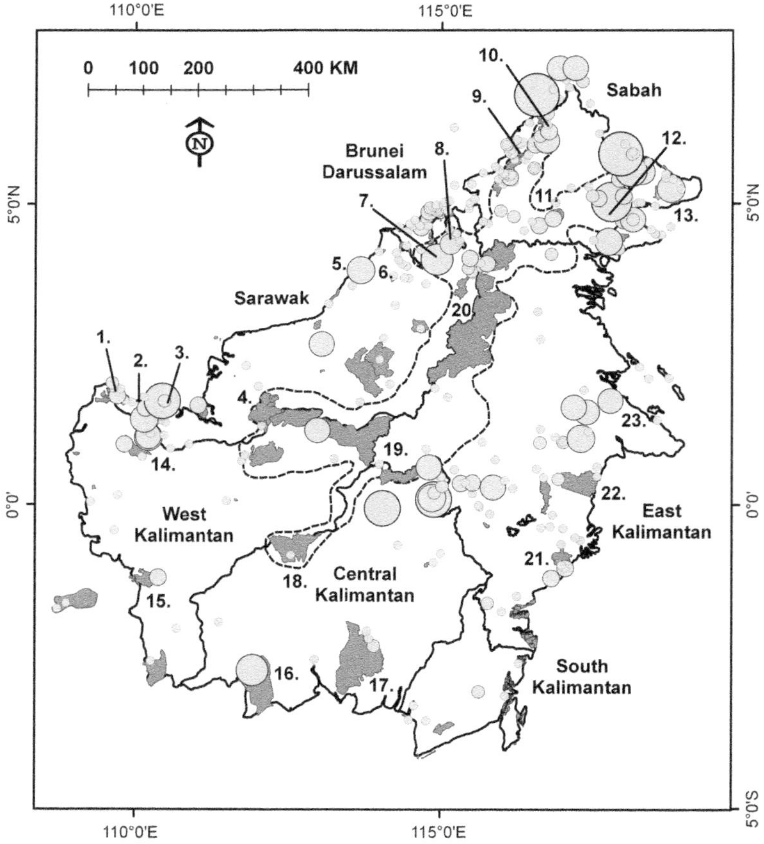

Fig. 1 Locality records of bats on Borneo overlaid with the current protected area network (dark shading), and the proposed Heart of Borneo boundaries (broken line, according to WWF in 2007). Bubble size is proportional to the number of bat species reported at the site, and numbers refer to the protected areas used in analyses and listed in Table 2. Lines indicate state/province boundaries

within and outside of the Heart of Borneo boundaries. Three estimates of species composition within each area were derived, which gave lower and upper bounds of species richness.

The first estimate was derived from the actual species records within each area and generated a lower bound of species composition from confirmed inventories. For this lower bound estimate all bat-locality records were assigned a latitude and longitude co-ordinate, entered into Arcview GIS (ESRI 1982–1998), and overlaid with a thematic layer of protected areas (based on WCMC data, but updated with local government data if applicable) to derive species lists.

The second estimate determined an upper bound to species composition in an area by identifying species potentially present using a simple gap analysis. We overlaid the species distribution polygons provided by the SAMD with protected area boundaries; all species that had a distribution overlapping with a boundary, regardless of the extent, were included in an inventory.

The third estimate used a similar gap analysis, but this time based on species distributions updated or corrected by our bat locality record database. We overlaid our bat locality records with the SAMD species distributions and digitised novel distribution polygons that encompassed new records if appropriate. We also corrected the distributions of species that predominantly roost in caves to exclude them from the coastal wetlands on the South coast. This represents an area of peat swamps and mangroves between the Barito and Kumai Rivers (see Langner et al. 2007 for delineation of these forest types) in which there are no known caves, and in which intensive surveys have not captured cave-roosting bat species (Struebig et al. 2006b).

We used the Bray–Curtis dissimilarity index to determine overall differences in bat species composition in protected areas, and hence assess the extent to which each reserve protected a unique subset of the bat species pool. Dissimilarities between reserve inventories based on each of the three estimates were represented using Non-Metric Multidimensional Scaling (NMDS) in PC-ORD (McCune and Meffor 1999). Ordinations were conducted on Bray–Curtis dissimilarity coefficients derived from presence–absence data at each protected area using 500 iterations and 250 runs of both real and randomised data. Two dimensions were chosen based on the reduction of stress from additional axes. We included the Heart of Borneo inventory for inspection of dissimilarity in bat species composition between this area and reserves outside of the proposed boundaries.

Species richness predictions from abundance data

We checked the consistency of our gap-derived estimates of bat species richness by further predicting richness using abundance data from seven standardised harp-trap inventories that provided this information. Given that harp-traps best target forest interior insectivorous species, we restricted our analyses to this bat ensemble (maximum 37 species). We used two multinomial prediction models introduced by Solow and Polasky (1999) and Shen et al. (2003) to predict species richness after an additional sampling effort. Although other methods for predicting species richness exist, they can become unstable or do not behave as expected at high predicted sample sizes (Shen et al. 2003). Furthermore, predictors and estimators have already been tested in Malaysian harp-trap inventories by Kingston (2009), and all but two led to substantial overestimation of bat species richness. Hence we used the two reliable predictors to create upper and lower bounds of species richness. We used a total predicted sample effort of 1,000 individuals because most species in this ensemble were represented by this effort in Malaysia (Kingston 2009).

Results

We amassed 1,811 independent bat-locality records of 96 species representing all families in Borneo, and at least 634 of these were additional to those available to the SAMD. Records per species ranged from only one for *Hypsugo impricatus* (previously *Pipistrellus impricatus*) to 89 records for *Cynopterus brachyotis* (Table 1). The majority of records represented the 17 frugivorous species within the Pteropodidae (576 records or 31.8% of total), while a further 774 records (42.7%) represented the 37 forest interior insectivorous species. The remainder (452 records, 25.5%) were of patchily recorded and/or poorly known species of the families Vespertilionidae, Emballonuridae or Molossidae (42 species). With 662 bat-locality records, Sabah remained the most studied part of Borneo, but by accounting for area the overall coverage in Brunei was more complete (0.87 records per 100 km^2 in Sabah vs. 2.17 for Brunei; Table 1; Fig. 1). Geographic coverage in Kalimantan remained poor compared to Sabah, Sarawak and Brunei: the least studied state was South Kalimantan with only 0.05 records per 100 km^2, followed by West Kalimantan with 0.06 records per 100 km^2.

Bat species composition in protected areas

Our three estimates of bat species composition yielded strikingly different reserve inventories (Table 2). Lower bound estimates derived from actual survey data in each reserve produced largely incomplete inventories compared to the bat species pool. The most species-rich protected area (according to the subset of 54 species) was Gunung Kinabalu in Sabah with 32 recorded species (59% of total). Danum Valley in Sabah and Sangkulirang in East Kalimantan also had relatively high inventories (29 species; 54% of total), followed by Ulu Temburong in Brunei (27 species, 50% of total). Conversely we could find only one species record (<2% total) for two of the reserves, Gunung Gading and Loagan Bunut in Sarawak, and the lists for Bukit Baka-Bukit Raya in West/Central Kalimantan and Kutai in East Kalimantan were limited to two and three species respectively (<6% total). Species poor inventories were often characterised by depauperate lists of forest interior insectivorous bats, particularly in Kalimantan and Sarawak where the majority of inventories listed <30% of the species pool. The high contrast between species-poor and species-rich inventories resulted in large differences between the protected areas in their representation of bat diversity (Fig. 2a). Wide scatter of protected areas in the NMDS plot highlighted large differences between the reserves, and also between the Heart of Borneo and reserves far outside the boundaries (e.g. No. 15 in Fig. 1; Table 2: Gunung Palung). By this evaluation the Heart of Borneo would support 81% of the bat species we considered (44 species), but complementarity between this area and reserves outside the boundaries would be low because these reserves represented depauperate faunas.

Upper bound estimates from SAMD-derived species distributions resulted in larger bat inventories in all protected areas (Table 2). According to this estimate, Kayan-Mentarang and Maliau Basin were predicted to contain almost 75% of the bat species pool (40 species each; Table 2), though the species composition of each reserve was markedly different (Fig. 2b). Much of the variation in species composition among reserves was accounted for by these protected areas and others in Sabah and Sarawak included within the Heart of Borneo boundaries, suggesting that this conservation area would protect substantial bat diversity. Indeed the Heart of Borneo would protect almost 95% of the pool of bat species (51 species; Table 2) according to this estimate. However, large distances on the NMDS plot between reserves within the Heart of Borneo and those in the south of the island

Table 1 Characteristics of the 54 bat species used in analyses and the number of independent locality records for these taxa in each state/province of Borneo

Family/species (previous species name)	Foraging strategy[a]	Red list status[b]	Range[c]	Roost ecology[d]	No. independent locality records							
					Sabah	Saraw	Brunei	W.Kal	C.Kal	S.Kal	E.Kal	∑
Pteropodidae												
Aethalops aequalis (*alecto*)	Bf		1	f?	15	9	2				1	27
Balionycteris maculata	Bf	NT	3	f	16	18	6	2	8		8	58
Chironax melanocephalus	Bf		4	f, c	2	2	1	1	3		3	12
Cynopterus brachyotis	Bf		6	f, c	36	23	3	5	10		12	89
C. horsfieldii	Bf		4/5	c, f	10	12		1	1		5	29
C. minutus	Bf		3	f			5				2	7
Dyacopterus spadiceus	Cf	NT	3	c, f?	4	10	2	2	1		3	22
Eonycteris major	Cf		1	C	8	10	2			1	1	22
E. spelaea	Cf		6	C	13	13	2	1		1	4	34
Macroglossus minimus	Bf		6	f	33	14	6	5	4	1	6	69
Megaerops ecaudatus	Bf		3	f	10	10	4	2	3		3	32
M. wetmorei	Bf	VU	2	f			4		2		1	7
Penthetor lucasi	Bf/Cf		3	C, f	13	19	5	4	4		7	52
Pteropus hypomelanus	Cf		5	F	5	3					2	10
P. vampyrus	Cf	VU	5	F	16	13	7	2	4		38	80
Rousettus amplexicaudatus	Cf		6	C	8	5	2	4			1	20
R. spinalatus	Cf	VU	2	C	2	3			1			6
Hipposideridae												
Coelops robinsoni	Ni		3	f, c		1			3		1	5
Hipposideros ater	Ni		6	c	8	5		1	3		3	20
H. bicolor	Ni		6	C	3	5	1		2		3	14

Table 1 continued

Family/species (previous species name)	Foraging strategy[a]	Red list status[b]	Range[c]	Roost ecology[d]	No. independent locality records							
					Sabah	Saraw	Brunei	W.Kal	C.Kal	S.Kal	E.Kal	∑
H. cervinus	Ni/Ei		6	C	17	13	1	3	13		11	58
H. cineraceus	Ni		6	C	7	2		2	1		5	17
H. coxi	Ni	DD	1	c		4			3			7
H. diadema	Ni/Ei		6	C, f	14	6	5	2	8	1	5	41
H. doriae (*sabanus*)	Ni	NT	3	f?	3	1	1		2		1	8
H. dyacorum	Ni		2	C, f	8	7	1	1	1		5	23
H. galeritus	Ni/Ei		6	c	11	7	1	2	2	1	7	32
H. larvatus	Ni/Ei		6	C		5		4		2	8	19
H. ridleyi	Ni	VU	2	f	4	2	2		1		1	10
Megadermatidae												
Megaderma spasma	Ni		6	f, c	16	5	1	2	2	1	5	32
Nycteridae												
Nycteris tragata (*javanica*)	Ni		4/5	f, c	6	10	1		5	1	3	26
Rhinolophidae												
Rhinolophus acuminatus	Ni		5	?	13	1			1		1	16
R. affinis	Ni/Ei		5	C	1	5		2	3	1	7	19
R. arcuatus	Ni		3	C?		2			2		1	5
R. borneensis	Ni		5	C	18	11	1	5	10	1	13	59
R. creaghi	Ni		4	C	11	1			1		9	22
R. luctus	Ni		6	f, c	9	8		3	7	1	1	29
R. philippinensis	Ni		6	C	4	7		1	2		1	15
R. pusillus	Ni		3	C	2						6	8

Table 1 continued

Family/species (previous species name)	Foraging strategy[a]	Red list status[b]	Range[c]	Roost ecology[d]	No. independent locality records							
					Sabah	Saraw	Brunei	W.Kal	C.Kal	S.Kal	E.Kal	∑
R. sedulus	Ni	NT	2	f	11	5	4	2	13		2	37
R. trifoliatus	Ni		6	f	16	4	5	2	10		7	44
Vespertilionidae (Kerivoulininae)												
Kerivoula hardwickii	Ni		6	f	7	3	1	1	5		4	22
K. intermedia	Ni	NT	2	f	10	5	1		4		3	23
K. minuta	Ni	NT	2	f	12	2	1		2		8	25
K. papillosa[e]	Ni		6	f	12	10	6		7		5	40
K. pellucida	Ni		3/4	f	8	3	2	1	4		1	19
K. whiteheadii	Ni		3	f	1	2						3
Phoniscus atrox	Ni	NT	3	f	5				3		1	9
P. jagorii	Ni		4	f	1				2			3
Vespertilionidae (Murininae)												
Harpiocephalus harpia (mordax)	Ni	VU	6	f	3							3
Murina aenea	Ni	VU	2	f	4				3			7
Murina cyclotis	Ni		6	f	6	2	3		7			18
Murina rozendaali	Ni	VU	2	f	7	1			1		1	10
Murina suilla	Ni		3	f	10	3			8		5	26
Total records					459	312	89	63	182	12	231	1,348
Other bat records					203	77	36	26	69	5	47	463

Table 1 continued

Family/species (previous species name)	Foraging strategy[a]	Red list status[b]	Range[c]	Roost ecology[d]	No. independent locality records							
					Sabah	Saraw	Brunei	W.Kal	C.Kal	S.Kal	E.Kal	∑
Coverage (records per 100 km^2)					0.87	0.31	2.17	0.13	0.16	0.05	0.06	0.23

Taxonomy follows Simmons (2005)

[a] Foraging strategies assigned based on wing morphology (Kingston et al. 2003; M. Struebig, Unpublished data) following those described in McKenzie et al. (1995): Ni, insectivorous species that forage in narrow-spaces or clutter (Strategy 1); Ei, insectivorous species that forage in partially cluttered edges and/or canopy edges (Strategy 2); Bf, frugivorous or nectarivorous species that forage in clutter below forest canopy (Strategy 4); Cf, frugivorous or nectarivorous species that forage in open areas and over large distances (Strategy 5)

[b] According to the SAMD (2006): DD data deficient; NT near threatened; VU vulnerable

[c] Species distribution according to the SAMD (2006) and categorised as (1) endemic to Borneo; (2) restricted to Borneo and one neighbouring land mass (e.g. Peninsular Malaysia, Sumatra or Philippines); (3) restricted to Borneo and two neighbouring land masses; (4) patchy distribution in three or more land masses; (5) large and covers majority or all of Southeast Asia; (6) large and extends beyond Southeast Asia

[d] Known roost types classified as f, tree cavities or foliage; and c, caves or other large permanent structures such as buildings and mines. Capitalisation indicates colony size: lower case, typically 1–100 individuals; upper case, >100 individuals. ? indicates that limited roost information is available

[e] The taxonomy of *Kerivoula papillosa* is being revised (K. Helgen, Personal Communication), and some individuals currently assigned to this species in Borneo may represent *K. lenis*, *K. flora*, or an undescribed cryptic species. Thus, for the purposes of consistency we assigned all records of these taxa to *K. papillosa*

Table 2 Bat species richness in protected areas in Borneo according to the observed species richness from inventories, and two upper bound estimates from gap analyses

No.[a]	Protected area	Habitats[b]	Approx. area (km^2)	Species richness, S[c]														
				Pteropodidae (max. 17 spp.)			Forest interior (max. 37 spp.)			Rhinolophidae (max. 10 spp.)			Hipposideridae (max. 12 spp.)			Kerivoulinae/ Murinae (max. 13 spp.)		
				Obs	SAMD	New	Obs	SAMD	New	Obs	SAMD	New	Obs	SAMD	New	Obs	SAMD	New
1	Gunung Gading NP, Sarawak	H	41	1	11	14	0	22	25	0	5	7	0	10	9	0	5	7
2	Kubah NP, Sarawak	L, H	22	7	12	13	11	22	26	4	5	7	4	10	10	1	5	7
3	Bako NP, Sarawak	L, M	27	6	11	13	17	22	26	4	5	7	8	10	10	3	5	7
4	Batang Ai NP, Sarawak	H, U	1,688	2	12	14	0	21	23	0	6	6	0	8	8	0	5	7
5	Niah NP, Sarawak	K	31	7	12	14	8	20	23	2	5	6	3	8	8	1	5	7
6	Loagan Bunut NP, Sarawak	S, L	107	1	13	15	0	21	24	0	5	6	0	8	8	0	6	8
7	Gunung Mulu NP, Sarawak	K	528	8	15	16	13	23	24	3	6	6	6	8	8	3	7	8
8	Ulu Temburong NP, Brunei	L, H	489	11	15	16	16	20	24	3	5	6	7	5	8	5	8	8
9	Crocker Range NP, Sabah	H	1,399	8	13	14	7	23	27	3	5	5	2	9	9	1	7	11
10	Gunung Kinabalu NP, Sabah	L, H, U	754	11	12	14	21	25	25	5	6	5	7	8	9	7	9	9
11	Malinau Basin CF, Sabah	H, L	390	4	13	15	7	27	29	4	8	8	2	8	8	1	9	11
12	Danum Valley CF, Sabah	L, H	427	7	11	14	22	28	29	5	7	7	6	9	9	10	10	11
13	Tabin WR, Sabah	L	1,120	5	10	13	15	26	27	4	7	7	5	9	9	5	8	9
14	Gunung Nyiut NP, West Kal.	H, U	1,800	6	11	14	4	22	25	2	5	7	2	10	10	0	5	6
15	Gunung Palung NP, West Kal.	M, S, L, H	900	3	11	13	6	20	22	3	5	5	3	9	9	0	4	6
16	Tanjung Puting NP, Cent. Kal.	M, S, L	3,550	6	12	11	17	20	19	3	6	4	2	10	3	10	4	10
17	Sebangau NP, Cent. Kal.	M, S	5,678	4	12	11	7	21	18	2	5	3	0	9	5	5	5	8
18	Bukit Baka-Raya NP, West/Cent. Kal.	H, U, L	1,810	1	10	12	1	20	24	1	5	5	0	8	9	0	5	8

Table 2 continued

No.[a]	Protected area	Habitats[b]	Approx. area (km^2)	Species richness, S^c														
				Pteropodidae (max. 17 spp.)			Forest interior (max. 37 spp.)			Rhinolophidae (max. 10 spp.)			Hipposideridae (max. 12 spp.)			Kerivoulinae/ Murinae (max. 13 spp.)		
				Obs	SAMD	New	Obs	SAMD	New	Obs	SAMD	New	Obs	SAMD	New	Obs	SAMD	New
19	Betung Kerihun NP, West Kal.	H, U, L	8,000	5	12	15	6	21	25	3	5	6	3	9	10	0	5	7
20	Kayan Mentarang NP, East Kal.	H, U, L	16,000	7	15	16	1	25	28	1	6	8	0	10	9	0	7	9
21	Bukit Soeharto WR, East Kal.	L, H	618	4	11	14	4	20	25	1	5	7	0	9	9	3	4	7
22	Kutai NP, East Kal.	L, H	1,986	2	12	15	1	20	25	1	7	7	0	8	9	0	3	7
23	Sangkulirang [proposed], East Kal.	K	2,000	7	11	14	22	22	27	5	6	8	8	9	9	7	5	8
24	HEART OF BORNEO	H, L, U, K	240,000	13	17	17	31	34	34	7	9	9	10	11	11	12	12	12

[a] Protected status: NP, national park; WR, wildlife reserve (*Cagar Alam*); CF, conservation forest

[b] Habitat types: M, mangrove; S, peat or riverine swamp forest; L, lowland Dipterocarp or heath forest; H, hill dipterocarp; U, montane forest; K, karst

[c] Three estimates of species richness: Obs., a lower bound from species-locality records in a reserve; SAMD, an upper bound derived from gap analysis using species distributions from the Southeast Asian Mammal Databank (SAMD); New, an upper bound derived from gap analysis using revised species distributions based on new species-locality records. Each estimate is partitioned for the main families/sub-families with forest interior insectivorous bats (Hipposideridae, Megadermatidae, Nycteridae, Rhinolophidae and vespertilionids of the sub-families Kerivoulinae and Murininae) shown separately. The total species richness can be derived by summing Pteropodidae and forest interior bats (max. = 54 species)

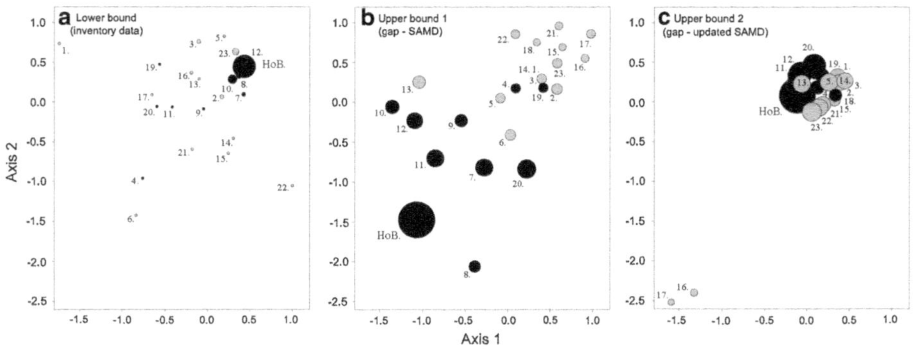

Fig. 2 Bat assemblage composition in protected areas in Borneo according to lower and upper bound inventories: **a** lower bound using observed data from biodiversity studies [stress = 0.23]; **b** upper bound prediction using overlays of species distribution polygons from the SAMD [stress = 0.13]; **c** upper bound prediction using distribution polygons revised as part of this study [stress = 0.09]. Each point is a non-metric multidimensional scaling (NMDS) representation of bat species composition (maximum 54 species) in a protected area, and distances between points reflect dissimilarity using the Bray–Curtis coefficient. Stress is a measure of the poorness-of-fit of the NMDS, with values <0.20 considered to indicate more reliable ordinations (McCune and Meffor 1999). Numbers on the plots refer to the protected areas in Table 2 (not all labelled for clarity), and bubble size is proportional to species richness. Black bubbles indicate protected areas that are within the current Heart of Borneo boundaries

suggested that species composition was still markedly different within and outside the Heart of Borneo boundaries (Fig. 2b).

After reviewing species-locality records we updated the SAMD data for 29 bat species (54% of total), and the majority of these changes were to extend distributions southwards following new records from Kalimantan. The subsequent upper bound estimates resulted in bat inventories that were higher than those from both other estimates in all but two of the protected areas—Tanjung Puting and Sebangau in Central Kalimantan (Table 2). The two distribution-based estimates identified the same reserves as the most species rich, but other areas far outside of the Heart of Borneo boundaries were shown to also exhibit high bat species richness when the updated distributions were used (e.g. Sangkulirang, East Kalimantan, 41 species; Gunung Nyiut, West Kalimantan, 39 species). Bat species representation in the Heart of Borneo remained high using this estimate (51 species, 95%). Furthermore, the scatter of points on the NMDS plot was greatly reduced when using inventories based on this estimate (Fig. 2c) suggesting that protected areas complemented each other in bat species they contained. Most reserves, both inside and outside of the Heart of Borneo were clustered together on the plot indicating strong similarities in bat species composition. However, the bat species composition of the two parks with depauperate bat faunas, was considerably different to that of other sites (Fig. 2c).

Species richness predictions from abundance data

Predicting bat species richness using species abundance data revealed that all but two harp-trap inventories (Tanjung Puting in Kalimantan, and Bau in Sarawak) were incomplete, and that the number of trappable species represented in the predicted inventories exhibited much variation between sites (Table 3). Bat inventories in the karst forests around Bau and the oligotrophic forests of Tanjung Puting remained depauperate subsets of the larger faunas of other sites. Barito Ulu, at the geographic centre of Borneo, supported the most

Table 3 Observed and predicted number of insectivorous bat species susceptible to capture by four-bank harp-trap at seven sites in Borneo, based on two multinomial species richness prediction models used on abundance data

	n	Species richness, S						
		Obs.	Solow and Polasky (1999) predictor			Shen et al. (2003) predictor		
			Predicted S[a]	Inventory completeness (%)[b]	Trappable species (%, max. = 37)[c]	Predicted S[a]	Inventory completeness (%)[b]	Trappable species (%, max. = 37)[c]
Bau, Sarawak (Mohd-Azlan et al. 2005)	83	13	14	93	35–38	14	93	35–38
Danum Valley, Sabah (T. Kingston, Unpublished data)	147	17	23	74	46–62	19	90	46–51
Sepilok, Sabah (Francis 1990)	442	17	24	71	46–65	23	74	46–62
Barito Ulu, Cent. Kalimantan (D. Pio, Unpublished data)	636	26	31	84	70–84	30	87	70–81
Tanjung Puting, Cent. Kalimantan (Struebig et al. 2006b)	575	17	18	94	46–49	18	94	46–49
Lampanut, Cent. Kalimantan (I. Maryanto and M. Struebig, Unpublished data)	219	14	22	64	38–59	21	67	38–57
Lesan, East Kalimantan (M. Struebig, Unpublished data)	310	18	25	72	49–68	24	75	46–65

[a] Species richness was predicted to a total sample size of 1,000 individuals with a cut-off for rare species set at 10 individuals. Decimal fractions were rounded

[b] Proportion of predicted species richness present in that observed

[c] Proportion of all insectivorous species susceptible to capture by harp-trap (max. = 37 species) that are represented in the observed and predicted inventories

forest-interior bat species with up to 85% of the maximum fauna represented in the predicted inventory. Forests of Lesan in East Kalimantan (68% complete), Sepilok (65%) and Danum Valley (62%) in Sabah also potentially supported a substantial number of species.

Predicted species richness varied considerably between adjacent sites, and also at the same site when compared to the distribution-based estimates. While some neighbouring sites were predicted to have quite similar species richness (Danum Valley, 19–23 species; Sepilok, 23–24; 90 km apart in Sabah), assemblages at other adjacent sites were predicted to be quite different (Barito Ulu, 30–31 species; Lampanut, 21–22 species; 95 km apart in Central Kalimantan). Species richness derived from prediction models was also similar to that estimated from revised species distribution estimates at Tanjung Puting (18 vs. 19 species), but was much lower than the estimates for Danum Valley (19–23 vs. 29 species).

Discussion

Representation of bat species in Borneo's protected areas

Our conclusions of how bat species are represented in protected areas depended greatly on which species composition data we used in the assessment. By using actual species lists, representation was highly irregular over the Bornean protected area network, and even if some reserves were included in the Heart of Borneo, the protected area network would still exhibit low complementarity. By inferring species composition from distribution ranges, the representation of bat species in reserves was significantly improved, but differences between reserves inside and outside of the Heart of Borneo were still apparent. Finally, when species distributions were revised to include up-to-date records, the reserve network achieved high complementarity in representing bat species richness, and species composition between reserves inside and outside the Heart of Borneo was much more similar.

Species inventories are often used by conservation practitioners to identify priorities (e.g. Meijaard and Nijman 2003; Sujatnika et al. 1995) or to promote the value of an area for protection (e.g. WWF 2005), yet the efficacy of these assessments depends critically on the quality of the inventory data available. By focusing on bats we demonstrate that using incomplete inventories is likely to grossly underestimate the faunal composition of most protected areas in Borneo. Inferring composition from species distributions suggests that species representation is more uniform than is currently considered, at least at the macro-scale of this analysis. The quality of the distribution data is important, and conservation biologists and practitioners should continue to work together to improve our knowledge of this. Our simple gap analyses should be equally as valuable in assessing potential species representation in protected areas of other large islands, which in Southeast Asia include Java, Sumatra, Sulawesi and Papua.

Despite our positive forecast of the efficacy of the reserve system and the Heart of Borneo in protecting biodiversity, there are important limitations to our approach that should be considered.

Firstly, our assessment was restricted to a subset of bat species for which distribution information is available, and biogeographic patterns of these taxa may not reflect those of other, poorly known, bat species, or indeed other vertebrates. Some bat species excluded from our analyses are considered endemic to a few localities in Borneo (e.g. *Myotis gomantongensis*), which gives the protection of these sites clear conservation priority. However, since the species we excluded are likely to forage in edges and open areas and have considerable dispersal capabilities (Meijaard et al. 2005), they are more likely to persist in disturbed habitats and exhibit wider ranges than currently recognised. Because these species are notoriously difficult to capture (Kingston et al. 2003), this paucity of records in Borneo most likely reflects poor sampling effort rather than actual rarity or endemism.

Secondly, our distribution based estimates should be considered as upper bounds to species composition in an area, and in reality some species are likely to be absent from inventories because habitats may not meet their ecological needs. Our abundance-based species richness predictions exhibited much variation between sites, reflecting landscape and region-wide heterogeneity of assemblage composition. While we could confidently account for the lack of cave-roosting species in oligotrophic forests of Southern Borneo, our overall knowledge of habitats over the island is quite limited, and as a result distributions may overestimate species composition in some areas. A simple gap analysis overlooks variation in the availability and quality of habitats in the protected areas assessed, and it is likely that much of the habitat throughout Borneo is unsuitable for all species. Many forest-interior insectivorous species are likely to be restricted to forest, and some of these species (e.g. *Murina aenea*, *M. rozendaali*, *Phoniscus jagorii*), have mostly been recorded in undisturbed habitats. Predictive modelling of species distributions has the potential to improve the accuracy of our estimates (Rodríguez et al. 2007), but our bat database for Borneo is still quite limited; sampling bias, unstandardised sampling effort, and low geographic and environmental coverage, all limit the use of biodiversity databases for detailed analyses (Hortal et al. 2007). Therefore, although our findings imply high complementarity within Borneo's protected area system, we suggest that the Heart of Borneo should not divert attention away from other areas of conservation importance until a better understanding of biodiversity value can be used to assess all areas.

Major threats to bats in Borneo

Of the five main categories of threat to island bat populations identified by Wiles and Brooke (2008), and the four prioritised for Southeast Asia by Kingston (2008), three are major concerns in Borneo: habitat alteration and loss, cave disturbance, and hunting.

A large number of this island's bat species depend on forests for food and shelter and deforestation can therefore have major impacts by reducing foraging habitats and roosting sites. Deforestation rates in Indonesia and Malaysia continue to be high (Langner et al. 2007), but forest loss in Brunei has slowed since 1990 (Koh 2007). Uncontrolled forest fires have destroyed vast areas of Kalimantan, and millions of hectares continue to be lost by illegal logging and land clearance (Fuller et al. 2004). Of most concern is that many reserves, in Kalimantan at least, are not fulfilling their protective roles. Illegal logging has been rampant in national parks such as Bukit Baka-Bukit Raya, Tanjung Puting and Gunung Palung (Curran et al. 2004; Jepson et al. 2001). Over 50% of reserves in East Kalimantan have already been significantly degraded by logging, encroachment and fire, and at least five, including Kutai National Park and Bukit Soeharto Reserve, have almost completely been destroyed (Jepson et al. 2002). A large proportion of deforested areas are converted to plantations of timber or commodities such as rubber, cocoa and palm oil. Although the bat diversity value of these plantations is far from clear, preliminary studies across vertebrate groups (including bats) suggest that species diversity is lower than in natural forests, and that biodiversity value depends on plantation type, with rubber and cocoa plantations sometimes supporting more species than oil palm (Fitzherbert et al. 2008). Borneo's remaining forests are becoming increasingly fragmented and isolated

(Curran et al. 2004), a process that leaves bats that roost in tree cavities or foliage particularly vulnerable to population declines and local extinction (Struebig et al. 2008). Most forest has been disturbed to some extent, and the loss of tree hollows and exfoliating bark in logged over forest is predicted to have significant impacts on some forest-roosting species (Meijaard et al. 2005).

Sustained disturbance to caves has had long-term negative impacts on bat populations worldwide (Hutson et al. 2001), and Borneo is likely to be no exception to this trend. At least 45% of Borneo's bat species use caves to some extent (Payne et al. 2000), and some formations have historically hosted very large populations—Bau, Mulu and Niah in Sarawak; Gomantong and Madai in Sabah; and the Sangkulirang peninsula in East Kalimantan (Abdullah et al. 2007; Hutson et al. 2001; Mohd-Azlan et al. 2005; Suyanto and Struebig 2007). Since cave formations are patchily distributed over Borneo (Mackinnon et al. 1996), the protection of these landscapes is of immediate importance to bat conservation efforts. The Meratus mountains and Sangkulirang peninsula in Kalimantan represent the largest areas of karst limestone in Borneo, and indeed much of Southeast Asia (Mackinnon et al. 1996), yet they remain understudied and are protected only for watershed management, not for biodiversity. Efforts to include Sangkulirang as a world heritage site have so far been unsuccessful, but Mulu has achieved this status and is protected as a national park; as is Niah. However, land clearance and fires around caves, quarrying of limestone for cement, and the collection of guano and swiftlet nests all significantly threaten resident bat colonies. Oil palm and cocoa plantations now surround most limestone karst outcrops in Sabah and Sarawak (Hobbs 2004), and fire has caused visible damage to formations at Sangkulirang (Suyanto and Struebig 2007). Anecdotal reports have implicated the activities of guano and swiftlet nest collectors to huge bat population declines at caves in Sabah and Sarawak (Hutson et al. 2001), and populations in caves with evidence of nest collection at Sangkulirang were small or non-existent (Suyanto and Struebig 2007). Nest collection rates in large caves in Sabah and Sarawak remain unsustainable, and issues of ownership, corruption and outside control of the trade need to be addressed if activities are to be regulated and cave visitation minimised (Hobbs 2004).

Hunting is a major threat to bat populations on tropical islands, especially the larger fruit bats, and Borneo is no exception (Wiles and Brooke 2008). On Borneo a few taxa are the target of opportunistic collection, for example those species that can be collected easily from bamboo culms (possibly *Tylonycteris* spp. or *Glischropus tylopus*), furled banana leaves (*Myotis muricola*) (Mohd-Azlan et al. 2003); or under palm fronds in village gardens (*Cynopterus* spp., M. Struebig, Unpublished data). These are typically common and widespread species, which are unlikely to undergo significant long-term population declines. The hunting of *Pteropus vampyrus*, however, is much more extensive, and in some places could be causing catastrophic population declines. Hunting is unlikely to have a major impact in areas where no commercial market exists, such as remote settlements (e.g. Bulungan in East Kalimantan, Puri 2001) or those with predominantly muslim inhabitants. However, preliminary studies of *Pteropus* hunting in Sarawak (Gumal 2000) and in the protected Sebangau catchment in Kalimantan (Struebig et al. 2007) suggest intensive harvests exist to meet market demand, and populations show no signs of recovery. The potential long distance movements of flying-foxes (Smith 2007) obscure the true impact of hunting and the efficacy of reserves in protecting *Pteropus* populations, but in the meantime the listing of all *Pteropus* species on Appendix 2 of CITES does little to curb domestic trade (Mohd-Azlan et al. 2001), and the quotas introduced in Indonesia (A. Suyanto, Personal Communication) are likely to be ignored.

Improving bat conservation research in Borneo

Bats are relatively understudied compared to most other animals groups and considerable geographic and methodological biases are evident in mammal research on Borneo. Our review of species-locality records highlights the paucity of research effort in Kalimantan compared to Malaysia and Brunei, and illustrates how this geographic bias has resulted in a distorted perception of species distributions on the island and hence species representation in protected areas. Although Sabah has hosted most bat studies (Payne et al. 2000), research effort in Sarawak and Brunei has increased in recent years (e.g. Abdullah et al. 2007; Kofron 2002), though a focus on a single sampling technique (mist-nets) has resulted in low inventories in some reserves compared to better studied sites. Fruit bats dominate species records (Table 1), and many forest-interior taxa (Kerivoulinae and Murininae in particular) are missing from protected area inventories (Table 2). This is particularly interesting for Brunei, which has the greatest coverage of bat-locality records overall, but most of these records are for fruit bats and hipposiderids (Table 1).

Using a range of sampling techniques in surveys should improve skewed bat inventories, and standardising effort across studies will allow for reliable comparisons between sites. It is commendable that more researchers are now using harp-traps (T. Abdullah et al. Personal Communication), though the species with the fewest locality records are those that forage in open areas where most capture techniques are likely to be ineffective. Using bat detectors may help to detect these species by their echolocation calls, but only if methodological biases are considered in survey design, data is interpreted appropriately, and there is sufficient information regarding intraspecific call variation (Fenton 2003). Unfortunately, there is still a long way to go in Southeast Asia before reliable data will be available. Our upper bound estimates of species compositions may prove valuable in evaluating the completeness of biodiversity inventories providing distribution information is accurate and up-to-date. Immediate priorities for bat diversity studies identified by our colleagues in Borneo include the limestone karsts of the Meratus mountains and Sangkulirang peninsula in Kalimantan; islands, particularly those off the coasts of Sabah, East, South and West Kalimantan, and Western Sarawak; disturbed landscapes throughout the island; and previously unsurveyed areas within the proposed Heart of Borneo.

As researchers we need to collaborate more in our biodiversity studies, standardise our efforts, and share our findings to a global audience, but we also need to apply more of our efforts to the conservation threats that our study species face. With a plethora of high profile mammals consuming the activities of Borneo's conservation community, improving public and governmental support for bats on this island is a big challenge, and one that bat researchers should tackle united. Research questions that need to be addressed to help promote and undertake bat conservation are similar to those for other animal groups in Southeast Asia (Meijaard and Sheil 2007), but specific topics for bats include: quantifying prey consumption by large colonies of insectivorous bats to advocate their protection in agricultural areas by plantation managers; describing roost types available to forest bats so that mitigation strategies can be developed by timber companies; defining a standardised roost count protocol to describe and monitor Borneo's bat colonies and determine where conservation intervention is needed; and identifying areas where intensive *Pteropus* hunting is taking place so that relationships between the movements of colonies over the island, hunting pressures and land-use changes can be quantified.

Finally, we note that protected area management will only be able to safeguard ca. 15% of Borneo's remaining forests, with an additional 20% included in forests that are protected for watershed protection. The remainder of Borneo's forest area is likely to be under extractive forest management. On Borneo, this generally means that a limited number of trees (mostly <10/ha) are extracted every 25–30 years, theoretically allowing the forest to regenerate. Such forests have considerable value for wildlife (Meijaard et al. 2005). Both Indonesia and Malaysia are making slow progress towards more sustainable forest management (Dennis et al. 2008), and we expect that the importance of these managed forests for wildlife conservation will increasingly be recognised. It is important to further promote the role of well managed timber concessions in wildlife conservation, because these timber concessions can not only provide important habitats for forest-dependent bats, but they can also provide links between strictly protected areas thereby retaining the overall ecological connectivity of Borneo's forest habitats. The near future will tell whether present trends of clearing degraded forest areas for monocultural plantations (Fitzherbert et al. 2008) will continue, or whether the value of forests will be recognised and further loss, both within and outside protected areas, will be prevented.

Acknowledgements We are most grateful to all the people who kindly contributed bat records to our database, and/or provided useful discussion on bat conservation issues in Borneo and elsewhere in Southeast Asia. These include Mohd. Tajuddin Abdullah, Mohd. Azlan J. Abd. Gulam Azad, Stephen Brend, Anne Brooke, Susan Cheyne, Louise Craig, Rona Dennis, Fanasudi, Charles Francis, Marty Fujita, Antonia Gorog, Melvin Gumal, Les Hall, Elery Hamilton-Smith, Mark Harrison, Kris Helgen, Robert Hodgkison, Jason Hon, Simon Husson, Faisal Ali Bin Anwarali Khan, Tigga Kingston, Christopher Kofron, David Lane, Suwido Limin, Ibnu Maryanto, Ellen McArthur, Kim McKonkey, Helen Morrogh-Bernard, Cahyo Rahmadi, Juliana Senawi, Indrawati Sendow, Martua Sinaga, Tony Start, Agustinus Suyanto, Gary Wiles and Shigeki Yasuma. Thanks also to Markus Radday for sharing the digitised boundaries of the Heart of Borneo. Bat surveys in Kalimantan were funded by The Nature Conservancy (in part through a grant from the Sall Foundation) and grants awarded to either MS or DP for The Kalimantan Bat Conservation Project from Bat Conservation International, British Ecological Society, Gilchrist Educational Trust, Percy Sladen Memorial Fund, Lubee Bat Conservancy, People's Trust for Endangered Species, Project Barito Ulu, Royal Geographical Society (Institute of Biology and Rio Tinto Inc.), The Orangutan Foundation and University of East Anglia. Permission to conduct bat research in Kalimantan to MS and DP was kindly granted by the Indonesian Institute of Sciences (LIPI).

References

Abdullah MT, Hall LH, Tissen OB, Tuuga A, Earl of Cranbrook (2007) The large bat caves of Malaysian Borneo. Bat Res News 48:99–100

Anwarali FA, Piksin SNS, Ketol B, Japning JRR, Julaihi AM, Hall LS et al (2007) Survey of bats in the tropical lowland dipterocarp forest of Bako National Park, Sarawak, Malaysian Borneo. Sarawak Mus J 83:1–29

Boitani L, Catullo G, Marzetti I, Masi M, Rulli M, Savini S (2006) The Southeast Asian mammal databank. A tool for conservation and monitoring of mammal diversity in Southeast Asia. Instituto di Ecologia Applicata, Rome. http://www.ieaitaly.org/samd/. Cited 31 March 2008

Bruner AG, Gullison RE, Rice RE, da Fonseca G (2001) Effectiveness of parks in protecting tropical biodiversity. Science 291:125–128. doi:10.1126/science.291.5501.125

Curran LM, Trigg SN, McDonald AK, Astiani D, Hardiono YM, Siregar P et al (2004) Lowland forest loss in protected areas of Indonesian Borneo. Science 303:1000–1003. doi:10.1126/science.1091714

Danielsen F, Beukema H, Burgess N, Parish F, Bruhl CA, Donald PF et al (2008) Biofuel plantations on forested lands: double jeopardy for biodiversity and climate. Conserv Biol (in press)

Dennis RA, Meijaard E, Nasi R, Gustafsson L (2008) Biodiversity conservation in SE Asian timber concessions: an overview of the implementation of guidelines and recommendations. Ecol Soc 13. http://www.ecologyandsociety.org/vol13/iss1/art25/

du-Toit JT, Walker BH, Campbell BM (2004) Conserving tropical nature: current challenges for ecologists. Trends Ecol Evol 19:12–17

Fenton MB (2003) Science and the conservation of bats: where to next? Wildl Soc Bull 31:6–15

Fitzherbert EB, Struebig M, Morel A, Danielsen F, Brühl C, Donald PF et al (2008) How will oil palm expansion affect biodiversity? Trends Ecol Evol 23:538–545. doi:10.1016/j.tree.2008.06.012

Francis CM (1990) Trophic structure of bat communities in the understorey of lowland dipteropocarp rain forest in Malaysia. J Trop Ecol 6:421–431
Fujita MS, Tuttle MD (1991) Flying foxes (Chiroptera: Pteropodidae): threatened animals of key ecological importance. Conserv Biol 5:455–463. doi:10.1111/j.1523-1739.1991.tb00352.x
Fuller DO, Jessup TC, Salim A (2004) Loss of forest cover in Kalimantan, Indonesia, since the 1997–1998 El Niño. Conserv Biol 18:249–254. doi:10.1111/j.1523-1739.2004.00018.x
Gumal MT (2000) Ecology and conservation of a fruit bat in Sarawak, Malaysia. Dissertation, University of Cambridge, p 234
Hobbs JJ (2004) Problems in the harvest of edible birds' nests in Sarawak and Sabah, Malaysian Borneo. Biodivers Conserv 13:2209–2226. doi:10.1023/B:BIOC.0000047905.79709.7f
Hortal J, Lobo JM, Jimenez-Valverde A (2007) Limitations of biodiversity databases: case study on seed-plant diversity in Tenerife, Canary Islands. Conserv Biol 21:853–863. doi:10.1111/j.1523-1739.2007.00686.x
Hutson AM, Mickleburgh SP, Racey PA (2001) Microchiropteran bats: global status survey and conservation action plan. IUCN/SSC Chiroptera Specialist Group. IUCN, Gland, Switzerland and Cambridge, UK
IUCN (1994) Guidelines for protected area management categories. IUCN Commission on National Parks/World Conservation Monitoring Centre, The World Conservation Union, Gland
Jayaraj VK, Ketol B, Khan FAA, Hall LS, Abdullah MT (2006) Bat survey of Mount Penrisen and notes on the rare *Kerivoula minuta, Kerivoula intermedia* and *Hipposideros coxi* in Sarawak, Borneo. J Biol Sci 6:711–716
Jennings MD (2000) Gap analysis: concepts, methods, and recent results. Landsc Ecol 15:5–20. doi:10.1023/A:1008184408300
Jepson P, Jarvie JK, MacKinnon K, Monk KA (2001) The end of Indonesia's lowland forests? Science 292:859–861. doi:10.1126/science.1061727
Jepson P, Momberg F, van-Noord H (2002) A review of the efficacy of the protected area system of East Kalimantan province, Indonesia. Nat Areas J 22:28–42
Kingston T (2008) Research priorities for bat conservation in Southeast Asia: a consensus approach. Biodivers Conserv. doi:10.1007/s10531-008-9458-5
Kingston T (2009) Analysis of species diversity of bat assemblages. In: Kunz TH, Parsons S (eds) Behavioral and ecological methods for the study of bats. Smithsonian Institution, Washington (in press)
Kingston T, Francis CM, Zubaid A, Kunz TH (2003) Species richness in an insectivorous bat assemblage from Malaysia. J Trop Ecol 19:67–79. doi:10.1017/S0266467403003080
Kofron CP (2002) The bats of Brunei Darussalam, Borneo. Mammalia 66:259–274
Koh LP (2007) Impending disaster or sliver of hope for Southeast Asian forests? The devil may lie in the details. Biodivers Conserv 16:3935–3938. doi:10.1007/s10531-007-9177-3
Lane DJW, Kingston T, Lee BPY-H (2006) Dramatic decline in bat species richness in Singapore, with implications for Southeast Asia. Biol Conserv 131:584–593. doi:10.1016/j.biocon.2006.03.005
Langner A, Miettinen J, Siegert F (2007) Land cover change 2002–2005 in Borneo and the role of fire derived from MODIS imagery. Glob Chang Biol 13:2329–2340. doi:10.1111/j.1365-2486.2007.01442.x
Mackinnon K, Hatta G, Halim H, Mangalik A (1996) The ecology of Kalimantan. Periplus Editions, Singapore
McCune B, Meffor MJ (1999) Multivariate analysis of ecological data. PcORD version 4.25. MJM Software Designs, Gleneden Beach, OR
McKenzie NL, Gunnell AC, Yani M, Williams MR (1995) Correspondence between flight morphology and foraging ecology in some Paleotropical bats. Aust J Zool 43:241–257. doi:10.1071/ZO9950241
Meijaard E, Nijman V (2003) Primate hotspots on Borneo: predictive value for general biodiversity and the effects of taxonomy. Conserv Biol 17:725–732. doi:10.1046/j.1523-1739.2003.01547.x
Meijaard E, Sheil D (2007) Is wildlife research useful for wildlife conservation in the tropics? A review for Borneo with global implications. Biodivers Conserv 16:3053–3065. doi:10.1007/s10531-007-9161-y
Meijaard E, Sheil D, Nasi R, Augeri D, Rosenbaum B, Iskandar D et al (eds) (2005) Life after logging: reconciling wildlife conservation and production forestry in Indonesian Borneo. Center for International Forestry Research, Bogor
Mohd-Azlan J, Zubaid A, Kunz TH (2001) Distribution, relative abundance and conservation status of the large flying fox: *Pteropus vampyrus*, in peninsular Malaysia: a preliminary assessment. Acta Chiropt 3:149–162

Mohd-Azlan J, Maryanto I, Kartono AP, Abdullah MT (2003) Diversity, relative abundance and conservation of chiropterans in Kayan Mentarang National Park, East Kalimantan, Indonesia. Sarawak Mus J 58:251–265

Mohd-Azlan J, Neuchlos J, Abdullah MT (2005) Diversity of chiropterans in limestone forest area, Bau, Sarawak. Malaysian. Appl Biol 34:59–64

Payne J, Francis CM, Phillips K, Kartikasari SN (2000) Mamalia di Kalimantan, Sabah, Sarawak dan Brunei Darussalam. Wildlife Conservation Society Indonesia Program, Jakarta

Puri RK (2001) Bulungan ethnobiology handbook. A field manual for biological and social science research on the knowledge and use of plants and animals amongst 18 indigenous groups in northern East Kalimantan. Centre for International Forestry Research (CIFOR), Bogor

Rodrigues ASL, Andelman SJ, Bakarr ML, Boitani L, Brooks TM, Cowling RM et al (2004) Effectiveness of the global protected area network in representing species diversity. Nature 428:640–643. doi:10.1038/nature02422

Rodríguez JP, Brotons L, Bustamante J, Seoane J (2007) The application of predictive modelling of species distribution to biodiversity conservation. Divers Distrib 13:243–251

Shen TJ, Chao A, Lin JF (2003) Predicting the number of new species in further taxonomic sampling. Ecology 84:798–804. doi:10.1890/0012-9658(2003)084[0798:PTNONS]2.0.CO;2

Simmons N (2005) Chiroptera. In: Wilson DE, Reeder DM (eds) Mammal species of the world: a taxonomic and geographic reference. John Hopkins University Press, Baltimore, pp 312–529

Smith C (2007) Use of satellite telemetry to study movement of the Malayan flying fox (*Pteropus vampyrus*): implications for conservation and public health. In: Wildlife health in a shrinking world: ecology, management and conservation. Proceedings of the wildlife disease association international conference, Cairns, Australia, 26 June–1 July 2005

Sodhi NS, Koh LP, Brook BW, Ng PKL (2004) Southeast Asian biodiversity: an impending disaster. Trends Ecol Evol 19:654–659. doi:10.1016/j.tree.2004.09.006

Solow AR, Polasky S (1999) A quick estimator for taxonomic surveys. Ecology 80:2799–2803

Struebig M, Benton-Browne A, Rachmad A, Yusliati N, Atmoko T, Rustam W, Fredriksson G, Meijaard E (2006a) A bat survey in Sungai wain protection Forest, East Kalimantan, Indonesia. Malay Nat J 59:189–196

Struebig MJ, Galdikas BMF, Suatma (2006b) Bat diversity in oligotrophic forests of Southern Borneo. Oryx 40:447–455. doi:10.1017/S0030605306001190

Struebig MJ, Harrison ME, Cheyne SM, Limin S (2007) Intensive hunting of large flying-foxes *Pteropus vampyrus* natunae in Central Kalimantan, Indonesian Borneo. Oryx 41:390–393. doi:10.1017/S0030605307000310

Struebig MJ, Kingston T, Zubaid A, Mohd-Adnan A, Rossiter SJ (2008) Conservation value of forest fragments to Palaeotropical bats. Biol Conserv 141:2112–2126. doi:10.1016/j.biocon.2008.06.009

Sujatnika JP, Soehartono T, Crosby MJ, Mardiastuti A (1995) Melestarikan keanekaragaman hayati Indonesia: pendekatan daerah burung endemik (Conserving Indonesian biodiversity: the endemic bird area approach). Perlindungan Hutan dan Pelestarian Alam and Birdlife International Indonesia Programme, Jakarta

Suyanto A, Struebig MJ (2007) Bats of the Sangkulirang limestone karst formations, East Kalimantan—a priority region for Bornean bat conservation. Acta Chiropt 9:67–95. doi:10.3161/1733-5329(2007)9[67:BOTSLK]2.0.CO;2

Wiles GJ, Brooke AP (2008) Conservation threats to bats in the tropical pacific islands and insular Southeast Asia. In: Fleming TH, Racey P (eds) Island bats: evolution, ecology and conservation. University of Chicago, Chicago (in press)

WWF (2005) Borneo—treasure island at risk. WWF Germany, Frankfurt am Main

Zubaid A (1993) A comparison of the bat fauna between a primary and fragmented secondary forest in Peninsular Malaysia. Mammalia 57:201–206

Biodivers Conserv (2010) 19:471–484
DOI 10.1007/s10531-008-9458-5

ORIGINAL PAPER

Research priorities for bat conservation in Southeast Asia: a consensus approach

Tigga Kingston

Received: 7 April 2008 / Accepted: 28 August 2008 / Published online: 12 September 2008

Abstract Southeast Asia is a critical area for biodiversity conservation; levels of species richness and endemism are among the highest in the world, but rapid land-use changes endanger much of the region's fauna. Bats are a critical component of this diversity, comprising nearly a third of Southeast Asia's mammal species and providing vital ecological and economic services. However, nearly half the species are of conservation concern and as many as 40% of bat species are predicted to be extinct by the end of this century if current deforestation rates persist. Conservation efforts are urgently needed, and the taxonomic continuity of Southeast Asia and the prevalence of major threats throughout suggest that a region-wide initiative could be an effective approach. The Southeast Asian Bat Conservation Research Unit (SEABCRU) was established in 2007 to provide an organizational framework to both accelerate the advancement of bat research, and to coordinate conservation efforts. The SEABCRU is an informal collaboration among institutes, NGOs and individuals and provides a web-based forum for the growing number of researchers and outreach workers to interact and coordinate activities. It was launched at the 1st International Southeast Asian Bat Conference in Thailand (May 2007), during which a forum was held to derive conservation research priorities using a consensus approach. Four priorities were identified by participants: flying fox conservation and monitoring, taxonomy, conservation of cave-dependent bats, and conservation of forest-dependent bats. Here I provide an overview of the rationales behind these priorities and list the specific recommendations for the actions identified.

Keywords Chiroptera · Caves · Conservation priority setting · Deforestation · Hunting · Flying foxes · Southeast Asia · Taxonomy

Introduction

The geopolitical region of Southeast Asia comprises 11 countries (Myanmar, Vietnam, Laos, Thailand, Cambodia, Malaysia, Singapore, Indonesia, Brunei, Philippines and East

T. Kingston (✉)
Department of Biological Sciences, Texas Tech University, Lubbock, TX 79409-3131, USA
e-mail: tigga.kingston@ttu.edu

Timor) and is one of the most biologically rich regions in the world, yet rapid land-use changes in recent decades have put much of its biodiversity at risk of extinction (Sodhi et al. 2004). As a consequence, the region encompasses four of the world's biodiversity hotspots (Myers et al. 2000) and seven of the 20 hotspots of latent extinction risk (Cardillo et al. 2006). Bats are a critical component of Southeast Asia's threatened fauna; they constitute approximately 30% of the region's mammal species, and can comprise as many as half of all mammal species in the tropical rainforest ecoregions (Kingston et al. 2006a). Moreover, the region is pivotal for international bat conservation as it supports nearly 30% of world's bat fauna; 320 species are listed by Simmons (2005) and a further ten described since (*Kerivoula kachinensis* (Bates et al. 2004); *K. krauensis* (Francis et al. 2007a); *K. titania* (Bates et al. 2007a); *Murina harrisoni* (Csorba and Bates 2005); *M. tiensa* (Csorba et al. 2007); *Rhinolophus chiewkweeae* (Yoshiyuki and Lim 2005); *Hipposideros khaokhaouayensis* (Guillén-Servent and Francis 2006); *H. boeadii* (Bates et al. 2007b); *Dyacopterus rickarti* (Helgen et al. 2007); and *Styloctenium mindorensis* (Esselstyn 2007). Species diversity is distributed across ten families, and 67 genera (Fig. 1), and of the 330 species identified, 197 are endemic to the region.

Like much of the region's fauna, the conservation status of many Southeast Asian bat species is a major concern. The most recent assessments of IUCN Red List status were conducted by the Southeast Asian Mammal Databank (SAMD) as part of the Global Mammal Assessment. Although yet to be ratified, these provide the most comprehensive evaluation of status to date (Biotani et al. 2006) (Fig. 1). Only 53% of species are currently classified as Least Concern, with the remaining species falling into one of four categories of conservation concern (Near Threatened, Vulnerable, Endangered, and Critically Endangered), considered Data Deficient, or have yet to be evaluated. Nearly 20% of species are Data Deficient, with fewest data available for the Rhinolophidae (14% species), Hipposideridae (24%), Vespertilionidae (27%) and Molossidae (42%) (Fig. 1). The status of the speciose Pteropodidae is better known, with only nine species considered Data Deficient, but 37 of the 96 species are threatened or nearly so (Near Threatened, 8 species; Vulnerable, 19 species; Endangered, 7; Critically Endangered, 3 species) (Fig. 1).

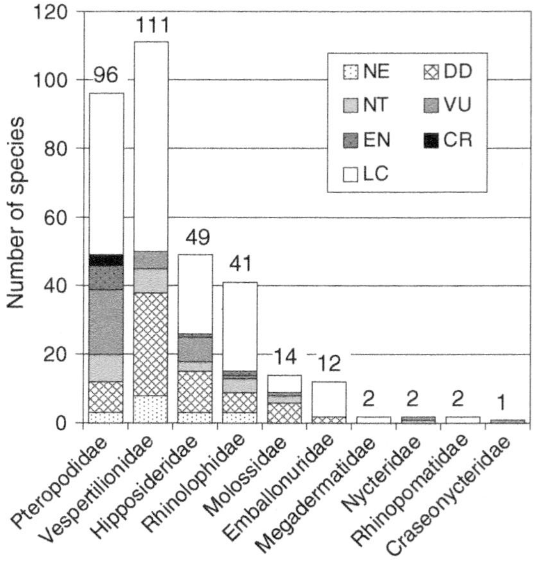

Fig. 1 Distribution of IUCN Red list status among bat families from the eleven countries of Southeast Asia (Myanmar, Vietnam, Laos, Thailand, Cambodia, Malaysia, Singapore, Indonesia, Brunei, Philippines, East Timor). Status from SAMD assessment (Biotani et al. 2006). NE, not evaluated, DD, data deficient, NT, near threatened, VU, vulnerable, EN, endangered, CR, critically endangered, LC, least concern. *Number* at top of *bars* is the total number of species recorded for the family, following Simmons (2005), Biotani et al. (2006) and new discoveries listed in the text

One of the most significant threats to the region's bat diversity is habitat loss. Relative deforestation rates in Southeast Asia are the highest of any tropical region (Achard et al. 2002; Sodhi et al. 2004; Sodhi and Brook 2006) and are expected to lead to extinction of many bat taxa. Upper bound estimates of regional species losses exceed 40% and global extirpation is anticipated for at least 23% of species by 2100 (Lane et al. 2006). Forest-dependent bats are particularly vulnerable (Kingston et al. 2003; Lane et al. 2006), but declines resulting from habitat loss are exacerbated by unregulated hunting of several species, most notably the plant-visiting flying foxes [*Pteropus* and *Acerodon* (Pteropodidae)]. Collection of bats for food and traditional medicine has led to dramatic population declines throughout the region (e.g., Mohd.-Azlan et al. 2001).

Not only does this represent a catastrophic decline in global vertebrate diversity, but the loss of the many ecosystem services that bats provide is likely to lead to an extinction cascade that will have profound ecological and economic impacts. The plant-visiting Pteropodidae provide essential ecosystem services as pollinators (Whittaker and Jones 1994; Corlett 2004; Srithongchuay and Bumrungsri 2007) and seed dispersers (Shilton et al. 1999; Corlett 1998) to more than 300 paleotropical plant species (Marshall 1983; Fujita and Tuttle 1991), and without bats both the regeneration of forest ecosystems (Marshall 1985; Cox et al. 1991; Rainey et al. 1995; Gumal 2001; McConkey and Drake 2006) and the success of bat-dependent fruit crops will be severely compromised. The economic value of these services is substantial. More than 500 products used by people are derived from bat-dependent plants (Fujita and Tuttle 1991); the global durian (*Durio zibethinus*) trade alone is worth in excess of US $1.5 billion per year (Lim 1998) and sales of petai, the beans of *Parkia speciosa*, generate approximately US $2.8 million annually in just the area around Kuala Lumpur (Malaysia) (Chuen et al. 1998). Recent pollination experiments assessing fruit set confirm nectarivorous bats as the most effective pollinators of both petai (Bumrungsri et al. 2007) and durian (S. Bumrungsri, personal communication).

The remaining bat families found in Southeast Asia comprise, with a few carnivorous exceptions, insectivorous species, and are the primary predators of nocturnal insects. Dietary analysis suggests that a single colony of 2.6 million wrinkle-lipped free-tailed bats (*Chaerephon plicata*) in Thailand may consume at least four tons of rice crop pests [white-backed planthoppers (*Sogatella* sp.)] each night (Leelapaibul et al. 2005). Such natural pest control agents help reduce the application of pesticides (Cleveland et al. 2006), and the consequent impacts on local biodiversity that result from their use. Large bat colonies are also a source of guano, which is widely used in Southeast Asia as an organic fertilizer. In Myanmar for example, over 260,000 kg per year are extracted, primarily for domestic use (Khin Maung Lwin 1995), and guano collection from the Khao Chong Pran Cave in Thailand generates an annual income for local villagers of US $135,000 (Leelapaibul et al. 2005). Consequently the conservation of Southeast Asia's bat fauna is a biodiversity priority from perspectives that range from the preservation of global species richness, through ecosystem maintenance to agricultural viability.

Research priorities for the conservation of Southeast Asian bats

Conservation research is urgently needed to reduce the impact of forest loss and hunting on the bat fauna of Southeast Asia, yet for historical, financial and logistical reasons, research in many countries is in its infancy. Nonetheless, there are a growing number of regional scientists and international collaborators dedicated to bat research and conservation in the region. Because of the taxonomic continuity of Southeast Asian bats, researchers share many common objectives and conservation issues yet to date there has been no mechanism for exchange of ideas and techniques, nor for discussion, synthesis and implementation of region-wide research and conservation objectives. To address this need, the Southeast Asian Bat Conservation Research Unit (SEABCRU) was established in 2007, and launched at the 1st International South East Asian Bat Conference in Thailand, May 2007. The SEABCRU is a collaboration between academic institutions and NGOs intended to provide an organizational framework to coordinate and implement research, capacity building, and outreach to promote the conservation of Southeast Asia's diverse but threatened bat fauna.

One of the first actions of the SEABCRU was to identify priorities for bat research, and to this end an open forum was held during the conference. Over seventy people attended from 20 countries, providing a unique opportunity for a diverse body of conservation stakeholders, ranging from outreach workers and first year students to researchers with 40 years experience, to define and shape bat conservation research for the next decades. Stakeholder involvement is commonly encompassed in applied conservation issues, but research directions are rarely derived from collective bottom-up consensus. The only constraint was that priorities center on activities or issues that are pertinent across the region and which would be best addressed by region-wide collaboration among researchers and conservation biologists. Consensus opinion identified four key areas of that fulfilled these criteria and are detailed below (in no particular order).

Flying fox conservation and monitoring

Based on current deforestation rates, as many as 24% of Southeast Asia's fruit bat (Pteropodidae) species are predicted to have become globally extinct by the end of this century (Lane et al. 2006). Within the Pteropodidae, the flying foxes [defined in this context as members of the genera *Pteropus* (25 species) or *Acerodon* (5 species)] are of particular concern. These large species roost in trees in aggregations of several thousand individuals (and historically in the tens of thousands; Lyon 1911) and are under intense hunting pressure for food and traditional medicine in many parts of Southeast Asia (Mickleburgh et al. 1992, 2002; Wiles and Brooke 2008).

Hunting and habitat loss were identified as major threats to nearly all flying fox species by both the SAMD assessors (Biotani et al. 2006), and participants of a flying fox forum at the 1st International South East Asian Bat Conference. In combination, these threats have led to dramatic, but unquantified, declines in flying fox populations throughout the region. As a consequence, more than half (18/30) of the region's flying fox species were Red Listed (Biotani et al. 2006) by the SAMD and all bar one species is listed by the Convention on International Trade in Endangered Species (CITES) as Appendix II, the exception being the Critically Endangered *Acerodon humilus* from the Taulad Islands in Indonesia which is listed as CITES Appendix I. Of the remaining species, five were considered to be Data Deficient, and only seven species were considered sufficiently secure in their populations to merit Least Concern status. As a family, the Pteropodidae are clearly more vulnerable to extinction than any other bat family, comprising at least 6 of the 14 probable global bat extinctions (Simmons 2005).

Conservation efforts are severely hampered by the lack of regional data on flying fox distributions and populations, and of quantified evaluations of the impact of hunting. Explicit population and hunting data are only available for a few localities, most notably *Pteropus vampyrus* populations in Peninsular Malaysia, which clearly illustrate the regional decline. Mohd.-Azlan et al. (2001) visited 115 sites known historically to have supported *P. vampyrus* colonies. There were no recent sightings at 40% of the sites and all local villagers reported dramatic declines. The authors attributed this to the loss of mangrove swamps and lowland forest, and the intensification of hunting pressure effected by the use of guns and greater access to remote roost sites. Moreover, although a hunting license is required, and limits the licensee to 50 individuals, the number of licenses issued is not restricted. Based on the number of licenses, harvests have been unsustainable for at least 15 years with 56,000 individuals taken between 1990 and 1996 (Mohd. Azlan et al. 2001) and 87,800 for the period 2002–2005 (Epstein et al. 2007). Similarly, Lee et al. (2005) monitored sales of four species (*P. hypomelanus, P. alecto, Acerodon celebensis, A. humilis*) from monthly surveys of markets in North Sulawesi, and estimated sales of c. 38,000 individuals over a 2 year period from just six markets. Intensive tower netting systems used in Central Kalimantan (Indonesian Borneo) harvested as many as 4,500 *P. vampyrus* from a single location in just a month, each selling for at most US $2.20 (Struebig et al. 2007).

Thus hunting is prevalent across Southeast Asia, but the need for regional cooperation is evinced further by satellite-tracking studies of *P. vampyrus* trans-national migration patterns. Individuals collared in Peninsular Malaysia have been recorded from Sumatra, Singapore and Thailand (Epstein et al. 2007), while those from the Democratic Republic of Timor-Leste cross into Indonesia West Timor (Breed et al. 2007). This not only complicates population assessments and long-term monitoring, but also means that threats must be addressed across migration routes; protection in one locality will not necessarily secure a population if threats persists in other parts of its home range.

As a consequence of the degree of threat and lack of population data, flying fox conservation was identified as key priority action for the SEABCRU. In a few instances, there are data to support immediate conservation action. For example, in those countries where hunting is regulated by the issuance of hunting licenses, there is an opportunity to prevent further population declines by reducing the number of licenses issued, limiting the number of individuals that can be taken per license, or halting the sale of licenses altogether. In countries without such a potential control mechanism, the survival of flying fox populations may ultimately only be secured by legal protection. The success of advocacy for protected status will likely rest on the availability of population data and quantified evaluations of the degree of threat faced. Consequently the research priorities centered on: (1) determining current populations and distributions of *Pteropus* and *Acerodon* species; (2) initiating and coordinating long-term monitoring of populations across the region; (3) assessing the impact of hunting and the bushmeat trade.

It is intended that the research output based on these priorities will identify those species most in need of protection and provide a mechanism to assess the success of conservation intervention.

Taxonomic network

The importance of taxonomy to conservation is increasingly being recognized (Dubois 2003; Golding and Timberlake 2003; Mace 2004). Accurate species identifications are fundamental to biodiversity assessments based on species richness estimates, levels of endemism, and complementarity (e.g., Struebig et al. 2008a, this issue), as well as to evaluations of threat to individual species (e.g., the IUCN Red List), both of which are central concepts guiding priorities for conservation action. Taxonomic studies of Southeast Asian bats began modestly in the mid-eighteenth century, with nine species described by 1800. The nineteenth century witnessed the most dramatic increase in species discoveries, leveling off somewhat during the first half of the twentieth century. Of note is the increase in rate of discovery in the last few decades (Fig. 2), indicating that true species richness may be

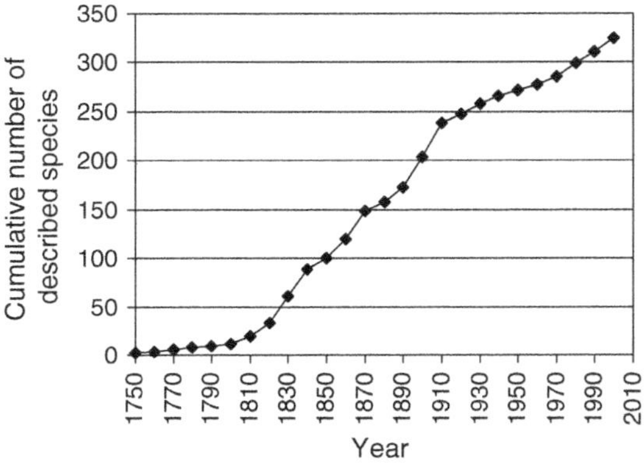

Fig. 2 Species description curve for bats from the eleven countries of Southeast Asia (Myanmar, Vietnam, Laos, Thailand, Cambodia, Malaysia, Singapore, Indonesia, Brunei, Philippines, East Timor). Species included and dates of first description are those currently recognized by Simmons (2005) or described since (as listed in "Introduction")

substantially higher than the currently recognized 330 species. The description of 14 species in the last 7 years reflects renewed survey effort in the region (e.g., Bates et al. 2000; Hendrichsen et al. 2001; Matveev 2005) and multifaceted survey approaches that employ a range of trapping techniques including mist nets, harp traps, tunnel traps (e.g., Sedlock 2001; Kingston et al. 2003). Not only have new species been discovered in areas not previously or recently studied (Bates et al. 2004, 2007b; Matveev and Csorba 2007), but they have also been found in relatively well-known localities as a consequence of greater research intensity (e.g., Francis et al. 2007a). Detailed reviews of museum collections have identified further additions (e.g., Helgen et al. 2007; Bates et al. 2007a), and the increasing use of molecular techniques is proving invaluable in elucidating hitherto hidden species diversity within cryptic species complexes (e.g., Kingston et al. 2001; Francis et al. 2007a, b) and clarifying relationships among them (Campbell et al. 2004; Thabah et al. 2006).

Nonetheless, the taxonomy of many Southeast Asian bat species remains uncertain, and one of the biggest challenges to the advancement of bat taxonomy in the region has been the constraints imposed by political boundaries. Although approximately 126 species are currently thought to be endemic to a single country, the remaining species are distributed across 2–11 countries (Fig. 3). Taxonomic resolution consequently requires mechanisms for multi-national collaboration among researchers that can coordinate taxonomic effort across Southeast Asia. The development of a Taxonomic Network was thus identified as the second SEABCRU research priority, with the specific initial remit to: (1) collate data on collections across the region; (2) establish loan protocols.

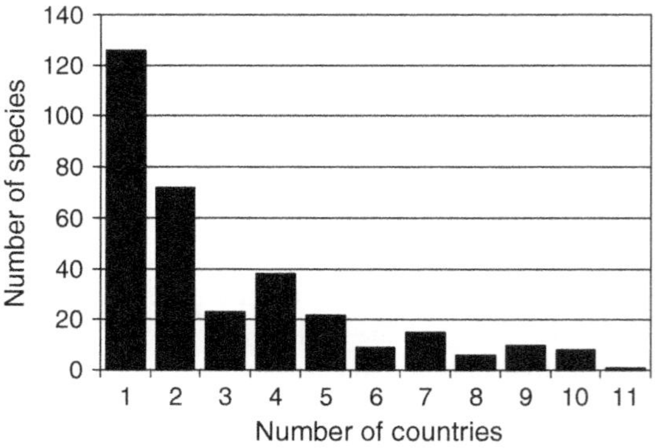

Fig. 3 Frequency of bat species occurring in one or more of the eleven countries of Southeast Asia (Myanmar, Vietnam, Laos, Thailand, Cambodia, Malaysia, Singapore, Indonesia, Brunei, Philippines, East Timor). Country occupancy derived from SAMD database (Biotani et al. 2006) and publications since SAMD assessment

Taxonomic efforts can be further advanced by recording and sharing molecular sequences of vouchered specimens. By contributing sequences to a central, public database (such as GenBank or the DNA Barcoding Initiative) researchers can greatly advance the delineation of taxonomic boundaries and enable ecologists and field workers to verify identifications for bats that are not being collected. The DNA Barcoding Initiative has a public database of sequences for over 165 Southeast Asian bat species, based on a standardized barcode of 655 base-pair segments of the Cytochrome C Oxidase subunit I gene. The barcode provides good discrimination of species and can thus be an effective tool for identifying known species, and for confirming the shared identity of specimens in different collections (Francis et al. 2007b). Small biopsies taken from the wing membrane can be readily obtained from both living and preserved individuals, but provision should be made to build local sequencing capacity to ensure the deposition of standard sequences in the database.

Conservation of cave-roosting bats

Bats spend over half their lives in roosts (Kunz 1982). Roosts provide shelter from the environment and predators, and a secure place to mate, rear young, and interact with other individuals (Kunz 1982). Caves are particularly attractive roost sites for many species because of their size, permanency and the stable microclimates they provide, and many species are highly dependent upon caves for roosting and form large aggregations. Free-tailed bats (Molossidae) roosting in the millions are known from caves around Southeast Asia [e.g., *Chaerephon plicata* in Gomantong Caves (Sabah), Mulu Cave (Sarawak), Khao Chong Pran Cave (Thailand)], but species of Rhinolophidae, Hipposideridae and Miniopterinae (Vespertilionidae) also congregate in numbers exceeding 100,000 individuals (e.g., *Miniopterus* spp., *Hipposideros larvatus*, *Rhinolophus creaghi*; Payne et al. 1985). The Pteropodidae tend to make less use of caves than do other families, presumably because species do not echolocate apart from the tongue-clicking *Rousettus*. However, several species are largely cave-dependent, most notably *Rousettus* spp., *Dobsonia* spp, *Eonycteris spelaea* and *Penthetor lucasi*.

Large complex cave systems provide an array of light and microclimate conditions and consequently harbor a great diversity of species. At least 12 species are known from Gomantong (Sabah), and Deer Cave (Mulu, Sarawak), and 19 from Niah Great Cave (Sarawak) (Abdullah et al. 2007). Although even small caves are likely to support a few bat species, it is the complex limestone karst systems that support the greatest diversity and abundance of bats (Racey 2007). These "Biodiversity Arks" (Clements et al. 2006) cover >400,000 km^2 in Southeast Asia, with particularly extensive systems in Indonesia, Thailand and Vietnam. Surveys in karst areas clearly illustrate their significance to bat diversity, with 35 species in the vicinity of the Sangkulirang karsts of Kalimantan, Indonesian Borneo (Suyanto and Struebig 2007), 33 around Kim Hy, Vietnam (Furey and Racey 2007), and 27 species around the karsts of Bohol, Philippines (Pamaong-Jose et al. 2007).

The threats to cave bats are many-fold. Direct hunting affects several species, most notably large or abundant species such as *Cheiromeles torquatus* in Borneo (Hutson et al. 2001), *Eoyncteris spelaea* in southern Palawan, Philippines (Shively 1997), *Dobsonia chapmani*, Philippines (Heaney and Heideman 1987), and *Chaerephon plicata*, in Lao PDR (Francis et al. 1999) and Cambodia (Yim Saksang, personal communication). More generally, human activities at caves for other purposes disturb bats and, if sustained, can have severe consequences for the long-term viability of the bat populations (Hutson et al. 2001). In Southeast Asia extractive visitations to collect guano or cave swiftlet (*Collocalia fuciphagus* and *C. maximus*) nests are major causes of disturbance (Suyanto and Struebig 2007; Wiles and Brooke 2008), and in several instances are compounded by visits by tourists (Hutson et al. 2001). Regionally, quarrying of karst areas for limestone and basement minerals probably represents the greatest threats to cave-roosting species. Southeast Asia has the highest annual quarrying rates in the tropics (178 million metric tons per year) and the rate is increasing by 5.7% per year (Clements et al. 2006).

The significance of caves to bat diversity and the multitude of threats to cave populations prompted the SEABCRU forum participants to identify cave bat conservation as the third priority for action. In this instance, it is the similarity of threats across Southeast Asia that supports a regional initiative. Foremost, there is a clear need for greater cave survey effort across Southeast Asia to identify and prioritize caves supporting the highest diversity of bats or populations performing key ecological functions in the local context. The specific resolutions reflect the importance of caves to local economies and are: (1) to develop cave management recommendations that recognize scientific, cultural/religious and economic values of caves AND provide for protection of bats; (2) instigate programs to monitor populations to assess effectiveness of management approaches.

Conservation of forest-dependent bats

Habitat loss is one of the main threats to bats worldwide (Hutson et al. 2001; Racey and Entwistle 2003), and in Southeast Asia, where as much as 74% of forest may be lost by the end of the century (Sodhi et al. 2004), it is a critical issue. A substantial component of the bat fauna appears to be highly dependent on intact stands of forest for foraging and roosting (Lane et al. 2006; Struebig et al. 2008b). Many species, particularly those in the families Hipposideridae, Rhinolophidae and the vespertilionid subfamilies Murininae and Kerivoulinae, combine wing morphologies and echolocation signal designs that confer efficient foraging in densely vegetated habitats, but which are less effective in the more open habitats typical of deforested landscapes (Kingston et al. 2003; Meijaard et al. 2005). The availability of appropriate cavity roosts in both standing and fallen trees is likely to further restrict species to forest habitats (Fletcher 2006). Similarly, although a number of pteropodids are readily captured in villages, secondary habitats, and small agricultural holdings (e.g., *Cynopterus* spp.), others, including over half of the Philippines' pteropodids (Heaney et al. 1998), are found primarily in forests and are highly vulnerable to habitat loss.

The most dramatic illustration of the consequences of forest loss to bat diversity comes from Singapore which has lost 95% of its primary forest since European colonization began in the 1800s (Corlett 1992). Of the 30 species documented for the island, only 20 have been recorded in recent surveys (Pottie et al. 2005), and this is likely a gross underestimate of true species losses because extensive land-use change was already underway before bat inventories began. Estimates of percentage species lost based on the likely presence of species before colonization fall between 67 and 72%, and it is the forest-dependent species that show greatest declines (Lane et al. 2006). The surviving species are generally either insectivorous species that forage in open and edge habitats or plant-visiting species able to forage in agricultural and secondary habitats.

Little is known of the ecological requirements and population dynamics of forest-dependent bats in undisturbed forest, nor the response of forest assemblages to different land-uses. Long-term monitoring of forest-dependent insectivorous bats is being conducted at five large study plots at Krau Wildlife Reserve, Peninsular Malaysia (Kingston et al. 2006a) with studies in the surrounding landscape investigating the persistence of species in forest fragments (Struebig et al. 2008b) and oil palm plantations (T. Wood, unpublished data).

However, there is a general lack of data in dramatic contrast to the growing number of bat assemblage disturbance studies from the Neotropics (reviewed in Kingston 2009). Thus the final SEABCRU resolution identified the need for: (1) a network of long-term study sites to elucidate the dynamics of bat assemblages in undisturbed forest; (2) studies focused on the response of forest-dependent bat species to land-use change. Such data are essential to identify both priority sites for protection and species most adversely affected by anthropogenic disturbance.

Discussion

Currently, bats are among the least-known vertebrates in Southeast Asia, yet their contribution to regional biodiversity and ecosystem functioning is critical, and the threats they face substantial. The adoption of a region-wide approach to conservation research could do much to accelerate the development of in-country programs, and is particularly well-timed. Research programs in many countries are just beginning, and in others transitioning from a few pioneers to a new generation of young researchers specializing in bat biology; a regional perspective can help focus individual efforts, and collaborations enhance their success. The probability that the SEABCRU will attain its objectives is further improved by the current level and rates of development within Southeast Asia. Although internet access in the field is of course limited, most researchers have adequate coverage at their place of work, paving the way for web-based collaborations and interactions through the SEABCRU portal (http://www.seabcru.org). Physical connections among countries are also increasing with development of trans-national highways, railways (e.g., Singapore-China Rail Link) and low-cost airlines. Collaborative efforts are also facilitated by the wide-spread and increasing use of English. As the only official language of ASEAN (Association of Southeast Asian Nations) it has even been adopted as the official second language in nations with a French colonial history (Cambodia, Laos and Vietnam), and initiatives to improve proficiency are common throughout the area.

Conservation resources are limited and it is essential to prioritize activities to maximize return on effort. Typically, conservation priority-setting has fallen into two broad categories. In the first, geographical or political units with high biodiversity value are identified, based on measures that include species richness, levels of endemism (e.g., Stattersfield et al. 1998; Myers et al. 2000), extinction threat (e.g., Ricketts et al. 2005), complementarity (e.g., Struebig et al. 2008a, this issue), and phylogenetic diversity (e.g., Isaac et al. 2007). In the second, individual species are prioritized, and again prioritization can be based on a variety of measures that include degree of threat (IUCN Red Lists), phylogenetic uniqueness (e.g., McGoogan et al. 2007), and the surrogacy role of the taxon as an umbrella, flagship or indicator species.

Although these approaches provide quantitative evaluations of priority that are much-preferred by policy makers, there are a number of limitations of such "top-down" approaches that can confound their application. First, the different biodiversity measures (e.g., species richness, endemism) may not generate congruent hotspots within a taxon (Orme et al. 2005; Ceballos and Ehrlich 2006), cross-taxon congruence may be limited or scale-dependent (Grenyer et al. 2006), and species selected as surrogates may not provide adequate coverage for other biodiversity components (Andelman and Fagan 2000; Roberge and Angelstam 2004). Second, they require substantial datasets that are unavailable for many lesser-known taxa; minimally accurate distribution maps for geographical analyses and some indication of population trends or ecological role for species-centric approaches.

Third, they describe pattern rather than process, and do not typically identify systemic threats. Finally, they are limited in the extent to which they can incorporate local expert knowledge. Researchers and conservation biologists working in situ may be only too aware of the local immediate threats to their taxa, but such "general knowledge" is typically unpublished, or poorly documented, and is difficult to synthesize in meta-analyses.

The bottom-up approach taken by the SEABCRU represents an attempt to synthesize local expert knowledge and identify consensus research priorities that are both relevant across Southeast Asia and best addressed through region-wide collaboration. Expert consensus has been used to identify conservation units and priorities for the jaguar (Sanderson et al. 2002) and American crocodile (Thorbjarnarson et al. 2006) but there are few examples of consensus thematic prioritization. During the SEABCRU forum consensus was relatively easily reached because each priority proposed was generally acknowledged by forum participants to be a major problem, regardless of the specifics of individual systems. None of the suggestions focused on single-species actions nor geographical localities; rather concern centered on systemic threats to bats (hunting) and key habitats (caves, forests), and major knowledge gaps that preclude action (taxonomy of all species and distribution and population data for flying foxes). It is doubtful that conventional priority setting would identify these as key issues, yet the bottom-up consensus reached here clearly indicated that they are central to successful bat conservation in Southeast Asia.

Identifying research priorities is a much-needed first step for bat conservation in Southeast Asia, providing foci for collaborative action and fund raising. To facilitate implementation of the priorities, forum participants requested an interactive website to promote international networking and coordination of efforts, and to enable easier exchange of research and outreach materials. The SEABCRU website (http://www.seabcru.org) was launched in 2008, and provides an interactive platform for all interested parties. Outreach activities are also a critical component in bat conservation (Kingston et al. 2006b), but the cultural and linguistic diversity of Southeast Asia are such that a regional program was not deemed appropriate or practical. Rather, materials developed by individual initiatives from member countries are made available on the SEABCRU website. The similarities among taxa and the key conservation issues across the region are such that outreach workers should have relatively little difficulty adapting materials to their specific locality.

Acknowledgments The establishment of the Southeast Asian Bat Conservation Research Unit was only made possible by the generous support of the British American Tobacco Biodiversity Partnership, and I would like to express my deepest thanks to the Partnership and their biodiversity partners Earthwatch Institute. Consensus priority setting is by nature a group effort and I would like to thank all the participants of the SEABCRU Forum held at the 1st International South East Asian Bat Conference for their contribution to the process, as well as Chutamas Satasook and the students and staff of the Prince of Songkla University for making the conference and forum such a success. Although the origins of the SEABCRU were inspired by discussions with many colleagues, the following have been instrumental in the founding and running of the SEABCRU: Paul Bates, James Burton, David Kingston, Paul Laird, Tammy Mildenstein, Paul Racey, Matthew Struebig, Danny Squire, and Allyson Walsh.

References

Abdullah MT, Halls LS, Tissen OB et al (2007) The large bat caves of Malaysian Borneo. Bat Res News 48:99–100

Achard F, Eva HD, Stibig H-J et al (2002) Determination of deforestation rates of the world's humid tropical forests. Science 297:999–1002. doi:10.1126/science.1070656

Andelman S, Fagan WF (2000) Umbrellas and flagships: efficient conservation surrogates or expensive mistakes? Proc Natl Acad Sci USA 97:5954–5959. doi:10.1073/pnas.100126797

Bates PJJ, Tin Nwe, Pearch MJ et al (2000) A review of bat research in Myanmar (Burma) and results of a recent survey. Acta Chiropt 2:53–82
Bates PJJ, Struebig MJ, Rossiter SJ et al (2004) A new species of *Kerivoula* (Chiroptera: Vespertilionidae) from Myanmar (Burma). Acta Chiropt 6:219–226
Bates PJJ, Struebig MJ, Hayes BD et al (2007a) A new species of *Kerivoula* (Chiropteridae:Vespertilioinidae) from Southeast Asia. Acta Chiropt 9:323–337. doi:10.3161/1733-5329(2007)9[323:ANSOKC]2.0.CO;2
Bates PJJ, Rossiter SJ, Suyanto A, Kingston T (2007b) A new species of *Hipposideros* (Chiroptera: Hipposideridae) from Sulawesi. Acta Chiropt 9:13–26. doi:10.3161/1733-5329(2007)9[13:ANSOHC]2.0.CO;2
Biotani L, Catullo G, Marzetti I et al. (2006) The Southeast Asian Mammal Databank. A tool for conservation and monitoring of mammal diversity in Southeast Asia. Istituto di Ecologia Applicata, Rome. http://www.ieaitaly.org/samd. Cited 31 March 2008
Breed A, Smith CS, Meers J, Field HE (2007) Transmigrating pteropids: international movements of flying foxes: implications for conservation and introduction of disease. Bat Res News 48:133–134
Bumrungsri S, Harbit A, Benzie C et al (2007) The cave nectar bat (*Eonycteris spelaea*), is the major pollinator of Petai (*Parkia speciosa* Hassk. and *P. timoriana* Merr.) in southern Thailand. Bat Res News 48:113
Campbell P, Schneider CJ, Adnan AM et al (2004) Phylogeny and phylogeography of old world fruit bats in the *Cynopterus brachyotis* complex. Mol Phylogenet Evol 33:764–781. doi:10.1016/j.ympev.2004.06.019
Cardillo M, Mace GM, Gittleman JL, Purvis A (2006) Latent extinction risk and the future battlegrounds of mammal conservation. Proc Natl Acad Sci USA 103:4157–4161. doi:10.1073/pnas.0510541103
Ceballos G, Ehrlich PR (2006) Global mammal distributions, biodiversity hotspots, and conservation. Proc Natl Acad Sci USA 103:19374–19379. doi:10.1073/pnas.0609334103
Chuen WW, Eaton D, Dan YM (1998) Estimation of petai consumption in the Klang Valley in Peninsular Malaysia. In: Lee SS, Dan, YM, Gauld I, Bishop J (eds) Conservation, management and development of forest resources. Proceedings of the Malaysia–UK Programme Workshop 21–24 October 1996. Forest Research Institute Malaysia, Kepong, pp 212–226
Clements R, Sodhi NS, Schilthuizen M, Ng PKL (2006) Limestone karst of Southeast Asia: imperiled arks of biodiversity. Bioscience 56:733–742. doi:10.1641/0006-3568(2006)56[733:LKOSAI]2.0.CO;2
Cleveland CJ, Betke M, Federico P et al (2006) Economic value of the pest control service provided by Brazilian free-tailed bats in south-central Texas. Front Ecol Environ 4:238–243. doi:10.1890/1540-9295(2006)004[0238:EVOTPC]2.0.CO;2
Corlett RT (1992) The ecological transformation of Singapore, 1819–1900. J Biogeogr 19:411–420. doi:10.2307/2845569
Corlett RT (1998) Frugivory and seed dispersal by vertebrates in the oriental (Indomalayan) region. Biol Rev Camb Philos Soc 73:413–448. doi:10.1017/S0006323198005234
Corlett RT (2004) Flower visitors and pollination in the oriental (Indomalayan) region. Biol Rev Camb Philos Soc 79:497–532. doi:10.1017/S1464793103006341
Cox PA, Elmqvist T, Pierson ED, Rainey WR (1991) Flying foxes as strong interactors in south Pacific island ecosystems: a conservation hypothesis. Conserv Biol 5:448–454. doi:10.1111/j.1523-1739.1991.tb00351.x
Csorba G, Bates PJJ (2005) Description of a new species of *Murina* from Cambodia (Chiroptera: Vespertilionidae: Murininae). Acta Chiropt 7:147–163. doi:10.3161/1733-5329(2005)7[1:DOANSO]2.0.CO;2
Csorba G, Thong VD, Bates PJJ, Furey NM (2007) Description of a new species of *Murina* from Vietnam (Chiroptera: Vespertilionidae: Murininae). Museum of Texas Tech University, Occasional Papers No. 268
Dubois A (2003) The relationships between taxonomy and conservation biology in the century of extinctions. C R Biol 326(Suppl 1):S9–S21
Epstein JH, Olival KJ, Smith C et al (2007) Hunting, international movement, and the need for regional conservation of flying foxes (*Pteropus* spp.). Bat Res News 48:104–105
Esselstyn JA (2007) A new species of stripe-faced fruit bat (Chiroptera: Pteropodidae: *Styloctenium*) from the Philippines. J Mammal 88:951–958. doi:10.1644/06-MAMM-A-294R.1
Fletcher CD (2006) Roosting ecology of insectivorous bats in Krau Wildlife Reserve, Pahang, Peninsular Malaysia. Dissertation, Universiti Kebangsaan Malaysia
Francis CM, Guillén A, Robinson MF (comps) (1999) Order Chiroptera: Bats. In: Duckworth JW, Salter RE, Khounboline K (eds) Wildlife in Lao PDR: 1999 Status Report. IUCN, WCS and CPAWM, Vientiane, Lao PDR pp 225–235
Francis CM, Kingston T, Zubaid A (2007a) A new species of *Kerivoula* (Chiroptera: Vespertilionidae) from Peninsular Malaysia. Acta Chiropt 9:1–12. doi:10.3161/1733-5329(2007)9[1:ANSOKC]2.0.CO;2
Francis CM, Guillén-Servent A, Borisekno AV et al (2007b) Use of DNA barcodes to understand Southeast Asian bat diversity. Bat Res News 48:103
Fujita MS, Tuttle MD (1991) Flying foxes (Chiroptera: Pteropodidae): threatened animals of key ecological and economic importance. Conserv Biol 5:455–463. doi:10.1111/j.1523-1739.1991.tb00352.x

Furey NM, Racey PA (2007) Community composition of bat assemblages in a Vietnamese karst protected area. Bat Res News 48:98–99

Golding JS, Timberlake J (2003) How taxonomists can bridge the gap between taxonomy and conservation science. Conserv Biol 17:1177–1178. doi:10.1046/j.1523-1739.2003.02404.x

Grenyer R, Orme DL, Jackson SF et al (2006) Global distribution and conservation of rare and threatened vertebrates. Nature 444:93–96. doi:10.1038/nature05237

Guillén-Servent A, Francis CM (2006) A new species of bat of the *Hipposideros bicolor* group (Chiroptera: Hipposideridae) from Central Laos, with evidence of convergent evolution with Sundaic taxa. Acta Chiropt 8:39–61. doi:10.3161/1733-5329(2006)8[39:ANSOBO]2.0.CO;2

Gumal MT (2001) Ecology and conservation of a fruit bat in Sarawak. Dissertation, University of Cambridge, England

Heaney LR, Heideman PD (1987) Philippine fruit bats: endangered and extinct. Bats 5:3–5

Heaney LR, Balete DS, Dolar ML et al (1998) A synopsis of the mammalian fauna of the Philippine Islands. Fieldiana Zool N Ser 88:1–61

Helgen KM, Kock D, Comez RKSC et al (2007) Taxonomy and natural history of the Southeast Asian fruit bat genus *Dyacopterus*. J Mammal 88:302–318. doi:10.1644/06-MAMM-A-276R.1

Hendrichsen DK, Bates PJJ, Hayes BD, Walston JL (2001) Recent records of bats (Mammalia: Chiroptera) from Vietnam with six species new to the country. Myotis 39:35–122

Hutson AM, Mickleburgh SP, Racey PA (2001) Microchiropteran bats: global status survey and action plan. IUCN/SSC Chiroptera Specialist Group, IUCN, Gland, Switzerland (comps)

Isaac NJB, Turvey ST, Collen B, Waterman C, Baillie JEM (2007) Mammals on the EDGE: conservation priorities based on threat and phylogeny. PLoS ONE 2:e296. doi:10.1371/journal.pone.0000296

Khin Maung Lwin (1995) Non-wood forest products in Myanmar. In: Durst PB, Bishop A (eds) Beyond timber: social, economic and cultural dimensions of nonwood forest products in Asia and the Pacific. Proceedings of the Regional Expert Consultation, Bangkok 28 November to 2 December 1994. FAO, RAP Publication 1995/13, Bangkok. http://www.fao.org/docrep/X5336E/X5336E00.HTM. Cited 31 Mar 2008

Kingston T (2009) Analysis of species diversity of bat assemblages. In: Kunz TH, Parsons S (eds) Behavioral and ecological methods for the study of bats. Smithsonian Institution Press, Washington (in press)

Kingston T, Lara MC, Jones G et al (2001) Acoustic divergence in two cryptic *Hipposideros* species: A role for social selection? Proc R Soc Lond B Biol Sci 268:1381–1386

Kingston T, Francis CM, Zubaid A, Kunz TH (2003) Species richness in an insectivorous bat assemblage from Malaysia. J Trop Ecol 19:67–79

Kingston T, Lim BL, Zubaid A (2006a) Bats of Krau Wildlife Reserve. University Kebangsaan Malaysia, Bangi

Kingston T, Zubaid A, Lim G, Hatta F (2006b) From research to outreach: environmental education materials for the bats of Malaysia. In: Proceedings of the best of both worlds international conference on environmental education for sustainable development 2005. FRIM, Kuala Lumpur

Kunz TH (1982) Roosting ecology. In: Kunz TH (ed) Ecology of bats. Plenum, New York, pp 1–55

Lane DJW, Kingston T, Lee BPY-H (2006) Dramatic decline in bat species richness in Singapore, with implications for Southeast Asia. Biol Conserv 131:584–593. doi:10.1016/j.biocon.2006.03.005

Lee RJ, Gorog AJ, Dwiyahreni A et al (2005) Wildlife trade and implications for law enforcement in Indonesia: a case study from North Sulawesi. Biol Conserv 123:477–488. doi:10.1016/j.biocon.2005.01.009

Leelapaibul W, Bumrungsri S, Pattanawiboon A (2005) Diet of wrinkle-lipped free-tailed bat (*Tadarida plicata* Buchannan, 1800) in central Thailand: insectivorous bats potentially act as biological pest control agents. Acta Chiropt 7:111–119. doi:10.3161/1733-5329(2005)7[111:DOWFBT]2.0.CO;2

Lim TK (1998) Durian. In: Hyde KW (ed) The new rural industries: a handbook for farmers and investors. Rural Industries Research and Development Corporation, Australia, pp 279–285

Lyon MW (1911) Mammals collected by Dr. W. L. Abbott on Borneo. Proc US Nat Mus 40:53–146

Mace GM (2004) The role of taxonomy in species conservation. Philos Trans R Soc Lond B Biol Sci 359:711–719. doi:10.1098/rstb.2003.1454

Marshall AG (1983) Bats, flowers and fruit: evolutionary relationships in the old world. Biol J Linn Soc Lond 20:115–135. doi:10.1111/j.1095-8312.1983.tb01593.x

Marshall AG (1985) Old world phytophagous bats (Megachiroptera) and their food plants: a survey. Zool J Linn Soc 83:351–369. doi:10.1111/j.1096-3642.1985.tb01181.x

Matveev VA (2005) Checklist of Cambodian bats (Chiroptera), with new records and remarks on taxonomy. Russ J Theriol 4:43–62

Matveev VA, Csorba G (2007) Bat fauna of Cambodia. Bat Res News 48:100

Meijaard E, Sheil D, Nasi R et al (2005) Life after logging: reconciling wildlife conservation and production forestry in Indonesian Borneo. Center for International Forestry Research, Bogor

McConkey KR, Drake DR (2006) Flying foxes cease to function as seed dispersers long before they become rare. Ecology 87:271–276. doi:10.1890/05-0386

McGoogan K, Kivell T, Hutchison M et al (2007) Phylogenetic diversity and the conservation biogeography of African primates. J Biogeogr 34:1962–1974. doi:10.1111/j.1365-2699.2007.01759.x
Mohd.-Azlan J, Zubaid A, Kunz TH (2001) The distribution, relative abundance, and conservation status of the large flying fox, *Pteropus vampyrus*, in Peninsular Malaysia: a preliminary assessment. Acta Chiropt 3:149–162
Mickleburgh SP, Hutson AM, Racey PA (1992) Old World Fruit Bats: An Action Plan for their Conservation. International Union for the Conservation of Nature and Natural Resources, Gland
Mickleburgh SP, Hutson AM, Racey PA (2002) A review of the global conservation status of bats. Oryx 36:18–34. doi:10.1017/S0030605302000054
Myers N, Mittermeier RA, Mittermeier CG et al (2000) Biodiversity hotspots for conservation priorities. Nature 403:853–858. doi:10.1038/35002501
Orme DL, Davies RG, Burgess M et al (2005) Global hotspots of species richness are not congruent with endemism or threat. Nature 436:1016–1019. doi:10.1038/nature03850
Pamaong-Jose R, Ramayla SP, Granert WG, Cariño AP (2007) Conservation of bats in Karst areas of Bohol Island, Philippines. Bat Res News 48:99
Payne J, Francis CM, Phillipps K (1985) A field guide to the mammals of Borneo. Sabah Society and World Wildlife Fund Malaysia, Kota Kinabalu
Pottie SA, Lane DJW, Kingston T, Lee BPYH (2005) The microchiropteran bat fauna of Singapore. Acta Chiropt 2:237–247. doi:10.3161/1733-5329(2005)7[237:TMBFOS]2.0.CO;2
Racey PA (2007) The role of karst in the conservation of bat biodiversity. Bat Res News 48:98
Racey PA, Entwistle AC (2003) Conservation ecology of bats. In: Kunz TH, Fenton MB (eds) Bat ecology. University of Chicago Press, Chicago, pp 680–743
Rainey WE, Pierson ED, Elmqvist T, Cox P (1995) The role of flying foxes (Pteropodidae) in oceanic island ecosystems of the Pacific. In: Racey PA, Swift S (eds) Ecology, evolution and behaviour of bats. Clarendon Press, Oxford, pp 47–59
Ricketts TH, Dinerstein E, Boucher T et al (2005) Pinpointing and preventing imminent extinctions. Proc Natl Acad Sci USA 102:18497–18501. doi:10.1073/pnas.0509060102
Roberge JM, Angelstam P (2004) Usefulness of the umbrella species concept as a conservation tool. Conserv Biol 18:76–85. doi:10.1111/j.1523-1739.2004.00450.x
Sanderson E, Redford KH, Chetkiewicz C-LB et al (2002) Planning to save a species, the jaguar as a model. Conserv Biol 16:58–72. doi:10.1046/j.1523-1739.2002.00352.x
Sedlock JL (2001) Inventory of insectivorous bats on Mount Makiling, Philippines using echolocation call signatures and a new tunnel trap. Acta Chiropt 3:163–178
Shilton LA, Altringham JD, Compton SG, Whittaker RJ (1999) Old world fruit bats can be long-distance seed dispersers through extended retention of viable seeds in the gut. Proc R Soc Lond B Biol Sci 266:219–223. doi:10.1098/rspb.1999.0625
Shively GE (1997) Poverty, technology, and wildlife hunting in Palawan. Environ Conserv 24:57–63. doi:10.1017/S0376892997000106
Simmons N (2005) Chiroptera. In: Wilson DE, Reeder DM (eds) Mammal species of the World: a taxonomic and geographic reference. John Hopkins University Press, Baltimore, pp 312–529
Sodhi NS, Brook BW (2006) Southeast Asian biodiversity in crisis. Cambridge University Press, Cambridge
Sodhi NS, Koh LP, Brook BW, Ng PKL (2004) Southeast Asian biodiversity: an impending disaster. Trends Ecol Evol 19:654–660. doi:10.1016/j.tree.2004.09.006
Srithongchuay T, Bumrungsri S (2007) Does pollination success of *Oroxylum indicum* Vent. (Bignoniaceae) depend solely on the nectarivorous bat, *Eonycteris spelaea* Dobson (Chiroptera: Pteropodidae)? Bat Res News 48:112–113
Stattersfield AJ, Crosby MJ, Long AJ, Wege DC (1998) Endemic bird areas of the world: priorities for biodiversity conservation. BirdLife International, Cambridge
Struebig MJ, Harrison ME, Cheyne SM, Limin S (2007) Intensive hunting of large flying-foxes *Pteropus vampyrus natunae* in Central Kalimantan, Indonesian Borneo. Oryx 41:390–393. doi:10.1017/S0030605307000310
Struebig MJ, Christy L, Pio D, Meijaard E (2008a) Bats of Borneo: diversity, distributions and representation in protected areas. Biol Conserv (Special Issue) (in review)
Struebig MJ, Kingston T, Zubaid A et al (2008b) Conservation value of forest fragments to palaeotropical bats. Biol Conserv 141:2112–2126. doi:10.1016/j.biocon.2008.06.009
Suyanto A, Struebig MJ (2007) Bats of the Sangkulirang limestone karst formations, East Kalimantan—a priority region for Bornean bat conservation. Acta Chiropt 9:67–95. doi:10.3161/1733-5329(2007)9[67:BOTSLK]2.0.CO;2

Thabah A, Rossiter SJ, Kingston T et al (2006) Genetic divergence and echolocation call frequency in cryptic species of *Hipposideros larvatus sensu lato* (Chiroptera: Hipposideridae) from the Indo-Malayan region. Biol J Linn Soc Lond 88:119–130. doi:10.1111/j.1095-8312.2006.00602.x
Thorbjarnarson J, Mazzotti F, Sanderson E et al (2006) Regional habitat conservation priorities for the American crocodile. Biol Conserv 128:25–36. doi:10.1016/j.biocon.2005.09.013
Whittaker RJ, Jones SH (1994) The role of frugivorous bats and birds in the rebuilding of a tropical forest ecosystem, Krakatau, Indonesia. J Biogoegr 245:245–258. doi:10.2307/2845528
Wiles GJ, Brooke AP (2008) Conservation threats to bats in the tropical Pacific islands and insular Southeast Asia. In: Fleming TH, Racey P (eds) Island bats: evolution, ecology and conservation. University of Chicago, Chicago (in press)
Yoshiyuki M, Lim BL (2005) A new horseshoe bat, *Rhinolophus chiewkweeae* (Chiroptera, Rhinolophidae), from Malaysia. Bull Natl Sci Mus Ser A 31:29–36

Biodivers Conserv (2010) 19:485–490
DOI 10.1007/s10531-009-9585-7

ORIGINAL PAPER

Insect extinctions on a small denuded Bornean island

Navjot S. Sodhi · David S. Wilcove · R. Subaraj · Ding Li Yong · Tien Ming Lee · Henry Bernard · Susan L. H. Lim

Received: 3 September 2008 / Accepted: 20 January 2009 / Published online: 4 February 2009

Abstract We report odonate and butterfly extinctions on Pulau Mengalum (Sabah) between 1928 and 2007. Pulau Mengalum has lost all of its closed-canopy forest; 55.5% of the odonates and 40% of the butterflies present in 1928 have likely been extirpated. Fourteen and five species of odonates and butterflies found by us were new records for the island, respectively. It is unclear if newly recorded species were missed by previous surveyors in 1928 or if they have colonized the island after that time. While our study indicates that deforestation is a serious threat to tropical insects, it remains unclear if deforested areas provide opportunities for new colonists.

Keywords Deforestation · Biodiversity crisis · Conservation · Odonata · Butterflies · Southeast Asia

N. S. Sodhi · R. Subaraj · D. L. Yong
Department of Biological Sciences, National University of Singapore,
14 Science Drive 4, Singapore 117543, Republic of Singapore

D. S. Wilcove
Department of Ecology and Evolutionary Biology, Princeton University,
Princeton, NJ 08544-1003, USA

T. M. Lee
Ecology, Behavior and Evolution Section, Division of Biological Sciences,
University of California, San Diego, 9500 Gilman Drive, La Jolla, CA 92093-0116, USA

H. Bernard
Institute for Tropical Biology and Conservation, Universiti Malaysia Sabah,
Locked Bag 2073, 88999 Kota Kinabalu, Sabah, Malaysia

S. L. H. Lim
Institute of Biological Sciences, University of Malaya,
50603 Kuala Lumpur, Malaysia

N. S. Sodhi (✉)
Department of Organismic and Evolutionary Biology, Harvard University,
Cambridge, MA 02138, USA
e-mail: nsodhi@oeb.harvard.edu; dbsns@nus.edu.sg

Introduction

Deforestation and forest degradation are endangering tropical biotas (Brook et al. 2003; Sodhi et al. 2007). Although the loss of birds, mammals, and other conspicuous groups is generally well documented, human-mediated extinctions of numerous insects may have been gone unnoticed (Mawdsley and Stork 1995; McKinney 1999; Dunn 2005). This is disconcerting considering that more than 57% described living species are insects (Stork 1997) and that 57,000 insect species per million, mostly in the tropics, could be extinguished over the next 50 years largely due to land-cover transformation (Pimm and Raven 2000; Dunn 2005). Specialized microhabitat requirements may have made numerous tropical insects vulnerable to habitat disturbance (Mawdsley and Stork 1995). However, a few extinctions of tropical insects have been recorded. In Singapore, 20 and 38% of phasmids and butterflies, respectively, were extirpated following the loss of more than 95% of the forest (Brook et al. 2003). Hanski et al. (2007) reported that Madagascar probably lost 43% of previously recorded forest dung beetles due to elimination of half of its forest. In contrast, Larsen (2008) reported that no butterfly species has been eradicated in all of West Africa despite 87% reduction in the forest. It is probable that a number of those butterfly species may already be committed to extinction and that further loss of protected forests may result in massive extinctions (Larsen 2008). A few insect extinctions due to land-cover transformation have also been documented in temperate areas (e.g., Korkeamäki and Suhonen 2002; Thomas et al. 2004; Samways 2006). Insects remain vulnerable to human-driven land-use transformation but more empirical data on insect extinctions are needed, particularly in the tropics, so that insects may gain more prominence in conservation initiatives.

Here we report local extinctions or extirpations (hereafter extinctions) of odonates (damselflies and dragonflies) and butterflies on an island, Pulau Mengalum (hereafter Mengalum), which was largely forested in 1928 but is now denuded (see below). Both study taxa provide important ecosystem functions (e.g., mosquito control or pollination; Korkeamäki and Suhonen 2002; Koh and Sodhi 2004; Suh and Samways 2005; Koh 2007) and can be good indicators of environmental quality.

Materials and methods

Mengalum (6°12′00″N, 115°35′00″E) is a sandy island of ~4.2 km^2 in size and lies offshore 56 km northwest of Kota Kinabalu, Sabah (Malaysian Borneo; Fig. 1). In 1928, most of the island was covered by native forest (Kloss 1930). All of the closed canopy forest has now been replaced with an open vegetation mosaic. Scrub (containing *Ficus* spp., *Instia bijuga*, *Chromolaena odorata*, etc.) covers about 40% of the island (estimated using GLC2000 landcover data in 2000; http://www-gem.jrc.it/glc2000/). A small number of Sulu fishing people inhabited the island until the 1960s, and the island was used as a quarantine facility for imported cattle up until 1997. This facility may have facilitated the spread of African star grass (*Cynodon* spp.) and elephant grass (*Pennisetum purpureum*), both of which were planted as fodder for cattle and now cover about 60% of the island (estimated using GLC2000 data). Currently there is a small resort on Mengalum to cater for recreational fishing and diving.

Insects on Mengalum were previously surveyed between 8 and 15 July 1928 (Pendlebury 1931a, b; Laidlaw 1931). We sampled odonates and butterflies on Mengalum on 4–5 June, 6–9 July, and 11–13 October 2007. Although we could completely survey the island

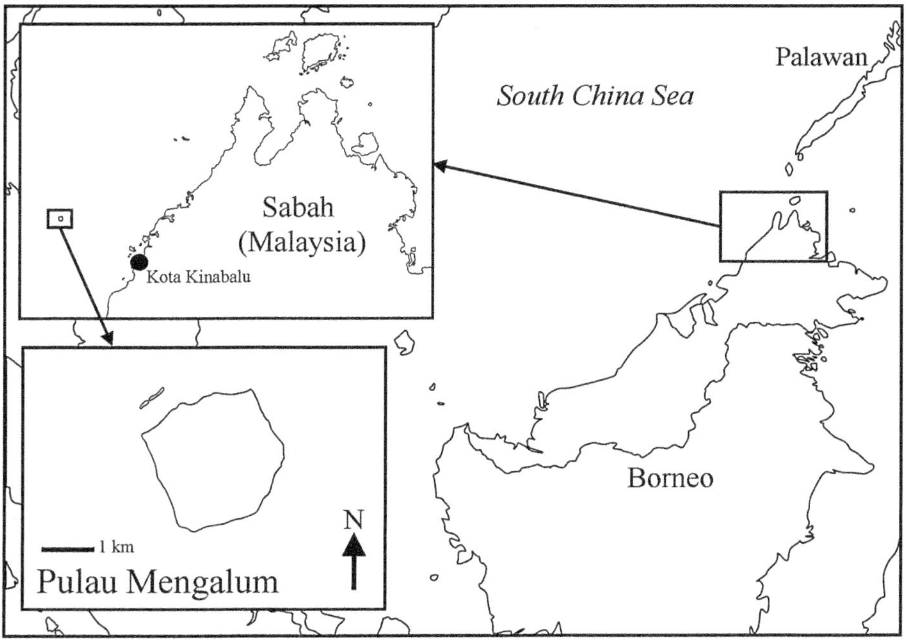

Fig. 1 Location of Pulau Mengalum

in 2 days, we decided to spread sampling over 5 months to minimize the chances of missing any previously recorded species. Each day, we surveyed the island between 09:00–13:00 h and/or 15:00–18:00 h. We conducted surveys only during fair weather (i.e., not during heavy rain and wind) and the whole island was surveyed during each of the three trips. All major waterbodies on the island were also surveyed during each trip. Each day, between 4–5 surveyors combed the area and captured all odonates and butterflies that were flying or flushed. The flat topography (only hill 12 m above the sea level) increased species detectability. Species that could be identified on the spot were released. However, a small number of individuals (n = 40 odonates and n = 42 butterflies) were kept to verify identification. Species were identified using Corbet and Pendlebury (1992) and Orr (2003, 2005). Any species that is predominantly found in primary and secondary forests based on aforementioned books was classified as a forest species. From the 1928 list, we excluded unidentified species or doubtful identifications (e.g., *Tramea limbata*). Species recorded in 1928 but not in 2007 were considered to be extinct.

Results and discussion

Mengalum has likely lost five (55.5%) of nine odonates since 1928 (*Archibasis melanocyana*, *Camacinia gigantea*, *Gynacantha dohrni*, *Lestes praemorsa*, and *Raphismia bispina*). Our last survey in October did not add any species to the previously recorded odonates suggesting that our sampling was adequate. Six (40%) of 15 butterflies recorded in 1928 are now probably extinct on Mengalum (*Bibasis harisa*, *Gangara thyrsis*, *Graphium empedovana*, *Libythea geoffroy*, *Nacaduba kurava*, and *Taractrocera ziclea*).

Our last survey in October added only one butterfly species (*Cupha arias*) to the ones recorded in 1928, indicating that the sampling was probably adequate. These insect extinctions are close to 25–50% predicted by species-area relationship with the 90% loss of habitat (Simberloff 1992). Due to factors such as high habitat specificity, endemic species are vulnerable to habitat loss (Posa and Sodhi 2006). As such, we determined whether disproportionately more species that are Southeast Asian endemics were extirpated in both groups but did not find any evidence for this (Fisher's exact test, $P > 0.08$), perhaps due to small sample sizes.

Both forest odonate species found in 1928 were extinct (*A. melanocyana* and *G. dohrni*). Of four forest butterfly species recorded in 1928, two were lost (*G. thyrsis*, and *L. geoffroy*). Previous studies on various taxa show that many forest species are unable to cope with deforestation due to erosion in habitat quantity and quality (e.g., increased predation and microhabitat alteration; Sodhi et al. 2007). Our study also shows that forest odonates may be more sensitive to habitat loss than butterflies. Extirpation of non-forest species may be due to loss of waterbodies or host plants (Koh and Sodhi 2004; Suh and Samways 2005). Some of these non-forest species may also have occupied mangroves or swamp forests, which have also been lost on Mengalum. However, since we used recent books to allocate habitats, it is possible that some of these non-forest dwelling species may in fact have been forest dwellers on the island in 1928. Habitat selection by a species can vary temporally and spatially (Cody 1985).

We recorded 14 new odonates in 2007 (*Archibasis viola*, *Agriocnemis femina*, *A. pygmaea*, *Brachydiplax chalybea*, *Diplacodes trivialis*, *Ischnura senegalensis*, *Lathrecista asiatica*, *Macrodiplax cora*, *Mortagrion falcatum*, *Orthetrum sabina*, *Pantala flavescens*, *Pseudagrion microcephalum*, *Rhyothemis phyllis*, and *Tramea transmarina*). One of these new records (*A. viola*) is a forest species. Similarly, we added five new butterfly species to the ones recorded in 1928 (*Charaxes solon*, *Danaus melanippus*, *Euploea eyndhovii*, *Hypolimnas anomla*, and *Catopsilia pomona*). Three of these are forest species (*C. solon*, *H. anomla*, and *E. eyndhovii*). Insects are known to migrate over 56 km (the distance between Borneo and Mengalum; see Wikelski et al. 2006), therefore it is possible that these species colonized the island following deforestation. We found significantly more widespread species (found in Southeast Asia and beyond) in newly recorded odonates (Fisher's exact test $P = 0.0001$) but not in the butterflies ($P = 0.2$). However, it is possible that some of our newly recorded species were missed by surveyors in 1928. Because some of the new records for butterflies and odonates are forest-dwelling species, we believe that this may be the case. We are also not certain if these new records were deliberately released by humans on the island.

A study such as ours is amenable to biases. One possibility is that we may have failed to record some species. We minimized this by visiting the island during three occasions and by using 4–5 surveyors. The flat topography of the island enhanced species detectability and further minimized this bias. The other bias is misidentification by previous surveyors, and we minimized this by critically inspecting the 1928 list and removing doubtful species (e.g., species not typically found in Southeast Asia).

Our study is yet another reminder that ongoing massive deforestation in Southeast Asia could result in considerable biotic extinctions (Sodhi et al. 2004; Sodhi and Brook 2006). Our study was conducted on a small island, and insular biotas are usually more vulnerable to disturbance because of factors such as small population sizes (Simberloff 2000). Therefore, it can be argued that our results may not be widely applicable. However, the percentages of butterfly extinctions reported by us are strikingly similar to those reported on the island of Singapore (540 km^2) with a similar level of deforestation (Brook et al.

2003). Additionally, another study also projected the loss of half of 151 recorded butterfly species in the Philippines due to deforestation (de Jong and Treadaway 1993). Reforestation activities could be carried out on Mengalum to preserve the residual forest biodiversity. For dragonflies, restoration of forests around waterbodies could be beneficial. On a wider scale, our study highlights that tropical insects are vulnerable to ongoing deforestation.

Acknowledgments Our study was supported by the National Geographic Society (8193-07) and Lady Y.P. McNeice through the Loke Wan Tho Memorial Foundation. Fieldwork was conducted with permission granted by the Economic Planning Unit of Malaysia. Sarah and Daniel Hrdy Fellowship provided support to NSS while this manuscript was prepared. Y.Takanami helped with butterfly identification. We thank them all.

References

Brook BW, Sodhi NS, Ng PKL (2003) Catastrophic extinctions follow deforestation in Singapore. Nature 424:420–423. doi:10.1038/nature01795

Cody ML (1985) Habitat selection in birds. Academic Press, London

Corbet AS, Pendlebury HM (1992) The butterflies of the Malay Peninsula. Malayan Nature Society, Kuala Lumpur

De Jong R, Treadaway CG (1993) The Hesperiidae (Lepidoptera) of the Philippines. Zool Verh Leiden 288:1–125

Dunn RR (2005) Modern insect extinctions, the neglected majority. Conserv Biol 19:1030–1036. doi:10.1111/j.1523-1739.2005.00078.x

Hanski I, Koivulehto H, Cameron A, Rahagalala P (2007) Deforestation and apparent extinctions of endemic forest beetles in Madagascar. Biol Lett 3:344–347. doi:10.1098/rsbl.2007.0043

Kloss CB (1930) The birds of Mangalum and Mantanani Islands off the west coast of British North Borneo. Bull Raffles Mus 4:117–123

Koh LP (2007) Impacts of land use change on South-East Asian forest butterflies: a review. J Appl Ecol 44:703–713. doi:10.1111/j.1365-2664.2007.01324.x

Koh LP, Sodhi NS (2004) Importance of reserves, fragments and parks for butterfly conservation in a tropical urban landscape. Ecol Appl 14:1695–1708. doi:10.1890/03-5269

Korkeamäki E, Suhonen J (2002) Distribution and habitat specialization of species affect local extinction in dragonfly Odonata populations. Ecography 25:459–465. doi:10.1034/j.1600-0587.2002.250408.x

Laidlaw FF (1931) Remarks on some Odonata (Dragonflies) from Mangalum Island, north-west Borneo. Bull Raffles Mus 5:90–92

Larsen T (2008) Forest butterflies in West Africa have resisted extinction…so far (Lepidoptera: Papilionoidea and Hesperioidea). Biol Conserv 17:2833–2847

Mawdsley NA, Stork NE (1995) Species extinctions in insects: ecological and biogeographical considerations. In: Harrington R, Stork NE (eds) Insects in a changing environment. Academic Press, London, pp 321–369

McKinney ML (1999) High rates of extinction and threat in poorly studied taxa. Conserv Biol 13:1273–1281. doi:10.1046/j.1523-1739.1999.97393.x

Orr AG (2003) A Guide to the Dragonflies of Borneo: their Identification and Biology. Natural History Publications (Borneo), Kota Kinabalu

Orr AG (2005) Dragonflies of Peninsular Malaysia and Singapore. Natural History Publications (Borneo), Kota Kinabalu

Pendlebury HM (1931a) Rhopalocera from Mangalum Island, north-west Borneo Bull. Raffles Mus 5:94–96

Pendlebury HM (1931b) A preliminary note on some insects from Mangalum Island, north-west Borneo. Bull Raffles Mus 5:97–101

Pimm SL, Raven P (2000) Biodiversity–extinction by numbers. Nature 403:843–845. doi:10.1038/35002708

Posa MRC, Sodhi NS (2006) Effects of anthropogenic land use on forest birds and butterflies in Subic Bay, Philippines. Biol Conserv 129:256–270. doi:10.1016/j.biocon.2005.10.041

Samways MJ (2006) Insect extinctions and insect survival. Conserv Biol 20:245–246. doi:10.1111/j.1523-1739.2006.00349.x

Simberloff D (1992) Do species-area curves predict extinction in fragmented forest? In: Whitmore TC, Sayer JA (eds) Tropical deforestation and species extinction. Chapman & Hall, London, pp 75–89

Simberloff D (2000) Extinction-proneness of island species–causes and management implications. Raff Bull Zool 48:1–9

Sodhi NS, Brook BW (2006) Southeast Asian biodiversity in crisis. Cambridge University Press, Cambridge

Sodhi NS, Koh LP, Brook BW, Ng PKL (2004) Southeast Asian biodiversity: an impending disaster. Trends Ecol Evol 19:654–660. doi:10.1016/j.tree.2004.09.006

Sodhi NS, Brook BW, Bradshaw CJA (2007) Tropical conservation biology. Blackwell Publishers Ltd, Oxford

Stork NE (1997) Measuring global biodiversity and its decline. In: Reaka-Kudla ML, Wilson DE, Wilson EO (eds) Biodiversity II. Joseph Henry Press, Washington, DC, pp 41–68

Suh A, Samways M (2005) Significance of temporal changes when designing a reservoir for conservation of dragonfly diversity. Biol Conserv 14:165–178

Thomas JA, Telfer MG, Roy DB, Preston CD, Greenwood JJD, Asher J, Fox R, Clarke RT, Lawton JH (2004) Comparative losses of British butterflies, birds, and plants and the global extinction crisis. Science 303:1879–1881. doi:10.1126/science.1095046

Wikelski M, Moskowitz D, Adelman JS, Cochran J, Wilcove DS, May ML (2006) Simple rules guide dragonfly migration. Biol Lett 2:325–329. doi:10.1098/rsbl.2006.0487

Biodivers Conserv (2010) 19:491–502
DOI 10.1007/s10531-009-9701-8

ORIGINAL PAPER

Vulnerable island carnivores: the endangered endemic dwarf procyonids from Cozumel Island

Katherine W. McFadden · Denise García-Vasco · Alfredo D. Cuarón · David Valenzuela-Galván · Rodrigo A. Medellín · Matthew E. Gompper

Received: 19 November 2008 / Accepted: 22 July 2009 / Published online: 15 August 2009

Abstract Insular carnivores represent some of the most critically threatened species, but also the least known. To evaluate the conservation status of these species, thorough abundance estimates are urgently required. To better understand the population biology and conservation status of the endemic and threatened pygmy raccoon (*Procyon pygmaeus*) and dwarf coati (*Nasua nelsoni*) on Cozumel Island, Mexico, we worked island-wide to identify the presence of these species, and for the pygmy raccoon we studied several

K. W. McFadden (✉)
Department of Ecology, Evolution, and Environmental Biology, Columbia University, New York, NY 10027, USA
e-mail: kwm6@columbia.edu

D. García-Vasco
Instituto de Ecología, A. C., 91070 Xalapa, Veracruz, Mexico
e-mail: denise.vasco@posgrado.inecol.edu.mx

D. García-Vasco · A. D. Cuarón
SACBÉ - Servicios Ambientales, Conservación Biológica y Educación, 77660 Cozumel, Quintana Roo, Mexico

A. D. Cuarón
Multicriteria SC, Unidad Independencia IMSS, Torre Uxmal 18, 10100 México, DF, México
e-mail: cuaron@gmail.com

D. Valenzuela-Galván
Departamento de Ecología y Conservación de los Recursos Naturales, Centro de Educación Ambiental e Investigación Sierra de Huautla, Universidad Autónoma del Estado de Morelos, 62209 Cuernavaca, Morelos, Mexico
e-mail: dvalen@buzon.uaem.mx

R. A. Medellín
Instituto de Ecología, A.C., Universidad Nacional Autónoma de Mexico, 04510 Ciudad Universitaria, DF, Mexico
e-mail: medellin@miranda.ecologia.unam.mx

M. E. Gompper
Department of Fisheries and Wildlife Sciences, University of Missouri, Columbia, MO 65211, USA
e-mail: gompperm@missouri.edu

populations in depth during 2001–2003. On Cozumel, trapping was conducted for >6,600 trap nights in 19 locations of varying habitat types. A total of 96 pygmy raccoons (47 males and 49 females) and a single adult, male dwarf coati (*N. nelsoni*) were captured. Estimated total annual pygmy raccoon population size ± SE was 80 ± 26.1, resulting in an average density of 22 ± 5.1 raccoons/km^2 for the three small sites where the animals persist. Based in part on the findings of this study that indicate these species have a restricted range and small population numbers, the IUCN recently changed the listing of the pygmy raccoon to Critically Endangered from Endangered. In contrast, the status of the dwarf coati (Endangered) has not been changed, although the taxon is in eminent danger of extinction and in need of immediate conservation action.

Keywords Carnivore conservation · Dwarf coati · Mark-recapture · *Nasua nelsoni* · *Procyon pygmaeus* · Pygmy raccoon · Population ecology

Introduction

Tropical and temperate islands typically lack terrestrial mammals, particularly mammalian carnivores, due to the poor over-water dispersal abilities and large body sizes of these taxa, as well as the low carrying capacities of islands (Williamson 1981; Alcover and McMinn 1994). Insular mammalian carnivores are particularly intriguing conservation subjects for several reasons. Species in the order Carnivora tend to exist at low densities, are often heavily harvested or persecuted by humans, are subject to epidemics via pathogen spillover from high-density domestic carnivore populations, and tend to have relatively large home-ranges which increase their likelihood of contact with habitat edges or altered landscape pressures (Gittleman et al. 2001). On islands, collectively these factors may place insular carnivore populations at particular risk relative to other mammalian guilds. Furthermore, because of the potential for strong top-down effects, the alteration of a carnivore community has the potential to result in strong and sometimes dramatic indirect effects—a scenario that has been documented repeatedly on island ecosystems (Roemer et al. 2009). Thus the status of island carnivores represents a significant conservation concern.

Cozumel Island, Mexico, is unusual in supporting several insular, endemic carnivores, of which the best described, are the pygmy raccoon (*Procyon pygmaeus*; Merriam 1901) and the dwarf coati (*Nasua nelsoni*; Thomas 1901). As is the trend for island fauna, both species are smaller, or dwarfed, compared to their mainland ancestors (Goldman 1950; Jones and Lawlor 1965). The pygmy raccoon's IUCN listing status has recently been changed from Endangered (EN) to Critically Endangered (CR) by the World Conservation Union (CR2a(i)b) (Cuarón et al. 2008). In contrast, the dwarf coati, whose status as a distinct species remains unclear (McFadden et al. 2008) has previously been listed as Endangered (EN C2a) despite recent work indicating these species should be considered Critically Endangered (CR A2c + C2a(i)b) according to the IUCN system (Cuarón et al. 2004; McFadden 2004). Given that conservation management of these species is urgently needed, information on the distribution, abundance, and population structure of these taxa across Cozumel is a priority to support the development of educated management strategies. Here we report the results of our work to address these topics.

Dwarf coatis and pygmy raccoons have existed on Cozumel Island for at least 3,500 years (Hamblin 1984) and phylogenetic analyses suggest divergence times from the mainland taxa that predate human colonization of the island (McFadden et al. 2008). A recent assessment of the status of these two species, however, suggests that populations of both have declined precipitously (Cuarón et al. 2004). The exact causes of their decline are unknown but both species are thought to be adversely affected by predators (dogs, *Canis familiaris*, and boas, *Boa constrictor*), habitat degradation, hunting, and possibly by resource competition and disease spill-over from free-ranging domestic carnivores (Navarro and Suárez 1989; Martínez-Morales and Cuarón 1999; Cuarón et al. 2004; McFadden 2004; García-Vasco 2005; McFadden et al. 2005; Bautista 2006; Mena 2007; Romero-Nájera et al. 2007). Natural threats such as hurricanes may also negatively impact these and other species given their already reduced population sizes (Navarro and Suárez 1989; Perdomo 2006; Copa-Álvaro 2007).

Little information exists on the abundance of either taxon. A brief survey carried out in 1987–1988 identified a few localized populations of pygmy raccoons and dwarf coatis in the mangrove forest of Chankanaab, the northwest tip of the island, and the adjacent Isla de la Pasión (Navarro and Suárez 1989). Martínez-Morales and Cuarón (1999) assessed the dwarf coati population in 1994–1995 based on line transect surveys, observing only two dwarf coatis in over 386 km of transects. Using these data, a preliminary estimate of the population density of dwarf coatis was 0.43 individuals/km^2 (Cuarón et al. 2004). Aside from these studies, however, no broad-based examination of either taxon has occurred. Therefore, we assessed the locations of pygmy raccoon and dwarf coati populations on the island and estimated the current population size and structure.

Materials and methods

Study sites, capture and handling

Cozumel, Mexico (20°16′18″ to 20°35′33″N; and 86°43′23″ to 87°01′31″W), is a c. 478 km^2 oceanic island located 17.5 km east of the Yucatan peninsula (Fig. 1). The island encompasses a variety of terrestrial habitats including semi-evergreen tropical forests, subdeciduous tropical forest, mangrove stands, sandy palm areas and other coastal vegetation (Téllez and Cabrera 1987; Cuarón 2009). Mean annual temperature is 25.5°C, and mean annual rainfall is 1,505 mm (Martínez-Morales 1996; Cuarón 2009), with a pronounced seasonality that influences the availability of foods such as fruit and insects (McFadden et al. 2006).

Trapping sites were selected based on pilot studies suggesting the possible presence of native carnivores, as well as an attempt to sample multiple habitats and regions of the island. An additional consideration was accessibility, as many areas on Cozumel are isolated and too wet (i.e., mangrove areas prone to flooding) to survey using a trapping-based approach. Ultimately we selected 19 sites throughout the island to sample with the majority of these sites sampled for at least 2 weeks (mean = 13 days, range = 4–22 days) per site over 3 years (2001–2003; Fig. 1). We attempted to trap throughout the year to examine if significant variation existed in capture probabilities between wet and dry seasons. The first sampling period was conducted from July through September 2001 (wet season), the second spanned from April (late dry season) to August 2002 (mid wet season), then every 2 weeks from September through December 2002 (late wet, early dry season), and the third season was February to March 2003 (dry season).

Three locations (sites 1–3; Fig. 1) in the northwestern mangrove-dominated habitats were further studied based on the identification of pygmy raccoon populations at these sites. Although these three sites are in relatively close proximity to each other (<4 km), we

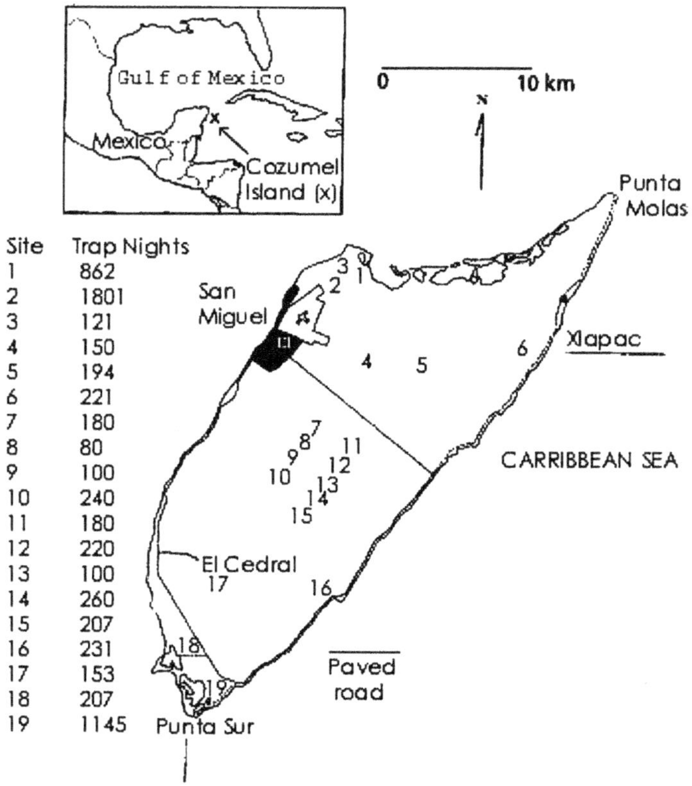

Fig. 1 Map of Cozumel Island, Mexico, and the locations of the 19 trapping sites and total sampling effort (number of trap nights) at each location from 2001 to 2003. Symbol *X* represents Cozumel Island

consider them to be independent populations because we did not detect movement of marked animals between sites.

During each sampling session, live traps (Tomahawk #207 box traps, Tomahawk Live Trap Co., Tomahawk, WI, USA or Havahart #1089, Havahart Live Trap Co., Lititz, PA, USA) were placed at 25–50 m intervals in each site. The placement and numbers of traps used differed between sites and depended on the habitat. Thus, we calculated trapping area differently depending on the configuration and shape of the trap site and placement of traps (i.e., polygon vs. linear). For sites 1 and 2 the trapping configuration was polygon shaped, so trapping area (T) was calculated as the area prescribed by a minimum convex polygon of the traps. Site 3 was configured as a strip transect, and traps were placed 0.02 km off both sides of a 0.40 km linear transect, and T was calculated as the total area the traps made up within this strip transect (0.04×0.40 km $= 0.016$ km^2).

Using GPS data, the mapping program TNTmips version 6.9 (MicroImages 2004), and a digitized satellite image of Cozumel Island (Landsat, Google Earth Professional v. 4.2 2007) the effective trapping area (A) was calculated by adding a buffer or boundary strip of width (W) to the area defined by the outer traps (T), so that the effective sampling area also includes areas covered by the outer traps beyond the outer polygon limits (Karanth and Nichols 1998). Based on Kenward (1985), we used a buffer width (W) of 1/2 the mean home range of pygmy raccoons (0.65 ± 0.26 km^2, based on 85% harmonic mean estimates, $n = 8$; García-Vasco et al., unpublished data).

Site 1 was primarily mangrove habitat, close to Isla de la Pasión. Site 2 was surrounding and within a golf course (i.e. not directly on golfing greens, but adjacent to), with trapping areas consisting of both semi-evergreen tropical forest and mangrove forests. Site 3, was located approximately 1.0 km north–northwest of site 2 (3.5 km from site 1) and consisted of mangrove forest close to the coast. For the three sites, $T = 0.02$–1.28 km^2, and $A = 0.69$–3.08 km^2 (Table 1).

Traps were baited with canned cat food, sardines, or a honey-banana mixture and checked at least once daily. We also explored additional bait types, including commercially available scented carnivore bait, various fruits, fish, and crabs. Trapped animals were immobilized with ketamine hydrochloride (10–12 mg/kg) and xylazine (2 mg/kg) prior to handling and ear tagging (Hasco Tag Co., Dayton, KY). Age was approximated as juvenile (0–12 months), subadult (13–21 months), or adult (>21 months) from tooth wear, body size, reproductive status, and recapture history (Grau et al. 1970; Sanderson and Nalbandov 1973). For males, canine size and the descent of the testes were used to differentiate subadult and adults. Females were considered to have bred based on nipple enlargement. Uncertainty about breeding status occurred in some cases, however, and these individuals were classified as non-breeding. All captured animals were released at the trap site after recovery from immobilization (within 2 h of checking the trap).

Population analysis

Capture frequencies were examined for variation by year, age, and sex, using chi-square tests. Using the CAPTURE feature in program MARK (White and Burnham 1999), chi-squares test were also used to assess time-related variation and behavioral response to capture. Specifically, the tests used to assess these factors were: Tests for heterogeneity of trapping probabilities in a population, tests for behavioral response after initial capture, tests for time-specific variation in trapping probabilities, goodness of fit tests of model. Results of these tests were used to select the appropriate model estimators. Due to low sample sizes, data sets for each site were pooled between years for each of these tests.

Using Pollock's (Pollock et al. 1990) robust population model in program MARK, annual capture ($\hat{p}$) and recapture ($\hat{c}$) probabilities, and abundance ($\hat{N}$) were calculated for Sites 1 and 2, for which we had trapped for three and two consecutive years, respectively. We used a Schnabel closed population model to estimate population size for Site 3, which was only sampled for one field season (2003) and which also had a relatively small population size. We used a robust design model, allowing closely-spaced trapping sessions to be viewed as a closed capture survey whereas longer intervals between sessions allows

Table 1 Effectively sampled areas calculated from mark recapture data and their resulting pygmy raccoon density estimates ($\hat{N}$) and standard error (SE)

Site	Sampled area size (km^2) outer trap polygon (T)	Effective sampled area (km^2) (A)	Mean $\hat{N}$	Resulting density estimate (raccoons/km^2)	±SE
1	0.030	0.74	23	31	3.72
2	1.280	3.08	41	13	2.40
3	0.016	0.69	16	23	0.97
Mean	0.45	1.26	27	22	1.13

for the possibility of temporary emigration from the trapping area and are viewed as an open capture survey. Thus, the primary sampling periods where gains (birth and immigration) and losses (deaths and emigration) are examined using a closed population model Cormack-Jolly-Seber (CJS) and the secondary sampling periods, or the day-to-day trapping sessions, are effectively closed to gains and losses (Kendall and Nichols 1995; Kendall et al. 1997). This model is powerful in that the probability that an animal is captured at least once in a trapping session can be estimated from just the data collected during the session using models developed for closed populations, and abundance and recruitment are less biased by heterogeneity in capture probability (Otis et al. 1978). The robust design model assumes: (1) the population is closed across all sampling days (i.e. within a field season); (2) samples are instantaneous and releases are made immediately after the sample; (3) animals do not lose marks between the sampling periods, (4) movement in and out of the area is random; (5) every marked animal in the population has an equal chance of survival and/or capture and (6) marking individuals does not affect their catchability. Assumptions 1–3 were met in our sampling design and marking procedures. Assumptions 4–6 were examined using the chi-square tests in CAPTURE as described above.

Average population size ($\hat{N}$) for each study site was calculated as the mean of the estimated abundance calculated by program CAPTURE over the 3 year study period. We estimated the total population size ± standard error of pygmy raccoons at our three main sites by combining multi-year averages from each site. Crude density estimates for each site were based on the average estimated population size over the entire study divided by the effective trapping area for each site (A).

Results

Variation in capture probability

We trapped at 19 sites for 6,652 trap nights, but captured pygmy raccoons or dwarf coatis at only three sites (Sites 1–3, 2,784 trap nights) in the northwestern tip of the island (Fig. 1). Throughout the study there were 175 captures of 96 (47 males and 49 females) pygmy raccoons (Appendix 1). Only a single, adult male dwarf coati was trapped in 2002 at Site 1. Therefore, the focus of our results refers to population estimates of the pygmy raccoon.

After our initial trapping program at Sites 1 and 3, traps were placed further inland (approximately 0.7 and 0. 3 km, respectively) from the edge of the coast, and near tropical forest mixed mangrove areas. However, at each of these sites, we did not capture any pygmy raccoons, despite regularly capturing individuals in close proximity to the coast.

At Sites 1 and 2 the distribution of the frequency of captures of males and females did not vary over the three-year study ($X^2 = 0.364$, $df = 1$, $P = 0.546$; Appendix 1). We captured more females at Site 1 in both 2002 and 2003 ($X^2 = 45.27$, $df = 1$, $P = 0.001$; $X^2 = 17.05$, $df = 1$, $P = 0.002$, respectively). The ratio of males and females in each of the three age classes (juvenile, subadult, adult) also significantly varied ($X^2 = 8.12$, $df = 2$, $P = 0.017$). Site 3 had nearly twice as many adult males captured as females ($X^2 = 3.84$, $df = 1$, $P = 0.050$), whereas all other age classes tended to have more females than males. The overall ratio of females to males for all individuals at all sites was equal (1:1). Adults made up the majority of captures (59%), followed by juveniles (22%), and subadults (19%). Although the proportion of each age class captured varied between sites ($X^2 = 12.60$, $df = 4$, $P = 0.013$), adults were the largest proportion of captures at all sites.

Program CAPTURE indicated the Zippin estimator (model M_b) as the most appropriate approach for population estimations for Site 1. Significant variation in capture probabilities may be due to behavioral responses to capture (trap happiness or shyness: $X^2 = 6.54$, $df = 1$, $P = 0.01$). The goodness of fit test for model M_b showed a reasonable fit ($X^2 = 45.53$, $df = 30$, $P = 0.03$). For Site 2, program CAPTURE recommended the jackknife estimator (model M_h). The goodness of fit test for model M_h showed a reasonable fit ($X^2 = 18.72$, $df = 1$, $P = 0.04$). No behavioral heterogeneity to capture was seen in Site 3 and we therefore used the null estimator (model M_o).

Average capture probability ($\hat{p}$) for the three pygmy raccoon populations (Sites 1–3) was 0.151. Probability of capture ($\hat{p} = 0.210$) was highest at Site 1 during the dry season of 2003, but this site also had the lowest probability of capture during the late dry and early wet season of 2002. There was no significant variation in capture or recapture probabilities among years or sites ($t = 1.28$, $P = 0.157$). The probability of recapture ($\hat{c}$) was highest at Site 3 (0.082) and lowest at Site 2 (0.004) (Table 2).

Population analysis

Average population estimates (±SE) for the three pygmy raccoon populations was 27.8 ± 5.5 individuals. Abundance estimates ranged from a low of 14 individuals at Site 1 (95% CI = 10–19) to a high of 48 individuals at Site 2 (95% CI = 39–79; Table 2). Site 2 was the largest trapping site ($A = 3.08$ km^2), resulting in the lowest density ($d = 13$). At Site 1, 34 individuals were captured in 2002; this was also one of our most comprehensive trapping efforts where after 1 month of intensive trapping only previously marked individuals were captured thereafter (3 year mean = 23). For Site 1 in 2002, population size was approximately 34 individuals (95% CI = 19–45), whereas in 2003, the estimate was 48 individuals (95% CI = 39–79; mean of 2002–2003 = 41). Closed capture models estimated the number of individuals at Site 3 to be 16 (range 13–20). Using the mean estimated population size calculated for each site over the 3 years (averaging abundance for site 1 for the 3 years, etc.), total population size at the three sites was 80 ± 26.1 individuals (95% CI = 72–103). Based on the average age structure (59% of all captures were adults), this translates to a population of approximately 47

Table 2 Pygmy raccoon population estimates based on Pollock's robust method of marking (*M*) and recapture (*C*)

Year	Site	Trapping occasions	*M*	*C*	$\hat{p}$	$\hat{c}$	$\hat{N}$	±SE	95% CI
2001	1	27	11	12	0.1042	0.0054	36	12.263	22–73
2002	1	31	19	34	0.1003	0.0312	19	0.763	19
	2	58	22	39	0.1451	0.0121	34	5.248	19–45
2003	1	16	8	13	0.2104	0.0208	14	4.677	10–19
	2	22	31	50	0.1566	0.0038	48	9.587	39–79
	3[a]	16	13	25	0.1919	0.0820	16	3.856	13–20

The number of nights, or trapping occasions, the probability of being captured ($\hat{p}$), probability of recapture ($\hat{c}$), estimated number of individuals ($\hat{N} \pm$ SE), and the range of population estimates based on 95% confidence intervals are noted for each year and site

[a] This site was only sampled for a single sampling session and therefore population estimates were based on a Schnabel closed population model

mature individuals. Average density for the three sites is 22 ± 5.1 raccoons/km^2 (range 13–31; Table 1).

Discussion

In assessing the abundance of pygmy raccoons on Cozumel Island, the most striking finding was that we did not capture any animals except in the most northwestern quadrant of the island. Although both species have been occasionally sighted in the northeastern, southern and central parts of Cozumel (Cuarón et al. 2004; personal observations), we only trapped a single coati throughout our three year study, and only captured pygmy raccoons at 3 of 19 sites. Mainland coatis (*N. narica*) and raccoons (*P. lotor*), and pygmy raccoons are fairly easy to capture when present (e.g., Gompper 1997; Valenzuela 1998; Carrillo et al. 2001), as is verified by our success at Sites 1–3. We tried numerous baits and worked in varying habitats, at many different times of the year. We trapped in areas where residents, or we, had previously seen these species on Cozumel (Cuarón et al. 2004), but if additional populations existed, they were too small to be detected through trapping. Thus, we feel confident our results reflect the true rarity of these insular species.

While variation in capture probabilities occurred between field seasons, even when we greatly increased trapping efforts capture probabilities slightly decreased. Small population estimates are therefore not likely an artifact of our trapping effort at these sites. Pygmy raccoons were caught primarily in mangrove habitats, or in semi-evergreen tropical forest adjacent to mangroves. Dietary studies have shown the reliance of pygmy raccoons on food items (i.e., crab) found in this type of habitat (McFadden et al. 2006). However, pygmy raccoons were not captured in mangrove areas on either the southern portion of the island (Fig. 1; Sites 18 and 19), where we spent considerable trapping effort (over 1,000 trap nights), or on the unpopulated northeastern side of the island (Site 6), although sightings of tracks and individuals have occasionally been made in both areas (Cuarón et al. 2004; personal observations). Even in the northwestern area of the island, where the most significant populations of pygmy raccoon exist, the range of the pygmy raccoon habitats was extremely patchy. For example, individuals trapped in mangrove habitat along the mangrove-bordered 2 km road leading to Isla de la Pasión, were only in the last 200 m of this habitat, closest to the ocean. Moreover, at Site 3, traps placed further inland by 300 m did not successfully trap individuals outside of the original trap area, even though this trap site had a high capture rate. It therefore appears that pygmy raccoon populations are locally clustered in areas closest to the shore. This may correspond to the availability of a preferred prey item, crabs (McFadden et al. 2006), and may also be associated with the distribution and abundance of potential predators such as *B. constrictor* and dogs, as the latter are less abundant in mangrove areas (Bautista 2006).

The number of pygmy raccoons captured at the three sites did not vary significantly over the course of our study. Some behavioral heterogeneity to capture occurred, which may in part reflect harassment by dogs, which was especially prominent at Site 3. Free-ranging dogs are abundant and widespread in Cozumel (Bautista 2006), and whereas we detected both pet and free-ranging dogs at all sites, Site 3 was the only locale that a raccoon died due to a dog attack (McFadden 2004; García-Vasco 2005). Also, dwarf coati remains have been identified in dog scats (Bautista 2006) indicating the threat this species plays to both carnivores.

Only a single dwarf coati was captured in over 6,600 trap nights, an alarming indication of the rarity of this species. Although we did observe dwarf coatis on several different occasions (once at Site 1, twice at Site 6, twice at Site 19, and several times at Sites 7–15, near San Gervasio, and between Xlapac and Punta Molas; Fig. 1), this species is clearly very rare on the island. These areas included semi-evergreen tropical forests, mangroves, and other coastal vegetation. Never were more than two individuals sighted together, providing further evidence of their rarity, as coatis generally travel in bands, and these tend to be quite notable when present. Although it is possible that small remnant populations exist elsewhere in the island, we conclude that the Cozumel coati is now extremely rare and in immediate danger of extinction.

Based in part on unpublished data from our population assessment, the IUCN recently changed the listing of the pygmy raccoon from Endangered to Critically Endangered (CR2a(i)b) (Cuarón et al. 2008). Although some propose that the dwarf coati be designated as a unique species (Glatston 1994; McFadden 2004; McFadden et al. 2008; Merriam 1901; Thomas 1901), the most recent Global Mammal Assessment (GMA) treated this taxon as a subspecies following Decker (1991). Because subspecies were not reassessed in the GMA, the IUCN status of the dwarf coati remains unresolved, as the taxon is currently listed as a subspecies of *N. narica*, which in turn is listed as a species of least concern (Samudio et al. 2008). However, based on our population estimates, as well as justification put forth by Cuarón et al. (2004), we propose that their designation be immediately changed to Critically Endangered (CR A2c + C2a(i)b). The requirements of the CR designation that both the pygmy raccoon and dwarf coati fulfill are (1) that population size is <250 mature individuals, and (2) no subpopulation estimated to contain >50 mature individuals (IUCN 2008). After having intensively trapped on Cozumel for over three field seasons, we feel the major populations have clearly been identified, and these populations are undoubtedly small, isolated, and patchily distributed. Both species also show extreme fluctuations in the number of mature individuals due to the effect of periodic major hurricanes. In addition to these criteria, there is evidence that the dwarf coati population has experienced significant reductions in their population size over the last 10 years. This species is extremely rare on Cozumel, while historically they were found throughout the interior of Cozumel (Glatston 1994). Therefore, both species clearly qualify as Critically Endangered and we consider their risk of immediate extinction to be extremely high.

Our work suggests that the habitat requirements of the Cozumel dwarf carnivores remain poorly understood. Population sizes are very small and heterogeneously distributed in mangrove habitat in the case of the pygmy raccoons. A conservation recommendation arising from this study is to minimize future reductions to the extent of mangrove habitat where the largest populations of pygmy raccoons persist in northern Cozumel and from semi-evergreen tropical forests (central Cozumel) and coastal areas in the northeast and eastern side of Cozumel where dwarf coatis have been recorded. Habitat protection measures, including the establishment of protected areas comprising key habitat areas for these Cozumel endemic species, are urgently needed. The spirit of the new (2008) Ecological Ordinances (*Programa de Ordenamiento Ecológico Local*) for Cozumel is that a substantial area of these habitats should be set for protection and conservation (Secretaria de Desarrollo Urbano y Medio Ambiente 2008), so these protected areas should be decreed immediately.

Acknowledgments This research was approved by Columbia University's Institutional Animal Care and Use Committee (IACUC), and supported by funds from the Explorer's Club, Sigma Xi, Sigma Delta Epsilon Women in Science, the American Museum of Natural History, Columbia University, the University of Missouri, the National Science Foundation (#0118867), the Consejo Nacional de Ciencia y Tecnología grant (#33635-V), and the Fondo Sectorial de Investigación Ambiental (SEMARNAT-2002-C01-0571). Field support and local assistance was provided by J. Benavides, R. Chacón, I. Fortes, C. González-Baca,

H. González, A. Laborde, M.A. Martínez-Morales, C. Naredo, G. Navarro, PROFEPA, I. Romero-Nájera, R. Sheese, and E. Vivas (CAPA).

Appendix 1

See Table 3.

Table 3 Trapping effort and capture success for each sex (M, Males, F, Females) and age class of pygmy raccoon at three main study sites on Cozumel Island, Mexico, from 2001 to 2003

Year	Site	Trap nights	Juveniles		Subadults		Adults		Total captures	Capture per effort
			F	*M*	*F*	*M*	*F*	*M*		
2001	1	354	0	0	2	1	1	8	12	0.034
2002	1	428	4	4	13	1	6	6	34	0.079
	2	1,246	1	0	3	9	17	10	40	0.032
2003	1	80	7	1	1	0	4	3	15	0.187
	2	555	10	4	0	4	13	17	48	0.086
	3[a]	121	4	3	0	0	4	15	26	0.216
Total	1	862	11	5	16	2	11	17	61	0.071
	2	1,801	11	4	3	13	30	27	88	0.049
	3[a]	121	4	3	0	0	4	15	26	0.216
All sites and years		2,784	26	12	19	15	45	59	175	0.063

[a] Site 3 was only trapped at during one field season and therefore the number of trap nights and individuals captured are the total for this site

References

Alcover JA, McMinn M (1994) Predators of vertebrates on islands. Bioscience 44:12–18

Bautista SM (2006) Distribución, abundancia y dieta de perros y gatos ferales en la Isla Cozumel. M.Sc thesis, Instituto de Ecología, A.C

Carrillo E, Wong G, Rodriguez MA (2001) Feeding habits of the raccoon (*Procyon lotor*) (Carnivora: Procyonidae) in a coastal, tropical wet forest of Costa Rica. Rev Biol Trop 49:1193–1197

Copa-Álvaro ME (2007) Efectos de los huracanes Emily y Wilma en los mamíferos medianos de Cozumel. M.Sc thesis, Universidad Nacional Autónoma de México

Cuarón AD (2009) Cozumel. In: Gillespie R, Clague DA (eds) Encyclopedia of Islands. University of California Press, Berkeley, pp 203–206

Cuarón AD, Martinez-Morales MA, McFadden KW, Valenzuela D, Gompper ME (2004) The status of dwarf carnivores on Cozumel Island, Mexico. Biodivers Conserv 13:317–331

Cuarón AD, de Grammont PC, Vázquez-Domínguez E, Valenzuela-Galván D, García-Vasco D, Reid F, Helgen K (2008) *Procyon pygmaeus*. In: IUCN 2008 IUCN red list of threatened species. www.iucnredlist.org. Downloaded on 19 November 2008

Decker DM (1991) Systematics of the coatis, genus *Nasua* (Mammalia: *Procyonidae*). Proc Biol Soc Wash 104:370–386

García-Vasco D (2005) Distribución, abundancia y aspectos poblacionales del mapache enano (*Procyon pygmaeus*), un carnívoro insular endémico. Licenciatura (Advanced BSc) Thesis, Universidad Veracruzana

Gittleman JL, Funk S, MacDonald D, Wayne R (2001) Why 'carnivore conservation'? In: Gittleman JL et al (eds) Carnivore conservation. Cambridge University Press, Cambridge, pp 1–7

Glatston AR (1994) Threats to procyonids and ailurids. In: Glatston AR (ed) The red panda, olingos, coatis, and raccoons and their relatives. IUCN, Gland, pp 1–42

Goldman EA (1950) Raccoons of North and Middle America. North Am Fauna 60:1–153
Gompper ME (1997) Population ecology of the white-nosed coati (*Nasua narica*) on Barro Colorado Island, Panama. J Zool 24:441–455
Grau GA, Sanderson GC, Rogers JP (1970) Age determination of raccoons. J Wild Manag 34:364–372
Hamblin NL (1984) Animal use by the Cozumel Maya. The University of Arizona Press, Tuscon
IUCN (World Conservation Union) (2008) IUCN red list categories and criteria: version 3.1. Species survival commission. IUCN, Gland
Jones JK, Lawlor TE (1965) Mammals from Isla Cozumel, Mexico, with description of a new species of harvest mouse. Univ Kansas Publ Mus Nat Hist 16:409–419
Karanth KU, Nichols JD (1998) Estimation of tiger densities in India using photographic captures and recaptures. Ecology 79:2852–2862
Kendall WL, Nichols JD (1995) On the use of secondary capture-recapture samples to estimate temporary emigration and breeding proportions. J Appl Stat 22:751–762
Kendall WL, Nichols JD, Hines JE (1997) Estimating temporary emigration using capture-recapture data with Pollock's robust design. Ecology 78:563–578
Kenward RE (1985) Ranging behaviour and population dynamics in grey squirrels. In: Sibly RM, Smith RH (eds) Behavioural ecology—ecological consequences of adaptive behaviour. Blackwell, Oxford, pp 319–330
Martínez-Morales MA (1996) The Cozumel curassow: abundance, habitat preference and conservation. MPhil thesis, University of Cambridge, UK
Martínez-Morales MA, Cuarón AD (1999) *Boa constrictor*, an introduced predator threatening the endemic fauna on Cozumel Island, Mexico. Biodivers Conserv 8:957–963
McFadden KW (2004) The ecology, evolution, and natural history of the endangered carnivores of Cozumel Island, Mexico. Dissertation, Columbia University
McFadden KW, Wade SW, Dubovi EJ, Gompper ME (2005) A serological and fecal parasitologic survey of the critically endangered pygmy raccoon (*Procyon pygmaeus*). J Wildl Dis 41:615–617
McFadden KW, Sambrotto RN, Medellín RA, Gompper ME (2006) Feeding habits of endangered pygmy raccoons (*Procyon pygmaeus*) based on stable isotope and fecal analyses. J Mammal 87:501–509
McFadden KW, Gompper ME, Valenzuela DG, Morales JC (2008) Evolutionary history of the critically endangered Cozumel dwarf carnivores inferred from mitochondrial DNA analyses. J Zool 276:176–186
Mena H (2007) Presencia de *Leptospira* spp. y moquillo canino en poblaciones de perros y carnívoros silvestres en la Isla Cozumel. M.Sc thesis, Universidad Nacional Autónoma de México
Merriam CH (1901) Six new mammals from Cozumel Island, Yucatan. Proc Biol Soc Wash 14:99–104
Navarro D, Suárez M (1989) A survey of the pygmy raccoon (*Procyon pygmaeus*) of Cozumel, Mexico. Mammalia 53:458–461
Otis DL, Burnham KP, White GC, Anderson DR (1978) Statistical inference for capture data on closed animal populations. Wildl Monogr 62:1–24
Perdomo H (2006) Caminos como barreras para el movimiento de aves y efectos del huracán Wilma, en una comunidad de aves de sotobosque de la selva mediana subcaducifolia de la Isla Cozumel. M.Sc thesis, Universidad Nacional Autónoma de México
Pollock K, Nichols JD, Brownie C (1990) Statistical inference for capture-recapture experiments. Wildl Monogr 107:1–97
Roemer GW, Gompper ME, Van Valkenburgh B (2009) The ecological role of the mammalian mesocarnivore. Bioscience 59:165–173
Romero-Nájera I, Cuarón AD, González-Baca C (2007) Distribution, abundance, and habitat use of introduced *Boa constrictor* threatening the native biota of Cozumel Island, Mexico. Biodivers Conserv 16:1183–1195
Samudio R, Kays R, Cuarón AD, Pino JL, Helgen K (2008) *Nasua narica*. In: IUCN, 2008 IUCN red list of threatened species. www.iucnredlist.org. Downloaded on 19 November 2008
Sanderson GC, Nalbandov AV (1973) The reproductive cycle of the raccoon in Illinois. Illinois Nat Hist Surv Bull 31:29–85
Secretaria de Desarrollo Urbano y Medio Ambiente (2008) Programa de Ordenamiento Ecológico Local del Municipio de Cozumel. http://seduma.qroo.gob.mx/bitacora/index.php?option=com_docman&task=catview&gid=108&Itemid=33. Downloaded on 19 November 2008
Téllez VO, Cabrera EF (1987) Listados floristicos de Mexico-VI. Instituto de Biología, Universidad Nacional Autonóma de México, México City
Thomas O (1901) New insular forms of *Nasua* and *Dasyprocta*. Ann Mag Nat Hist 7:271–273

Valenzuela D (1998) Natural history of the white-nosed coati, *Nasua narica*, in a tropical dry forest of western Mexico. Rev Mex Mastozoo 3:26–44
White GC, Burnham KP (1999) Program MARK: survival estimation from populations of marked animals. Bird Study 46:120–138
Williamson M (1981) Island populations. Oxford University Press, Oxford

Biodivers Conserv (2010) 19:503–518
DOI 10.1007/s10531-009-9742-z

ORIGINAL PAPER

Distribution of endemic and threatened herpetofauna in Mt. Malindang, Mindanao, Philippines

Olga M. Nuñeza · Fritzie B. Ates · Apolinario A. Alicante

Received: 5 April 2008 / Accepted: 27 October 2009 / Published online: 12 November 2009

Abstract Mt. Malindang is one of the upland ranges where biodiversity has been severely threatened due to forest loss. Fieldwork was conducted from October 2003 to December 2004 in 14 sampling sites from an elevation of 120 to over 1,700 m above sea level to assess the distribution of endemic and threatened herpetofaunal species. Twenty-six species of amphibians and 33 species of reptiles were observed for all sampling sites. The level of endemism for amphibians was 42% where 7 of the 11 recorded species are found only in Mindanao. Nine species were in the threatened category, 8 vulnerable and 1 endangered. For the reptiles, 48% endemicity was observed. No threatened species was found. Field observations show that the major threat to the herpetofauna is habitat destruction, particularly the conversion of the forest to agricultural farms by the local people. It was also observed that endemic and threatened species were distributed in high elevation sites (submontane, dipterocarp, almaciga, and montane forests). Despite habitat loss in Mt. Malindang, 18% of the recorded herpetofaunal species recorded in the Philippines were found in Mt. Malindang, indicating the conservation importance of this mountain range.

Keywords Endemic · Herpetofauna · Mindanao · Mt. Malindang · Species · Threatened

O. M. Nuñeza (✉)
Department of Biological Sciences, College of Science and Mathematics, MSU-Iligan Institute of Technology, A. Bonifacio Drive, Iligan City, Philippines
e-mail: olgamnuneza@yahoo.com

F. B. Ates
Biology Department, University of the Philippines—Mindanao, Davao City, Philippines

A. A. Alicante
Biology Department, College of Natural Sciences and Mathematics, Mindanao State University, Marawi City, Philippines

Introduction

The herpetofauna are very good indicators of ecosystem health. A decrease or the absence of frogs, especially forest dwellers and endemic species would indicate a disturbance in their habitat. Range size, habitat loss, and more extreme seasonality in precipitation contributed to decline risk in the 2,454 amphibian species that declined between 1980 and 2004, compared to species that were stable ($n = 1{,}545$) or had increased ($n = 28$). These empirical results show that amphibian species with restricted ranges should be urgently targeted for conservation (Sodhi et al. 2008), and one of the Southeast Asian countries that needs immediate attention is the Philippines. The rapid destruction of tropical rainforest in Southeast Asia in the twentieth century drastically altered the environmental conditions where many forms of vertebrate life had been adapted (Sodhi et al. 2004). The Philippines is regarded by nearly every major international conservation organization as one of the top global priorities for conservation action (Heaney and Oliver 1997). An account of the country's herpetofaunal biodiversity lists down 101 amphibians with 79 endemic species and 258 reptiles with 170 endemics (Diesmos et al. 2002).

Mt. Malindang is one of the upland ranges in the Philippines where biodiversity has been severely threatened due to forest loss. It is the highest mountain (2,175 masl) in the Zamboanga peninsula, a western part of Mindanao. The Malindang range is part of the national park, the boundaries of which define the Important Bird Areas (IBA). While several bird species with a restricted-range and threatened status have been reported from Mt. Malindang, limited information has been reported for herpetofauna. A survey conducted by Tabaranza et al. (2001) in Mt. Malindang reported 26 herpetofaunal species belonging to nine families. This study presents additional information on the species richness as well as the distribution of the endemic and threatened species of herpetofauna in Mt. Malindang.

Methodology

Fourteen sampling sites in Mt. Malindang (Fig. 1) were established for herpetofaunal survey.

Sampling site 1. This site was a mossy forest (1,700 masl) with a plant community of *Ascarina philippinensis* and *Xanthomyrtus diplycosifolium.* The canopy layer was mainly composed of *Viburnum* sp., *Lithocarpus* sp., *Caldeluvia* sp., *Pometia* sp., *Macaranga deptirocarpifolia*, *Polyosma philippinensis*, *Clethra lancifolia*, *Podocarpus neriifolius*, *Syzygium*, and *Polyosma philippinensis*. Other vegetations were: canopy epiphytes {wild orchids, bird's nest fern (*Asplenium* sp.) and mosses}; the canopy vines {*Plagiogyria* sp., *Pteris tripertita*, *Smilax* sp. and climbing pandan (*Freycinetia* sp.)} and the understory plants {*Clethra lancifolia*, tree ferns (*Cyathea*), ferns and the ground cover plants, mosses, and lichens}. Fallen trees were moderately to commonly abundant. The distance to the nearest creek was 50–300 m for subsite 1; more than 300 m for subsite 2. Both subsites were at a distance of 500–2,000 m from the nearest anthropogenic clearing.

Sampling site 2. This montane forest (1,400–1,700 masl) had a plant community of *Clethra lancifolia* and *Impatiens montalbanica. Lithocarpus philippinensis, Lithocarpus mindanaensis, Polyosma philippinensis, Gordonia luzonica,* and *Clethra lancifolia* made up the canopy layer. Epiphytic ferns, wild orchids, lichen, and mosses were observed. Canopy vines included the *Smilax* sp., *Piper* sp., and *Freycinetia* sp. The forest understory was made up of saplings and shrubs. Bryophytes and ferns covered the ground. *Pandanus*

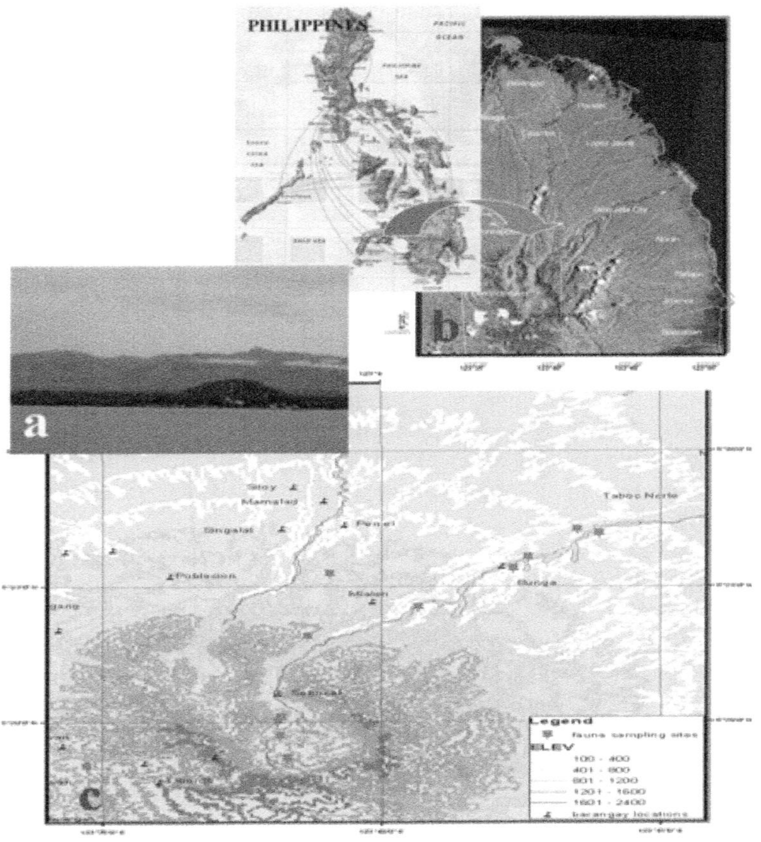

Fig. 1 **a** Panoramic view and **b** location map of Mt. Malindang showing **c** the 14 sampling sites

sp. was moderately abundant. The moss density was moderately high. Fallen trees and exposed rocks were common 50–500 m away from the creek to the site. The site was disturbed by timber extraction.

Sampling site 3. This almaciga forest (1,200–1,400 masl) was dominated by a community of *Agathis philippinensis* which ranged in height from 15 to 40 m, together with *Cinnamomum mercadoi.* Epiphytic plants such as wild orchids, bird's nest fern, and vines were observed. Few pitcher plants, *Musa*, and some fruit-bearing plants were recorded. Tree ferns, saplings, seedlings of almaciga trees, and ferns comprised the understory plants. Fallen trees were observed moderately frequently. The site was 100 m away from a river. Timber extraction was the main disturbance in the area.

Sampling site 4. This site was located at the mid elevation and characterized as the submontane dipterocarp forest (900–1,000 masl). It had a dominant plant community of *Ficus minahassae* and *Bischofia javanica. Shorea negrensis* was the emergent tree at 15–30 m in height. The canopy layer was at 10–15 m in height. Epiphytic ferns, wild orchids, mosses, pandan, and bamboo were found. Tree ferns, saplings, shrubs, and palm trees were some of the understory plants recorded. *Ficus* and pandan were moderately frequent. Slightly moist loam soil was covered with 1–2 cm humus and a moderate dense layer of leaf litter. The gathering of rattan was observed. The river was a distance of less than 1,000 m.

Sampling site 5. This site was the lowland dipterocarp forest (220–500 masl) just 5 m away from a creek in Barangay Mialen, with a plant community of *Ficus variegata* and *Selaginella moellendorfi.* Coconut trees dominated the lower elevation (subsite 1) with some fruit-bearing trees such as mango and guava. *Musa* was moderately common. The upper elevation (subsite 2) had *Lithocarpus* sp. with a height ranging from 10 to 15 m and DBH of 1 m; and *Dendrocnide* sp. with 6–8 m height. The ground was covered with sedges. Mosses, lichens, bird's nest fern were epiphytic to the canopy. Small-scale farming, clearings, and trails were the disturbance to the site. The upper elevation was less than 1.5 km away from the Layawan River while the lower elevation was about 200–500 m away.

Sampling site 6. This site was a mixed dipterocarp forest (450–900 masl). It had a plant community of *Syzgium* and *Sticherus laevigata.* The height of emergent tree ranged from 10 to 15 m while the canopy trees ranged from 3 to 5 m in height. Canopy epiphytes and only one species of canopy vine were found; understory plants were rarely observed. Fruit-bearing plants were moderately abundant while fallen trees were rarely observed. The site was 100–200 m away from Langaran River. Anthropogenic disturbances observed were the clearings, farms, and trails.

Sampling site 7. This site was a mixed lowland dipterocarp forest (220–450 masl). The dominant vegetation was composed of Diplodiscus *paniculatus* and *Lithocarpus* sp. Subsite 1 was 200 m away from Layawan River. The area was planted with coconuts and fruit trees. Cogon (*Imperata cylindrica*) and other grasses covered the ground. Subsite 2 was a hilly/mountainous area. The ground was covered with grasses and ferns. Several fruit-bearing trees were abundant. Exposed rocks were moderately abundant. Subsite 2 was about 100 m away from Layawan River.

Sampling site 8. The plantation forest site 1 (120–900 masl) was located in Barangay Peniel about 400 m away from a creek. This site had a plant community of *Acacia mangium, Cocos nucifera*, and other fruit trees.

Sampling site 9. The plantation forest site 2, in Barangay Bunga, had the same vegetation as plantation forest site 1. Subsite 1 was 1.5 km away from a water course while subsite 2 was 75–100 m away from the Layawan River.

Sampling site 10. This was Agroecosystem site 1 (1,400–1,550 masl) at Barangay Lake Duminagat. This site was located adjacent the montane forest. Fruiting plants like wild raspberries, *Medinilla*, and *Senna septemtrionalis* were commonly observed. Bare rocks were moderately abundant. Small-scale vegetable farms were observed throughout the cultivated area; cogon grass (*Imperata cylindrica*) made up most of the principal ground cover of grassland. The nearest creek was at 50–100 m.

Sampling site 11. Agroecosystem site 2 in Barangay Sebucal (1,235–1,550 masl) was adjacent to the almaciga forest in the same barangay. Ferns, *Imperata cylindrica* and grasses covered the ground. Together with mixed vegetables, corn and sugarcane, fruit-bearing plants like *Homolanthus, Tremma*, and *Everittia* were found. Exposed rocks were moderately common. The site was 50 m away from the Layawan River.

Sampling site 12. Agroecosystem site 3, Barangay Mansawan. This site was adjacent to a forest at a lowland at elevation of 600–985 masl.

Sampling site 13, Agroecosystem site 4 (Barangay Peniel, 600–610 masl) and Sampling site 14, Agroecosystem site 5 (Barangay Mamalad, 165–465 masl) were habitats in the lowland with high degree of disturbance due to farming activities.

Sampling of amphibians and reptiles was done using the cruising method for a total of 203 field days. Aquatic, subterranean, surface, and arboreal strata were searched for amphibians by day (0600 hours–0800 hours) and at night (1800 hours–2200 hours). Sampling for reptiles was done late in the morning to the early afternoon (1000 hours–1400 hours) and in the evening (1800 hours–2200 hours) to capture nocturnal reptiles. Reptiles were searched by digging and trenching using bolos and sticks, climbing arboreal strata of the forest, checking isolated pools, seepage areas, tree holes, burrows, rotten logs, vines, ant mounds, leaf litter, tree foliage, and other microhabitats.

Two to three voucher specimens of each species captured (the permit limit), particularly species not identified in the field, were prepared following standardized preservation techniques (Heyer et al. 1994). These specimens were deposited at the Natural Science Museum of the Mindanao State University-Iligan Institute of Technology, Iligan City. Reference works such as Inger (1954), Brown and Alcala (1978, 1980), Alcala (1986), and Alcala and Brown (1998) were used for species identification. Nomenclature followed Alcala and Brown (1998) for amphibians and Alcala (1986) for reptiles. A list of endemic and threatened species was generated based on published data and IUCN criteria.

Seriation using the Markov Monte Carlo method was used to determine assemblage of herpetofaunal species by placing the species into sequence so that the sampling sites with the most similar proportions of certain taxa are together. This was done using PAST software developed by Hammer et al. (2001).

Results

We recorded a total of 59 herpetofaunal species from the 14 sites explored within the Mount Malindang range (Tables 1, 2). These included 26 amphibians and 33 reptiles. Of these species, 27 are endemic to the Philippines including seven species, all amphibians, that have a range that is restricted to Mindanao (Table 3). The 59 species recorded in this study form 18% of the total number recorded for the Philippines as a whole ($n = 324$) (Table 4).

Two amphibian species *Limnonectes magnus* and *Rana grandocula* were found to be the most widely distributed being found in 11 of the 14 sampling sites. The Mindanao endemic and vulnerable *Philautus acutirostris* was also found to be widely distributed (nine sampling sites) from the higher to lower elevation and in almost all habitat types. The Mindanao endemic *Ansonia mcgregori* was found in eight sites while *Philautus surrufus*, and *Philautus worcesteri* were distributed in seven sampling sites. Interestingly, *Ichthyopis mindanaoensis*, a Mindanao endemic caecilian was found only in two sites. The Philippine endemic *Philautus surdus. was* noticeably found in the four forest sampling sites at higher elevation and in the lowland dipterocarp forest. Other endemic species were found only in one or two sampling sites mostly in the forest sampling sites at higher elevation.

Five of the six endemic species in the mossy forest are Mindanao endemic. These were *Ansonia mcgregori, Philautus acutirostris, Philautus poecilus,* and *Philautus worcesteri* which are of vulnerable status and the endangered *Philautus surrufus*. The Mindanao endemic species found in the mossy forest were also present in the montane and almaciga forests.

Among the reptiles, the Philippine endemic *Sphenomorphus fasciatus* was the most widely distributed. Another endemic species, *Rhabdophis auriculata* was found in five sampling sites while *Naja samarensis* was found in four sampling sites. Other endemic species had limited distribution in one to three sampling sites.

Among the sampling sites surveyed, the submontane dipterocarp forest had the highest species richness for amphibians and reptiles followed by the lowland dipterocarp forest.

Table 1 Species list of amphibians (nomenclature follows Alcala and Brown 1998)

Species	Sampling sites													
	A	B	C	D	E	F	G	H	I	J	K	L	M	N
*Ansonia mcgregori*** Taylor	4	23	5	7	0	0	0	0	2	2	5	0	0	1
Bufo marinus Linnaeus	0	0	0	0	2	0	1	0	2	1	2	1	1	0
Chaperina fusca Mocquard	0	0	2	3	0	6	1	0	0	0	0	3	0	0
*Ichthyophis mindanaoensis*** Taylor	0	0	0	4	1	0	0	0	0	0	0	0	0	0
Leptobrachium hasselti Muller	0	0	0	0	0	0	0	0	0	0	1	0	0	0
*Limnonectes leytensis** Boettger	0	0	0	0	1	3	0	0	0	0	0	0	0	0
*Limnonectes magnus** Stejneger	0	0	3	2	1	6	2	1	1	6	4	8	0	2
*Limnonectes parvus*** Taylor	0	0	0	1	0	0	0	0	0	0	0	0	0	0
Megophrys stejnegeri Kuhl and Van Hasselt	0	6	0	1	1	0	1	2	1	0	0	0	0	0
*Nyctixalus spinosus** Taylor	0	0	0	2	0	0	0	0	0	0	0	0	0	0
*Occidozyga laevis*Gunther	0	0	0	0	0	0	0	1	0	0	0	1	1	0
*Philautus acutirostris*** Peters	36	18	27	50	1	0	1	3	0	0	2	1	0	0
*Philautus poecilus*** Brown and Alcala	11	4	0	0	0	0	0	0	0	0	0	0	0	0
Philautus sp. A	22	0	0	0	1	0	0	0	0	0	0	0	0	0
Philautus sp. B	7	0	0	1	0	0	0	0	0	0	0	0	0	0
Philautus sp. C	0	0	0	0	2	0	0	0	0	0	0	0	0	0
*Philautus surdus** Peters	23	8	1	0	1	0	0	0	0	0	0	0	0	0
*Philautus surrufus*** Brown and Alcala	70	14	3	8	6	0	0	0	0	1	2	0	0	0
*Philautus worcesteri*** Stejneger	22	7	1	3	0	0	0	0	0	1	1	1	0	0
Platymantis dorsalis Dumeril	0	1	0	1	0	0	1	0	0	0	0	0	0	0
Platymantis rabori Brown	0	0	0	1	0	0	0	0	0	0	0	0	0	0
Polypedates leucomystax Gravenhorst	0	0	0	2	2	2	1	0	0	0	0	15	1	1
Rana grandocula	0	1	3	8	13	5	0	1	0	4	1	11	1	8
Rhacophorus bimaculatus Peters	0	0	6	10	1	0	0	0	0	0	0	0	0	0
Rhacophorus pardalis Gunther	0	0	0	1	0	0	0	0	0	0	2	1	0	0
Staurois natator Gunther	0	0	0	0	9	2	0	1	1	0	0	0	0	1

Legend: * Philippine endemic ** Mindanao endemic

Nos. indicate no. of individuals

A mossy forest, *B* montane forest, *C* almaciga forest, *D* submontane dipterocarp forest, *E* lowland dipterocarp forest, *F* mixed dipterocarp forest, *G* mixed lowland dipterocarp forest, *H* plantation (peniel 12), *I* plantation (Bunga), *J* agro ecosystem (Barangay lake), *K* agro ecosystem sebucal, *L* agro ecosystem mansawan, *M* agro ecosystem peniel, *N* agro ecosystem mamalad

High species richness for amphibians was also documented in the forest sampling sites at high elevation and in two agroecosystem areas. Reptiles were noticeably absent in the mossy and montane forest sites as well as in one plantation site and one agroecosystem site. Low species richness was noted in the mixed dipterocarp forests and in three agroecosystem sites.

High amphibian endemism was recorded in sampling sites at high elevation like the mossy forest, montane forest, almaciga forest, and agroecosystem sites 1 and 2. Reptilian

Table 2 Species list of reptiles (nomenclature follows Alcala 1986)

Species	Sampling sites													
	A	B	C	D	E	F	G	H	I	J	K	L	M	N
Ahaetulla prasina Reinwardt	0	0	1	2	3	0	0	2	1	0	0	0	0	0
Aplopeltura boa Boei	0	0	0	1	0	0	0	2	0	0	0	0	0	0
Boiga dendrophila Boei	0	0	0	0	1	0	0	0	0	0	0	0	0	0
Brachymeles boulengeri	0	0	0	0	0	0	0	0	0	0	2	0	0	0
*Calamaria gervaisi** Dumeril and Bibron	0	0	0	0	0	0	0	0	0	0	15	0	0	0
Calamaria lumbricoidea	0	0	0	0	0	0	0	0	0	0	0	1	0	0
Calotes sp.	0	0	0	0	0	1	0	0	0	0	0	0	0	0
*Cyclocorus lineatus** Reinwardt	0	0	0	0	1	0	0	1	0	0	0	0	0	0
*Cyrtodactylus annulatus** Taylor	0	0	0	0	2	0	1	0	0	0	0	1	0	0
*Dasia grisea**	0	0	0	0	0	0	0	2	0	0	1	0	0	0
Dendrelaphis caudolineatus Peters	0	0	0	0	0	0	0	0	0	0	0	0	0	2
Draco sp.	0	0	0	0	1	1	0	0	1	0	0	0	0	0
Elaphe erythrura Dumeril and Bibron	0	0	0	0	0	0	1	0	0	0	0	0	0	1
Gekko gecko Linnaeus	0	0	0	0	0	0	0	0	1	0	0	0	0	0
Gonocephalus sp.	0	0	0	1	0	0	0	0	0	0	0	0	0	0
Hemidactylus frenatus Dumeril and Bibron	0	0	0	0	0	0	0	0	0	0	0	0	0	2
Mabuya multicarinata Gray	0	0	0	0	1	0	0	1	0	0	0	3	0	0
Mabuya multifasciata Gray	0	0	0	0	1	0	2	0	0	0	5	8	0	1
*Naja samarensis** Linnaeus	0	0	1	1	1	0	1	0	0	0	0	0	0	0
Opisthotrophis alcalai	0	0	0	1	0	0	0	0	0	0	0	0	0	0
*Oxyrhabdium modestum** Gunther	0	0	0	0	0	0	0	0	0	0	0	2	0	0
Psammodynastes pulverulentus Boei	0	0	7	12	0	0	0	1	0	0	0	1	0	0
Pseudorabdion sp.	0	0	0	0	0	0	0	0	0	0	3	0	0	0
Python reticulatus	0	0	1	1	1	0	0	0	0	0	0	0	0	0
*Rhabdophis auriculata** Gunther	0	0	0	1	3	0	1	0	0	1	0	1	0	0
*Sphenomorphus abdictus**	0	0	0	0	0	0	1	0	0	0	0	0	0	0
*Sphenomorphus cumingii**	0	0	0	0	0	0	0	1	0	0	0	0	0	0
*Sphenomorphus fasciatus** Gray	0	0	1	3	1	0	1	0	0	0	4	7	0	0
*Sphenomorphus steerei** Stejneger	0	0	3	1	0	0	0	0	0	0	1	0	0	0
*Sphenomorphus variegates*** Peters	0	0	0	1	0	0	0	0	0	0	1	1	0	0
*Trimeresurus flavomaculatus** Gray	0	0	0	0	0	0	0	1	1	0	0	0	0	0
*Tropidonophis dendrophiops**	0	0	0	1	0	0	1	2	0	0	0	0	0	0
*Tropidophorus misaminus** Stejneger	0	0	0	1	2	0	0	0	0	0	0	0	0	1

Legend: * Philippine endemic ** Mindanao endemic

Nos. indicate no. of individuals

A mossy forest, *B* montane forest, *C* almaciga forest, *D* submontane dipterocarp forest, *E* lowland dipterocarp forest, *F* mixed dipterocarp forest, *G* mixed lowland dipterocarp forest, *H* plantation (peniel 12), *I* plantation (Bunga), *J* agro ecosystem (Barangay lake), *K* agro ecosystem sebucal, *L* agro ecosystem mansawan, *M* agro ecosystem peniel, *N* agro ecosystem mamalad

endemism was observed to be high in the submontane dipterocarp forest, lowland dipterocarp forest and in two agroecosystem sites at high elevation which are adjacent to a forest.

Table 3 Species richness and endemism of the herpetofauna in Mt. Malindang

Sampling sites	A	B	C	D	E	F	G	H	I
(1) Mossy forest (over 1,700 masl)	8	6 (5)	75	0	0	0	8	6	75
(2) Montane forest (1,400–700 masl)	9	6(5)	67	0	0	0	9	6	67
(3) Almaciga forest (1,200–1,400 masl)	10	6(4)	60	6	3(0)	50	16	9	56
(4) Submontane dipterocarp forest (900–1,100 masl)	17	8(6)	47	13	9(0)	69	30	17	57
(5) Lowland dipterocarp forest (220–500 masl)	14	6(3)	43	12	8(0)	67	26	14	54
(6) Mixed dipterocarp forest (450–900 masl)	7	2(0)	28	2	0	0	9	2	22
(7) Mixed lowland dipterocarp forest (220–450 masl)	7	2(1)	28	8	5(0)	62	15	7	47
(8) Plantation forest1 (120–900 masl)	8	3	38	8	4	50	16	7	44
(9) Plantation forest 2 (120–900 masl)	6	2	33	3	0	0	9	2	22
(10) Agroecosystem 1 (1,400–1,550 masl)	7	5	71	1	1	100	8	6	75
(11) Agroecosystem 2 (1,235–1,550 masl)	10	6	60	8	5	63	18	11	61
(12) Agroecosystem 3 (600–985 masl)	10	4	40	11	6	55	21	10	48
(13) Agroecosystem 4 (600–610 masl)	4	0	0	1	0	0	5	0	0
(14) Agroecosystem 5 (165–465 masl)	5	2	40	6	1	17	11	3	27
Total # of species	26			33			59		
Endemic species	11			16			27		
Mindanao island endemic	7			0			7		
% Endemism	42			48			46		
% of Mindanao endemics (total species)	27			0			12		
% of Mindanao endemics(phil. endemics)	64			0			26		

Legend: *A* no. of species amphibians, *B* phil. endemic (Mindanao endemic), *C* % endemic amphibians, *D* no. of species reptiles, *E* phil. endemic (Mindanao endemic), *F* % endemic reptiles, *G* total no. of species, *H* phil. endemic (Mindanao endemic), *I* % endemic herpetofauna

Table 4 Herpetofaunal species richness in Mt. Malindang

	Mt. Malindang this study	Mt. Malindang (Don Victoriano) (Tabaranza et al. 2001)	Philippines (Alcala et al. 2006)
Amphibians	26 (11)	16	110 (79)
Reptiles	33 (16)	10	197 land, ca 17 marine (170)
Total number of species	59 (27)	26	324 (249)

Legend: () Philippine endemic

More threatened species were found in the sampling sites at high elevation (Table 5) compared to the low elevation sampling sites. No threatened species was recorded for reptiles.

A high amphibian diversity was recorded in the lowland dipterocarp forest as well as in the agroecosystem site in Sebucal. Other sampling sites exhibited moderate diversity. Over-all amphibian diversity was high. A more or less even distribution was noted for amphibians except in the submontane forest which recorded low evenness. Pooled data,

Table 5 Distribution of threatened species of herpetofauna in Mt. Malindang (conservation status follows IUCN 2007)

Sampling sites	Amphibians	Reptiles
(1) Mossy forest	5 (1Endangered (E) 4Vulnerable (V)	0
(2) Montane forest	6 (1E,5 V)	0
(3) Almaciga forest	5 (1E,4 V)	0
(4) Submontane dipterocarp forest	8 (1E,7 V)	0
(5) Lowland dipterocarp forest	4 (1E,3 V)	0
(6) Mixed dipterocarp forest	0	0
(7) Mixed lowland dipterocarp forest	2 V	0
(8) Plantation forest 1 (120–900 masl)	2 V	0
(9) Plantation forest 2 (120–900 masl)	2 V	0
(10) Agroecosystem 1 (1,400–1,550 masl)	3 (2 V, 1E)	0
(11) Agroecosystem 2 (1,235–1,550 masl)	5 (4 V, 1E)	0
(12) Agroecosystem 3 (600–985 masl)	3 V	0
(13) Agroecosystem 4 (600–610 masl)	0	0
(14) Agroecosystem 5 (165–465 masl)	1 V	0
Threatened species	9	0
Threatened phil. endemic (Mindanao endemic)	7(6)	0
% Threatened species	35	0

Legend: *E* endangered, *V* vulnerable

however, showed high diversity and a more or less even distribution. A much higher diversity was recorded for reptiles but the distribution was less even (Table 6).

A graphical structure of the communities of herpetofauna in the fourteen sampling sites was done using seriation, a method typically applied to an association between taxa and sampling stations (Brower and Kile 1988) (Figs. 2, 3). The seriation reorganized the data matrix such that the presence of the taxa is concentrated along the diagonal. The results showed that the structures of the communities are closely associated with the elevation pattern of the sampling sites.

Rarefaction analysis was done on the abundance data using the algorithm from Krebs (1989) to estimate the number of species if the sample sizes for all sites were equal. The paleontological statistics software developed by Hammer et al. (2001) was used to compute for the estimates. Results showed that the estimated species richness matched with that of the observed (Table 7).

Discussion

Species distribution showed that only the endemic *Ansonia mcgregori* and the *Philautus* group were found in the mossy forest. Changes in the abundance of certain types of breeding sites with altitude may affect amphibians (Inger and Stuebing 1989). In the mossy forest of Mt. Malindang, the terrain was so steep that most breeding habitats formed by rainfall were shallow and swift, and diminished quickly in volume because of rapid runoff. Inger and Stuebing (1989) explained that the lack of various types of breeding microhabitats, specifically lentic microhabitats, at higher elevations might have favored the

Table 6 Diversity indices of the 14 sampling sites

Sampling sites	Amphibians			Reptiles			Herps		
	Species	Shannon	Evenness	Species	Shannon	Evenness	Species	Shannon	Evenness
Mossy forest	8	1.785	0.745	0	0	0	8	1.785	0.745
Montane forest	9	1.874	0.724	0	0	0	9	1.874	0.724
Almaciga forest	9	1.597	0.549	6	1.431	0.697	15	2.082	0.535
Submontane dipterocarp forest	17	1.97	0.422	13	2.018	0.579	30	2.486	0.401
Lowland dipterocarp forest	14	2.118	0.594	12	2.37	0.892	26	2.804	0.635
Mixed dipterocarp forest	6	1.694	0.907	2	0.693	1	8	1.888	0.826
Mixed lowland dipterocarp forest	7	1.906	0.961	8	2.043	0.964	15	2.67	0.963
Plantation 1 (peniel)	7	1.82	0.8818	9	2.138	0.943	16	2.678	0.9097
Plantation 2 (Bunga)	6	1.733	0.943	4	1.386	1	10	2.254	0.952
Agro ecosystem 1 (Barangay lake)	6	1.529	0.7691	1	0	1	7	1.667	0.757
Agro ecosystem 2 (sebucal)	9	2.039	0.8535	8	1.625	0.635	17	2.451	0.6821
Agro ecosystem 3 (mansawan)	9	1.668	0.589	9	1.821	0.687	18	2.386	0.6037
Agro ecosystem 4 (peniel)	4	1.386	1	1	0	1	4	1.386	1
Agro ecosystem 5 (mamalad)	5	1.179	70.65	5	1.55	0.9421	10	1.956	0.7071
For all sites (pooled data)	26	2.59	0.5127	33	3.037	0.6315	59	3.174	0.405

existence of species that do not require them. Considering the scarcity of such microhabitats at high elevations, evolving a reproductive mode that would not require them, like developing directly into froglets and skipping the aquatic larval stage, would prove to be advantageous to a species if it is to successfully colonize the high altitude areas. This explains the dominance in the mossy and montane forests of the genus *Philautus*, a group of amphibians in which the eggs develop directly into froglets (Alcala and Brown 1998). *Philautus acutirostris* which topped the list of abundant species was encountered in almost

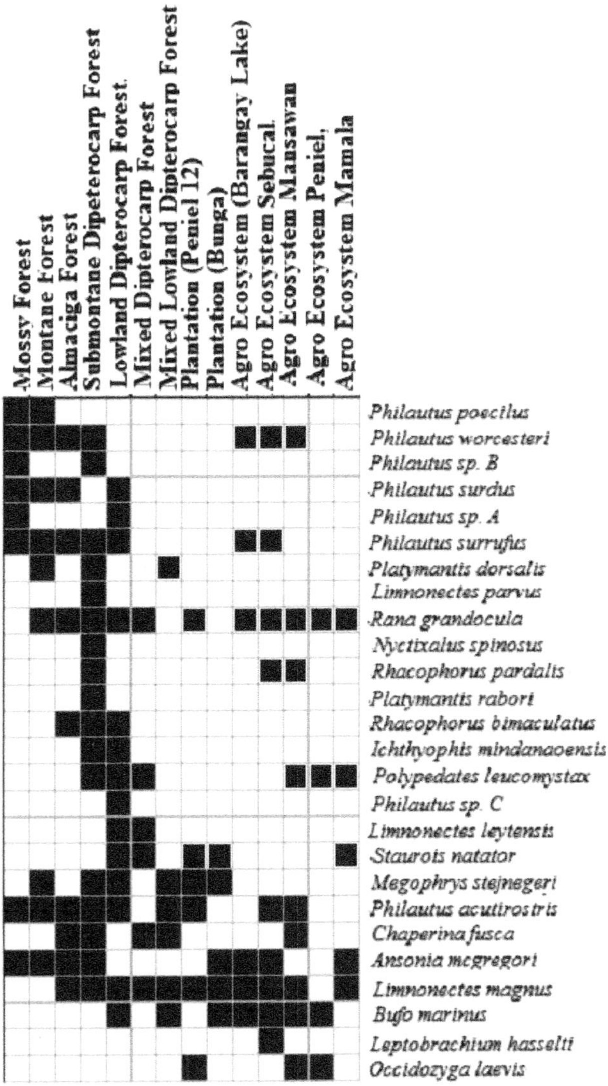

Fig. 2 Assemblage of amphibian species placed into sequence so that the sampling sites with the most similar proportions of certain taxa are together

all sampling sites in Mt. Malindang. This species with conservation status listed as vulnerable (IUCN 2007) was collected either from the ground or on the leaves of low-lying vegetation. *Philautus surrufus*, the second most abundant species takes the common name Malindang Tree Frog (Brown and Alcala 1994). This species was encountered in several vegetation types but was recorded to occur highest in the mossy forest. This species has been classified as endangered (IUCN 2007). *A. mcgregori* is a vulnerable species captured in various vegetation types including the highly disturbed agro ecosystems, but was recorded highest in abundance at the montane forest. This species was mostly collected

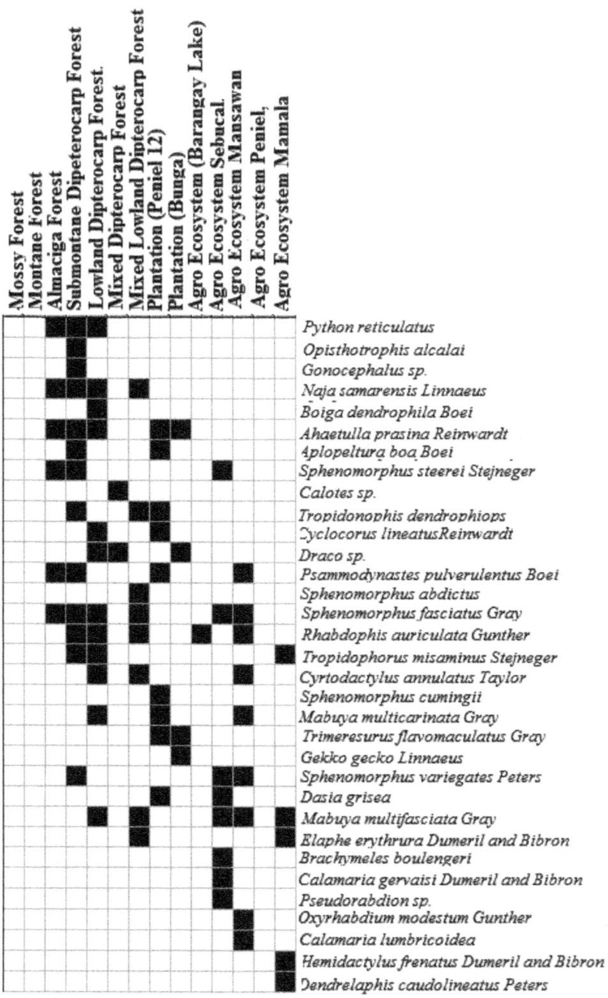

Fig. 3 Assemblage of reptilian species placed into sequence so that the sampling sites with the most similar proportions of certain taxa are together

from the forest floor. *Philautus worcesteri* is a Mindanao endemic that has also been classified as vulnerable by the IUCN (2007). Although it was also encountered in various vegetation types, the highest abundance recorded for this species was in the mossy forest.

In this study, three unidentified species of *Philautus* which are probably new species were found distributed in one to two sampling sites. Bossuyt and Dubois (2001) in their review of the specific taxonomy of the frog genus *Philautus* Gistel, 1848 found 143 types to be extant. They mentioned that from 1822 to 1999, 177 nominal species were either described as members of this genus, or of other genera but subsequently referred to this genus. Numerous species of this group remain to be described, particularly in Sri Lanka. Brown and Alcala (1994) reported that data on zonation and microhabitats in the mountains of the larger islands are available for eight Philippine species of *Philautus*. Most have

Table 7 Results of the rarefaction analysis on the abundance data

Site	Amphibians		Reptiles		Pooled	
	Observed	Expected (195)	Observed	Expected (32)	Observed	Expected (195)
A	8	8	0	0	8	8
B	9	9	0	0	9	9
C	9	9	6	6	15	15
D	17	17	13	13	30	30
E	14	14	12	12	26	26
F	6	6	2	2	8	8
G	7	7	8	8	15	15
H	6	6	9	9	15	15
I	5	5	4	4	9	9
J	6	6	1	1	7	7
K	9	9	8	8	17	17
L	9	9	9	9	18	18
M	4	4	0	0	4	4
N	5	5	5	5	10	10

Expected number of species was calculated against the largest sample size in parenthesis

been recorded from the mid-dipterocarp (300–500 m) through the montane forest that starts between about 800 and 2,000 m depending on the height of the mountain surveyed. For Philippine species of *Philautus*, density, based on sample sizes, is higher in the submontane and montane forest zones. This observation is in agreement with the reported zonal distribution of *Philautus* in Bornean forests, eight of 11 species of *Philautus* from Mt. Kinabalu, Mt. Mulu, Mt. Lumuko, and the Crocker range in Northwest Borneo being thus far recorded only from the montane zone, above the dipterocarp zone. The results of distribution of *Philautus* in this present study agree well with the above observations.

The endemic *Ichthyophis mindanaoensis* was found only in the submontane dipterocarp and lowland dipterocarp forests. This caecilian species is a lowland primary forest dweller and occurs at 600–850 masl (Alcala and Brown 1998; IUCN 2007). It is apparently common, but is very locally and patchily distributed. This species inhabits the rotting logs and decaying plant litter of the forest floor where it was captured during sampling. The adults and larvae were found in the mud in recently cleared agricultural fields at forest edge. IUCN has listed this species to be data deficient because there is still much to be learned about the status of this primitive amphibian.

Our data showed that except for the absence of reptiles in the montane and mossy forests, the majority of the amphibians and reptiles were found in more forested sites and in agroecosystem sites adjacent to forests. Herpetofauna flourished better in older forest than in young forest (Mitchell et al. 1997). The mossy, montane, almaciga, and submontane forests in this study were found to be old-growth forests. Likewise, the results analysed by seriation showed that the distribution of the herpetofauna seems to be correlated with elevation (Figs. 2, 3).

The present study agrees with the findings of Diesmos et al. (2003) who found that majority of the amphibians and reptiles in Balbalasang-Balbalan National Park (BBNP), Philippines are forest-restricted. They further reported that the distribution pattern of the herpetofauna of BBNP was positively related to elevation and the coinciding gradients of temperature and humidity. At lower elevations, forests are warmer and drier and support higher levels of reptile diversity and abundance in most species. At higher elevations, reptiles are scarcer and, when encountered, species are less abundant with fewer individuals encountered. This concurs with the absence of reptiles noted for the mossy and montane forests in the present study. Rahbek (1995) similarly reported that the trend of decreasing species richness with increasing elevation had been widely demonstrated for many taxa.

Since amphibians are moisture-sensitive (Heatwole 1982) and reptiles are moisture-selective organisms (Alcala 1976), an alteration of their suitable microclimate would have a negative impact on their populations. Ernst et al. (2006) found that South American anurans exhibited higher species richness in primary sites but in West Africa, anuran species richness did not differ between primary and exploited forest sites. In Mt. Malindang, except for two agroecosystem sites, amphibian species richness was high in primary and exploited forest sites but endemic amphibians especially the Mindanao endemics were mostly found in relatively intact forests.

Other sampling sites in this study have low to moderate numbers. For the reptiles, the more open canopy facilitates their basking behavior (Zug 1993). This was observed in this study where several individuals of *Mabuya multifasciata,* a non-endemic species that inhabits primary and secondary forests and can be found close to villages, were captured in agroecosystem sites while basking. Alcala et al. (2004) elucidated the effects of fragmentation and degradation of tropical rainforest on Negros Island (Philippines) on tropical herpetofauna. They found that the factors responsible for the loss of species appeared to be the removal of large trees where the decreased canopy cover caused physical changes, and forest fragmentation increased the edge areas.

It is interesting to note that the plantation forest site 1 had a herpetofaunal species richness of 16 while plantation forest site 2 had 9. This indicates that rehabilitating degraded areas through planting of trees could encourage wildlife in the area.

Results showed that the endemic herpetofauna in Mt. Malindang has to date been thriving relatively well. The relatively high level of endemism observed suggests that Mt. Malindang is favorable for amphibian speciation. The greater occurrence of endemics in the more highly forested sites indicates that forests indeed are important habitats for the herpetofauna, supporting several rare and restricted-range species.

In their study of the fauna of Mt. Malindang, Tabaranza et al. (2001) found fewer species with only 16 amphibians and 10 reptiles. They did not encounter the Malindang tree frog, *Philautus surrufus*. In this study, 17 species of amphibians and 31 reptiles found are additional records. Seven species of amphibians and eight reptiles earlier reported were not documented in this study. Considering the earlier record and the result of this study, Mt. Malindang has now an inventory of 33 species of amphibians and 41 reptiles. Since the earlier survey was done on the other side of the mountain range, it is possible that species not encountered in this study are not common in the present sites sampled but could be common elsewhere, or the present sampling might have simply missed them. Results of this study showed that Mt. Malindang has a higher total number of species compared with other mountains in Mindanao like Mt. Kitanglad (which still has patches of primary forest at 700 masl). A total of only 14 herpetofaunal species was recorded in Mt. Kitanglad (Heaney and Peterson 1992). The many sampling sites and representative elevations sampled may be contributing factors accounting for the higher number of species documented in this study. The sites with the highest number of threatened species were the forests at high elevation with varied layers of vegetation. It was also observed during the sampling that these were the sites which are less disturbed by humans due to their inaccessibility. Sexton et al. (1964) noted that the arrangement of vegetation in space and particularly its effect on the micro-environment is important in determining the local distribution of species. The structural complexity of forests provides the suitable micro-environmental conditions for the species. Alcala (1976) noted that the land vertebrates of the Philippines are generally distributed in areas covered by natural vegetation, especially forests. Amphibians due to their dependence on cutaneous respiration and sensitivity to dehydration require the general availability of continuous moisture, such as presented by tropical rainforests where the relative humidity is always at or near saturation. Zug (1993) reported that reptiles, even though not directly dependent on moisture, are affected by humidity, as shown by their frequent choice for moist or humid microhabitats in the tropical rain forests. Moisture conditions also affect the growth and differentiation of reptilian eggs and embryos. Forest sites with tall trees and high canopy cover assure the availability of environmental conditions required by a wide variety of land vertebrates.

Sites of community importance have higher habitat heterogeneity with wider surface areas and include higher number of species (Canova and Marchesi 2007). Our data demonstrate the importance of lush forests for the endemic and threatened fauna. The high diversity recorded for the herpetofauna is very encouraging considering that many areas in Mt. Malindang have already been converted for agricultural purposes. More extensive sampling in future studies could yield additional species.

Conclusion and recommendation

Mt. Malindang is an area of high herpetofaunal diversity. Given the high number of endemic and threatened species, Mt. Malindang is also a key area for conservation. The sampling sites at high elevation (submontane dipterocarp forest, almaciga forest, montane forest, mossy forest and two agroecosystem sites adjacent to forests) which showed high endemism, should be listed as priority areas for conservation. However, the fact that endemic and some threatened species are also found in the lowland forests (which are mostly threatened by habitat loss) shows the need to develop a conservation/management plan for the whole landscape.

Acknowledgment We thank three anonymous reviewers and the guest editor for helpful comments on our paper. Funding for this project was provided by the Netherlands Ministry for Development Cooperation (DGIS) through the Southeast Asian Regional Center for Graduate Study and Research in Agriculture (SEARCA).

References

Alcala AC (1976) Philippine land vertebrates. New Day Publishers, Quezon City, 176 pp

Alcala A (1986) Guide to Philippine flora and fauna. Amphibians and reptiles. Natural Resources Management Center, Ministry of Natural Resources and University of the Philippines, Philippines, 195 pp

Alcala AC, Brown WC (1998) Philippine amphibians. An illustrated field guide. Bookmark, Inc, Makati City, Philippines, 116 pp

Alcala E, Alcala A, Dolino C (2004) Amphibians and reptiles in tropical rainforest fragments on Negros Island, the Philippines. Environ Conserv 31(3):254–261

Alcala AC, Alcala EL, Buot LE, Diesmos A, Dolar M, Fernando ES, Gonzales J, Tabaranza B (2006) Philippine biodiversity: ecological roles uses and conservation status. Plenary paper presented during the 28th National Academy of Science and Technology annual scientific meeting, NAST, Manila, 12–13 July 2006

Bossuyt F, Dubois A (2001) A review of the frog genus *Philautus* Gistel, 1848 (Amphibia, Anura, Ranidae, Rhacophorinae). Zeylanica 6(1):1–112

Brower JC, Kile KM (1988) Seriation of an original data matrix as applied to palaeoecology. Lethaia 21:79–93

Brown WC, Alcala AC (1978) Philippine lizards of the family Gekkonidae. Silliman University. Nat Sci Monogr Ser 1, Dumaguete City, Philippines, 131 pp

Brown WC, Alcala AC (1980) Philippine lizards of the family Scincidae. Silliman University. Nat Sci Monogr Ser 1, Dumaguete City, Philippines, 263 pp

Brown WC, Alcala AC (1994) Philippine frogs of the family Rhacophoridae. Proc Calif Acad Sci 48(10):85–220

Canova L, Marchesi M (2007) Amphibian and reptile communities in eleven sites of community importance (SCI): relations between SCI area, heterogeneity and richness. Acta Herpetol 2(2):87–96

Diesmos A, Brown R, Alcala A, Sison R, Afuang L, Gee G (2002) Amphibians and reptiles. Philippine biodiversity conservation priorities. A second iteration of the national biodiversity strategy and action plan. DENR-PAWB, Quezon City, Philippines, p 26

Diesmos A, Brown R, Gee G (2003) Preliminary report on the amphibians and reptiles of Balbalasang-Balbalan National Park, Luzon Island, Philippines. Sylvatrop. Tech J Philipp Ecosyst Nat Resour 13(1 and 2):63–80

Ernst R, Linsenmair K, Rodel M (2006) Diversity erosion beyond the species level: dramatic loss of functional diversity after selective logging in two tropical amphibian communities. Biol Conserv 133:143–155

Hammer Ø, Harper DAT, Ryan PD (2001) PAST: paleontological statistics software package for education and data analysis. Palaeontol Electron 4(1):9 pp. http://palaeo-electronica.org/2001_1/past/issue1_01.htm

Heaney LR, Oliver W (1997) Biodiversity and conservation in the Philippines: an introduction to a global priority. Philippine Red Data Book, Bookmark, Manila

Heaney LR, Peterson AT (1992) Inventory of the vertebrates of Mt. Kitanglad Nature Park. Final report for 1992. Field Museum of Natural History, USA

Heatwole H (1982) A review of structuring in herpetofaunal assemblages. In: Scott NJ Jr (ed) Herpetofaunal communities: a symposium of the Society for the study of Amphibians and Reptiles and the Herpetologists' League, August 1997. Wildlife Research Report 13. U.S. Department of the Interior, Fish and Wildlife Service, Washington, DC, USA, pp 1–9

Heyer WR, Donnelly RW, McDiarmid RW, Hayek L, Foster MS (1994) Measuring and monitoring biological diversity: standard methods for amphibians. Smithsonian Institution Press, Washington, DC, 364 pp

Inger RF (1954) Systematics and zoogeography of Philippine amphibia. Fieldiana: zoology, vol 33, No. 4. Chicago Natural History Museum, Chicago, pp 185–531

Inger RF, Stuebing RB (1989) Frogs of Sabah. Sabah Parks. Publication No. 10. Kota Kinabalu, 120 pp

IUCN Red List of Threatened Species (2007) http://www.iucnredlist.org. Cited 15 Dec 2007

Krebs CJ (1989) Ecological methodology. Harper and Row, New York

Mitchell JC, Rinehart SC, Pagels JF, Buhlman KA, Pague CA (1997) Factors influencing amphibian and small mammal assemblages in central Appalachian forests. For Ecol Manag 96:65–76

Rahbek C (1995) The elevational gradient of species richness: a uniform pattern? Ecography 18:200–205

Sexton OJ, Heatwole H, Knight D (1964) Correlation of microdistribution of some Panamian reptiles and amphibians with structural organization of the habitat. Carrib J Sci 4:261–295

Sodhi NS, Koh LP, Brook BW, Ng PKL (2004) Southeast Asian biodiversity: an impending disaster. Trends Ecol Evol 19(12):654–660

Sodhi N, Bickford D, Diesmos A, Lee T, Koh L, Brook B, Sekercioglu C, Bradshaw C (2008) Measuring the meltdown: drivers of global amphibian extinction and decline. PloS ONE 3(2):e1636. www.plosone.org

Tabaranza BR, Tabaranza ACE, Bagaloyos AP, Dimapilis A (2001) Flora and fauna of Mt. Malindang Natural Park, Don Victoriano and Lopez Jaena, Misamis Occidental (final report). Haribon Foundation for the Conservation of Natural Resources, Inc., Diliman Quezon City, pp 34–36

Zug RG (1993) Herpetology: an Introductory biology of amphibians and reptiles. Academic Press, Inc, California, USA, 527 pp

Biodivers Conserv (2010) 19:519–529
DOI 10.1007/s10531-009-9596-4

ORIGINAL PAPER

Fuelling the biodiversity crisis: species loss of ground-dwelling forest ants in oil palm plantations in Sabah, Malaysia (Borneo)

Carsten A. Brühl · Thomas Eltz

Received: 26 March 2008 / Accepted: 5 February 2009 / Published online: 20 February 2009

Abstract Oil palm plantations today cover large areas of former tropical lowland rain forest in Southeast Asia and are rapidly expanding on the island of Borneo. Study of the community of ground-dwelling ants in different plantations in Sabah, Malaysia, over 2 years using tuna baiting, revealed that the oil palm plantation ground ant community was severely reduced in species richness in comparison to the forest interior, regardless of age, undergrowth cover, or proximity to neighbouring forest. The results indicate that oil palm plantation habitats, now covering more than 15% of Sabah's land area, can sustain only about 5% of the ground-dwelling ant species of the forest interior. Nine of the 23 ant species baited in the plantations were never recorded inside forest. All numerically dominant ants were non-forest species. The most common species was *Anoplolepis gracilipes*, an invasive species present at 70% of all bait sites and known to cause 'ecological meltdowns' in other situations. The low frequency and species number of forest ground ants indicates that oil palm plantations act as effective dispersal barriers leading to community isolation in rain forest remnants. The replacement of natural forests with oil palm plantations poses a serious threat to the conservation of biodiversity on Borneo if similar results are confirmed in other taxa.

Keywords *Anoplolepis gracilipes* · Ants · Biodiversity · Dispersal barrier · Fragmentation · Isolation · Malaysia · Oil palm

Introduction

With the growing global demand for agricultural products landscapes throughout the tropics are increasingly dominated by agriculture (Tilman et al. 2001; Achard et al. 2002).

C. A. Brühl (✉)
Institute for Environmental Sciences, University Koblenz-Landau, Fortstr. 7, 76829 Landau, Germany
e-mail: Bruehl@uni-landau.de

T. Eltz
Department of Neurobiology, Sensory Ecology Group, University of Düsseldorf, Universitätsstr. 1, 40225 Düsseldorf, Germany
e-mail: Eltz@uni-duesseldorf.de

The increase in oil palm (*Elaeis guineensis*) cultivation is cited as a major driver of deforestation and biodiversity loss in tropical countries (Donald 2004; Koh and Wilcove 2007, 2008) and has also major impacts on global carbon balance (Danielsen et al. 2008). Oil palm plantations expanded almost exponentially in recent decades (Donald 2004), now covering over 13 million ha, primarily in Southeast Asia (Danielsen et al. 2008). Palm oil is now the world's most traded oilseed crop with uses from cooking oil to biofuel (Carter et al. 2007) and with over 37 million metric tons produced in 2005 (Turner et al. 2008). In 2006, 85% of the global palm oil crop was produced by Indonesia (43%) and Malaysia (42%) (FAOSTAT 2007), two countries whose combined annual forest loss is around 2 million ha (Danielsen et al. 2008). Malaysia has the highest levels of palm oil production per unit area but also the highest relative number of endangered species, and therefore impacts on biodiversity are expected. However, to date studies addressing this topic are few (Fitzherbert et al. 2008; Koh and Wilcove 2008; Turner et al. 2008).

In the Malaysian state of Sabah, oil palm plantations have been established on a large scale in the lowland areas that had previously been degraded by selective logging of variable intensity (McMorrow and Talip 2001). This valuable crop is currently grown on more than 1.2 million ha in Sabah equalling around 30% of the total of Malaysia's plantation area (MOPB 2007; FAOSTAT 2007) or ca. 15% of the total land area of Sabah. Oil palm plantations today stretch all along the coastal plains in Eastern Sabah and most of the lowland rain forest has been converted into an extensive monoculture of immense dimensions.

Before the establishment of an oil palm plantation, the forest must be completely cleared of vegetation. The wooden debris and trees easily catch fire, especially in dry El Niño years. The extensive fires in Borneo in 1998 were mostly associated with agricultural and land clearing operations (Chazdon 1998; Cao 2000; Dennis et al. 2005), in Sabah often related to the establishment of oil palm plantations. After clearing, the ground is terraced before planting the young oil palms. A humus and leaf litter layer is lacking in the newly established plantations because of erosion of the top soil associated with high surface water runoff, which reaches more than twice the amount measured on land under forest cover (MacKinnon et al. 1996). Lack of organic matter is the main factor determining the unfavourable change in soil properties in oil palm plantations, leading to a marked deficiency in most nutrients (Yeboua and Ballo 2000).

Leaf litter ants are considered useful indicators of ecosystem disturbance (Agosti et al. 2000; Underwood and Fisher 2006) and they have been found to show congruent patterns with other taxa in their responses to environmental change in different habitats (Lawton et al. 1998; Alonso 2000). The functional importance of ground-dwelling ants in the process of nutrient cycling due to their role in bioturbation (Gunadi and Verhoef 1993; Wilson 2000) and enhancement of microbial activity (Dauber and Wolters 2000) established their role as ecosystem engineers (Jones et al. 1994; Folgarait 1998). They are also recognised as main predators of other ground arthropods (Carroll and Janzen 1973; Hölldobler and Wilson 1990).

Selective logging in Sabah has resulted in a pronounced reduction in forest leaf litter ant diversity and density, and communities did not recover even 25 years after the logging impact (Brühl 2001). Ant diversity was also reduced in smaller primary forest fragments (Brühl et al. 2003) where the genetic diversity of forest litter ants was affected as well (Bickel et al. 2006), implying that ant populations in forest remnants were effectively isolated. Here, the ground ant community in surrounding oil palm plantations was studied to evaluate if this agricultural matrix is a suitable habitat for native forest ant species.

Methods

Study sites

Ground-dwelling ants were collected in four oil palm plantations in Eastern Sabah (Fig. 1): In the Mayvin oil palm estate (MOP) and in plantations near Tawai Hills Forest Reserve (THFR), Deramakot Forest Reserve (DFR) and Sepilok Forest Reserve (SFR). MOP is situated in the core of a very large plantation area (15 × 10 km) and was ~8–10 years old with palm trees reaching 8 m in height. Undergrowth consisted of various grasses and herbs (e.g. *Panicum* sp., *Mimosa pudica*). In all but the MOP location sampling transects extended from the forest edge into the oil palm plantation. The THFR plantation was 5–7 years old, palm trees were about 3–5 m in height and the undergrowth was sparse with bare ground mostly visible. A similar plantation type was present at the DFR site, bordered by heavily logged forest. The SFR transects ran from the edge of a primary forest remnant (42.9 km^2) into an old abandoned oil palm plantation (ca. 15–20 years old). The palm trees reached up to 10 m in height, and various herbs, bushes and saplings were present, forming a comparatively dense undergrowth. The soil in the oil palm plantations was mostly bare and no substantial leaf litter layer was present. For comparative purposes ants were also collected in the Danum Valley Conservation Area (DVCA), a contiguous primary rain forest of 438 km^2 ~80 km south of the plantation area (Fig. 1). The transect in DVCA was situated on the West Trail Transect ~1.5 km from the field center.

Sampling

Sampling transects were initially established in 1999 at two sites (SFR1, THFR1). Subsequently, in 2000, except for the single site established at MOP and the control site at DVCA, paired transects were established (SFR, THFR, DFR; see Table 1). Since the ant fauna in the plantations was found to be impoverished, some of the transects were operated at more widely spaced intervals (100 m) to detect potential turnover (see Table 1). The longest transects covered almost 1 km, which was the maximum length manageable for monitoring and sampling the bait sites.

All transects were sampled once in the morning from 8:00–10:00 hour, at least 2 days after rain in dry conditions in March and April. Ground-dwelling ants were attracted with baits consisting of a teaspoon of canned tuna in oil placed on tissue paper. Three tuna baits were placed at every bait site at a distance of 5, 10 and 15 m from the access trail to avoid possible disturbance by humans or machinery. Baits were left for 2 h to attract ants and were checked every 30 min. All ant species present at a bait site were collected using forceps and were pooled for a bait site sample. During the 2 h of active baiting the area was also searched for ants by hand collecting. During this process stones, twigs or palm leaves on ground level were turned and searched for ants. Ants were stored separately and later compared with the ants collected at baits.

The ants were sorted to genus level, then mounted following the convention of Bolton (1994) and finally sorted to morphospecies. Morphospecies classification was based on external morphological characters of the worker class visible under a dissecting microscope (Leica MZ8 16 × 14). Morphospecies labels were replaced by valid species names whenever possible. Voucher specimens have been deposited with the Entomology Section, Forest Research Centre, Sandakan, Sabah, the Department of Animal Ecology and Tropical Biology, University of Würzburg, Germany and are integrated in a web based ant identification key (www.antbase.net).

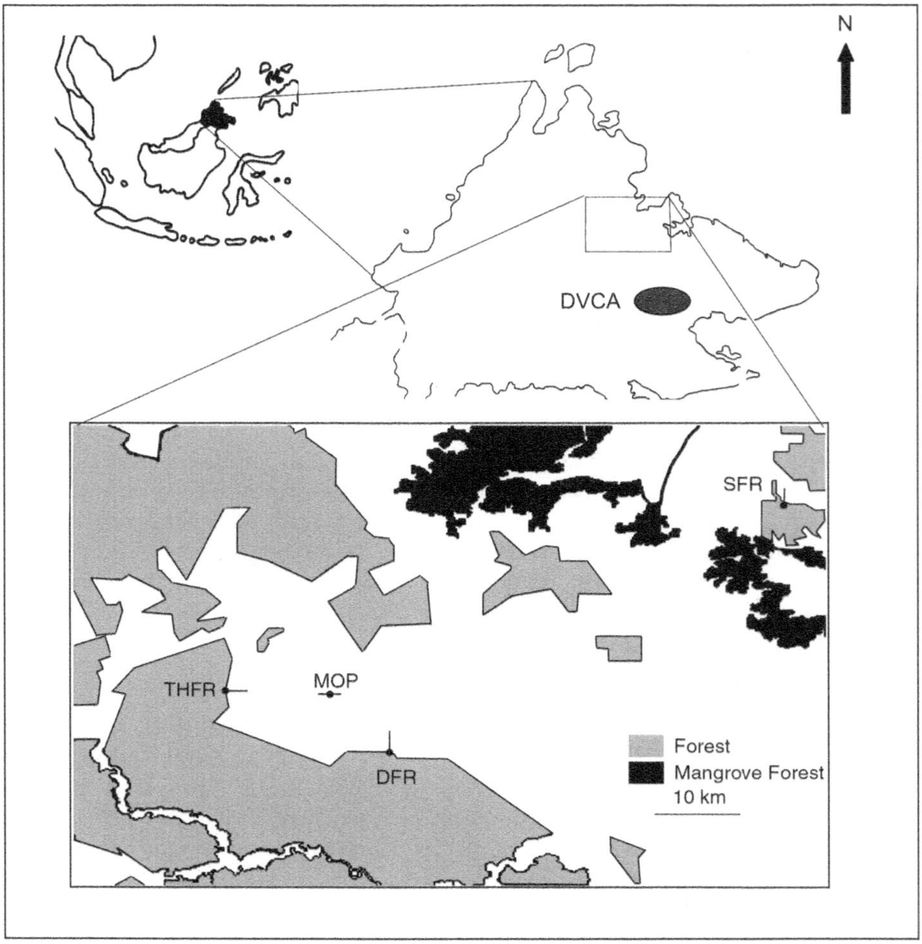

Fig. 1 Location of the four sampled oil palm plantations in East Sabah: plantation bordering the Tawai Hills Forest Reserve (*THFR*), the Deramakot Forest Reserve (*DFR*) and the Sepilok Forest Reserve (*SFR*). The Mayvin oil palm plantation (*MOP*) is situated in the center of a large plantation area. The primary forest site was located in the Danum Valley Conservation Area (*DVCA*)

Table 1 Transects used for tuna baiting of ground-dwelling ants in oil palm plantation (SFR, THFR, DFR, MOP) and primary forest (DVCA)

Transects	SFR1	SFR2	SFR3	THFR1	THFR2	THFR3	DFR1	DFR2	MOP1	DVCA
Sampling year	1999	2000	2000	1999	2000	2000	2000	2000	2000	1999
Number of bait sites	5	13	13	13	13	10	13	10	10	6
Distance between bait sites (m)	25	10	10	50	10	100	10	100	20	100
Transect length (m)	100	120	120	600	120	900	120	900	180	500

Statistical methods

Presence and absence of ant species at bait sites was analysed. Species accumulation curves for the different plantation transects were calculated using EstimateS 5.0 (Colwell 1997) randomising the data 100 times. Five species richness estimators were calculated for the total oil palm plantation ground ant community: ICE, Jackknife 1 and 2, Bootstrap, MMMean (Colwell and Coddington 1994; Colwell 1997). Sampling efficiency was obtained after dividing observed by estimated species number.

Results

In total 23 ant species in 14 genera were recorded along nine transects in oil palm plantations (Table 2). Most species belonged to the subfamily Myrmicinae (11 species) followed by the Dolichoderinae and Formicinae (five species each). Two Ponerinae species were also sampled. Hand collecting did not reveal any species not recorded at the tuna baits.

The randomised species accumulation curves of the oil palm plantation transects were much flatter than that of the primary forest transect, yielding lower species numbers (Fig. 2). The rate of increase in species with bait sites was low in all plantation transects. In the DFR1 transect only one species (*A. gracilipes*) was recorded at all 13 bait sites (Table 2). Observed species number in single transects was similar in the different plantation areas (Fig. 2), although longer transects (600–900 m, THFR1, THFR3, DFR2; Table 1) tended to yield more species.

When all samples were combined in order to estimate total projected ground ant species number in oil palm plantations the highest estimate was 48 (Jack2), the lowest 25 species (MMMean) with sampling efficiencies of 47.9 and 92.0%, respectively (Table 3).

Discussion

Loss of forest ants in oil palm plantations

The ground-dwelling ant communities of oil palm plantations were found to be highly impoverished with respect to forest taxa and, instead, were dominated by a small number of, partly invasive, non-forest species. Species richness and community composition did not vary much between the four plantations. This was unexpected since the SFR, THFR and DFR transects were directly connected to forest, in the case of SFR a primary forest fragment. Proximity to closed forest could theoretically promote an influx of forest ant species into the plantations (Laurance and Bierregaard 1996). Instead species numbers were similar to the MOP transect which was isolated in the core of a large oil palm plantation. Also, there seemed to be no effect of plantation age or undergrowth structure on ant species richness: The abandoned SFR plantation, with old palm trees, dense vegetation and no mechanical or chemical treatment, harboured as few ant species as the young plantations near THFR, where the ground had been completely cleared of forest just 4–5 years prior to sampling. Slightly more species were only recorded in the three longer transects THFR1, THFR3, and DFR2, indicating that monitoring of ants might need to consider larger spatial scales in monocultures.

The species accumulation curve of the tuna bait samples in the primary forest had a steep slope revealing that sampling of the ground ant community was far from complete.

Table 2 Occurrences of ground-dwelling ant species and cumulative species number at the different oil palm transects (maximum number of bait sites)

Species	SFR1 (5)	SFR2 (13)	SFR3 (13)	THFR1 (13)	THFR2 (13)	THFR3 (10)	DFR1 (13)	DFR2 (10)	MOP1 (10)	Total (100)
Pachycondyla b			1						1	2
Odontoponera denticulata		1				3		4	3	11
Dolichoderus sp. 1						1			2	3
Dolichoderus sp. 2				1	9	6		1		17
Technomyrmex sp. 1									1	1
Technomyrmex sp. 2				4		5		2		11
Technomyrmex sp. 1						1			2	3
Oecophylla smaragdina	1									1
Anoplolepis gracilipes	5	10	10	11	4	2	13	5	10	70
Paratrechina longicornis			3	1		1			2	7
Paratrechina sp. 1		1								1
Paratrechina sp. 2				1						1
Crematogaster sp. 1								1		1
Recurvidris sp. 1									1	1
Lophomyrmex bedoti		3	1	2						6
Tetramorium noratum				1						1
Monomorium floricola	1			2	1	3		1	2	10
Monomorium sp. 1				1	1					2
Monomorium sp. 2				1		3				4
Solenopsis sp. 1								1		1
Pheidole sarawakana			1							1
Pheidole sp. 1				4	2	1				7
Pheidole sp. 2			1							1
Number of species	3	4	6	11	5	10	1	7	9	23

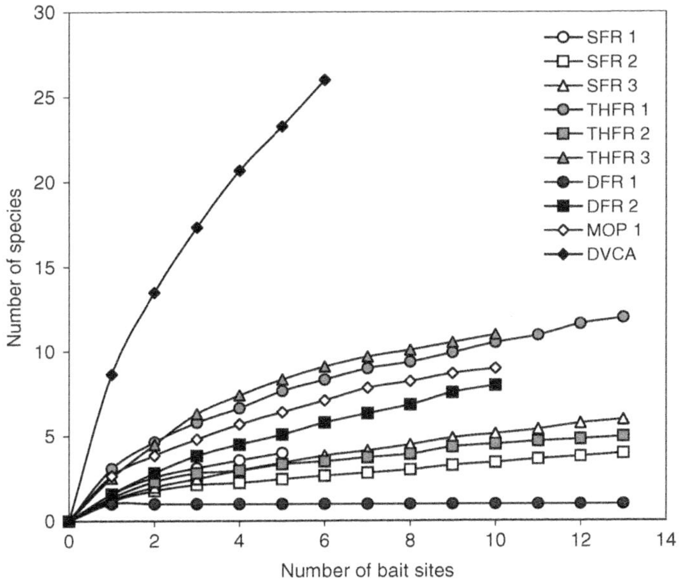

Fig. 2 Randomised species accumulation curves (100 runs) of the sampled transects in the different oil palm plantations (*SFR*, *THFR*, *DFR*, *MOP*) and the primary forest (*DVCA*)

Table 3 Observed and estimated species and sampling efficiency (%) of the ant communities for all bait sites ($N = 100$) in the oil palm plantations

	Observed species	ICE	Jack 1	Jack 2	Bootstrap	MMMean
Total oil palm	23	38 (60.5)	33 (69.7)	48 (47.9)	27 (85.2)	25 (92.0)

Nonetheless, species richness in the primary forest site was on average five times higher than that of the plantation sites. Furthermore, sampling with tuna baits is likely to underestimate the true diversity of ground-dwelling ants in forest habitats. Sampling with litter sifting, a technique that is more appropriate to sample habitats rich in leaf litter, revealed 187 ground-dwelling ant species at the same primary forest locality (Danum Valley, Brühl et al. 2003). More than 250 species were obtained by litter sifting along an altitudinal transect from primary lowland to subalpine forest in Northwestern Sabah (Brühl et al. 1999). In plantations, where no substantial leaf litter is available, litter sifting is not a feasible approach to collect ground-dwelling ants; instead tuna baiting is a very efficient alternative (Agosti et al. 2000). This is confirmed by our study. Sampling at tuna baits along the various transect was evidently efficient since intensive hand collecting did not reveal additional species at the respective sites. Thus, despite differences in the methodology that was used in the forest and plantation studies, we conclude that oil palm plantations retain only a small fraction (perhaps <5% of species) of the forest ground-dwelling ant community.

Two factors are likely responsible for the loss of ground-dwelling ants in oil palm plantations. First, the almost complete absence of leaf litter reduces the availability of nest sites for litter-nesting species (Carvalho and Vasconcelos 1999). Second, hot and dry conditions possibly prevent colony establishment and reduce colony survival of shade-adapted taxa. Microclimate is an influential factor on the distribution and competitive performance of ground-dwelling forest ants (Torres 1984; Perfecto and Vandermeer 2007; Brühl 2001). Forest ants have a lower tolerance to high temperature than ants from non-forested areas (Torres 1984). This may explain their absence in oil palm plantations, where conditions are warmer and drier and the unprotected soil layer can become extremely hot especially during the establishment phase of plantations. In addition to not providing a suitable habitat for forest ants, oil palm plantations likely form a dispersal barrier between forest populations. Queens of forest ants may be killed by adverse climatic conditions during dispersal between forest remnants. Forest ant populations that persist in small Virgin Jungle Reserves in Eastern Sabah may therefore be effectively isolated (Brühl et al. 2003). This was indicated by the lack of gene flow between fragmented populations of two forest ant species in Sabah (Bickel et al. 2006).

Our results on ground-dwelling ants are in good general agreement with findings concerning the arboreal ant community of oil palm plantations. Arboreal ants were studied by Pfeiffer et al. (2008), who intensively sampled from oil palm stems and freshly cut palm fronds. They recorded 53 species of ants with 17 species accounting for 95% of all species incidences. Thus, arboreal ant species richness in oil palm plantations was also dramatically lower than that of primary forest in Sabah, where 385 ant species were recorded in the lower vegetation and canopy (Brühl et al. 1998). Similar to our findings for ground-dwelling ants, almost 50% of all species occurrences represented tramp ant species, and at least four of these 12 species were invasive (Pfeiffer et al. 2008).

Ant community shift in oil palm plantations

Besides the evident reduction in species richness of forest ants, there was a pronounced shift in community composition towards non-forest taxa. Nine species of oil palm plantation ants were never observed in forest leaf litter in an extensive sampling program (Brühl 2001): *A. gracilipes*, *Dolichoderus* sp. 1 and 2, *Odontoponera denticulata*, *Monomorium* sp. 2, *Technomyrmex* sp. 2, *Oecophylla smaragdina*, *Paratrechina* sp. 2, and *Crematogaster* sp. 1. The 14 remaining species, which were also recorded in forest leaf litter (Brühl 2001; Brühl et al. 2003), were all rare in the plantations. Even *Lophomyrmex bedoti*, a common ant in highly disturbed logged-over forest in Sabah (Brühl 2001) was not frequently observed in oil palm plantations (six of 100 bait sites). The oil palm plantation ground ant community was numerically dominated by three species not occurring in forest habitats: *O. denticulata*, *Dolichoderus* sp. 2, and *A. gracilipes* (Table 2). The yellow crazy ant, *A. gracilipes*, was the most abundant species, collected at 70 out of 100 bait sites. It was present in all oil palm plantations and was the only species collected along the transect DFR1. *A. gracilipes* was also recognised as the dominant arboreal ant in oil palm plantations in Sabah and Peninsular Malaysia (Pfeiffer et al. 2008) and in cocoa agroforests in Sulawesi (Bos et al. 2008). The presence of *A. gracilipes* may in itself be responsible, by means of predation or competitive exclusion, for the very low ant diversity in some of the transects (especially DFR1). The yellow crazy ant has been recognised as an invasive tramp ant species (Wetterer 2005) and is documented as a dominant species in many ant communities in disturbed habitat all over the world (Majer 1993; Wetterer 2005; Bos et al. 2008). Its ecological impact was studied especially on islands (Lester and Tavite 2004) where it has revealed the potential to influence a wide variety of fauna and flora, negatively affecting ant communities but also other insects, spiders and, in the case of Christmas Island, the land crab *Gecaroidea natalis*. The reduction of the crab resulted in mutitrophic-level distortions in the ecosystem causing an invasional 'meltdown' with changes in forest structure and tree composition (O'Dowd et al. 2003; Abbott and Green 2007). *A. gracilipes* is also known to directly attack hatchling birds and reptiles and preys upon small livestock (Wetterer 2005). Reduced breeding success and behavioural changes were observed in birds on Christmas Island in the presence of *A. gracilipes* (Davis et al. 2008). *A. gracilipes* supercolonies reach densities of up to 2,250 ants per m^2 (O'Dowd et al. 2003; Abbott and Green 2007) and cover large areas. Supercolonies consist of interrelated nests that show no or very low levels of intraspecific aggression, which is a key feature for the ecological dominance of *A. gracilipes* (Hölldobler and Wilson 1990). The population structure of *A. gracilipes* was recently studied in Malaysian Borneo and a mosaic of supercolonies and expanding, newly introduced propagules was revealed (Drescher et al. 2007). Although the direct and indirect impacts of *A. gracilipes* on vertebrates have not been studied outside an island context, it seems possible that this species not only affects the ant community in oil palm plantations but also poses more wide-ranging threats to the residual amphibian, reptile, bird and mammal fauna. Another tramp species, *Monomorium floricola* (Schultz and McGlynn 2000) was also fairly common in the plantations and present at 10% of all bait sites. This species was also recorded in disturbed forests in Sabah (Brühl 2001).

Conclusion

The ground-dwelling ant communities of oil palm plantations were found to be highly impoverished with respect to forest taxa and, instead, were dominated by a small number of, partly invasive, non-forest species. The latter also included *A. gracilipes*, an invasive tramp species that is known for its adverse effects on biodiversity on tropical islands. Therefore it seems important to evaluate the ecological impact of this ant species in oil palm plantations, not only on other ants and insects but also on vertebrates.

The low abundance and species richness indicates that oil palm plantations cannot serve as a refuge for ground-dwelling forest ants. Furthermore, oil palm plantations are likely to act as dispersal barriers for forest taxa leading to community isolation in rain forest remnants. There is a clear need for more research to understand how animal and plant populations of the remaining forest fragments are affected by the surrounding plantation. A shift of the current focus in conservation biology from large mammals and birds towards arthropods that make up the bulk of the biodiversity in tropical rain forests would be desirable.

Acknowledgments We want to thank K. Eduard Linsenmair and Brigitte Fiala for discussions of many aspects of this research and Tobias Bickel for his help with tuna baits. We are also thankful to Chey Vun Khen, Arthur Chung, Robert Ong, Hubert Petol and Ruben Nilus of the Forest Research Center in Sandakan for all their support during our stay in Sabah. Permission to conduct research in Sabah was granted by the Economic Planning Unit, the Forestry Department Sabah and the Danum Valley Committee and is gratefully acknowledged. Three anonymous reviewers improved the quality of the paper. This study is project number DV163 of the Danum Valley Rainforest Research and Training Programme and was supported through a DAAD scholarship and DFG travel grant to CAB and by GTZ (TÖB) funding to TE.

References

Abbott KL, Green PT (2007) Collapse of an ant-scale mutualism in a rainforest on Christmas Island. Oikos 116:1238–1246. doi:10.1111/j.0030-1299.2007.15629.x

Achard F, Eva HD, Stibig H-J, Mayaux P, Callego J, Richards T, Malingreau J-P (2002) Determination of deforestation rates of the world's humid tropical forests. Science 297:999–1002. doi:10.1126/science.1070656

Agosti D, Majer J, Alonso E, Schultz TR (2000) Ants: standard methods for measuring and monitoring biodiversity. Biological diversity handbook series. Smithsonian Institution Press, Washington DC, 280 pp

Alonso LE (2000) Ants as indicators of diversity. In: Agosti D, Majer J, Alonso E, Schultz TR (eds) Ants: standard methods for measuring and monitoring biodiversity. Biological diversity handbook series. Smithsonian Institution Press, Washington DC, pp 80–88

Bickel TO, Brühl CA, Gadau JR, Hölldobler B, Linsenmair KE (2006) Influence of habitat fragmentation on the genetic variability in leaf litter ant populations in tropical rainforests of Sabah, Borneo. Biodivers Conserv 15:157–175. doi:10.1007/s10531-004-4248-1

Bolton B (1994) Identification guide to the ant genera of the world. Harvard University Press, London, 1689 pp

Bos MM, Tylianakis JM, Steffan-Dewenter I, Tscharntke T (2008) The invasive yellow crazy ant and the decline of forest ant diversity in Indonesian cacao agroforests. Biol Invasions 10:1399–1409. doi:10.1007/s10530-008-9215-4

Brühl CA (2001) Leaf litter ant communities in tropical lowland rain forests in Sabah, Malaysia: effects of forest disturbance and fragmentation. Dissertation, Julius-Maximilians-Universiät, Würzburg, Germany, 168 pp. Available from http://www.opus-bayern.de/uni-wuerzburg/

Brühl CA, Gunslam G, Linsenmair KE (1998) Stratification of ants (Hymenoptera, Formicidae) in a primary rain forest in Sabah, Borneo. J Trop Ecol 14:285–297. doi:10.1017/S0266467498000224

Brühl CA, Maryati M, Linsenmair KE (1999) Altitudinal distribution of leaf litter ants along a transect in primary forests on Mount Kinabalu, Sabah, Malaysia. J Trop Ecol 15:265–277. doi:10.1017/S0266467499000802

Brühl CA, Eltz T, Linsenmair KE (2003) Size does matter—fragmentation of tropical rain forest in Sabah Malaysia and the effects on the leaf litter ant community. Biodivers Conserv 12:1371–1389. doi:10.1023/A:1023621609102

Cao KF (2000) Water relations and gas exchange of tropical saplings during a prolonged drought in a Bornean heath forest, with reference to root architecture. J Trop Ecol 16:101–116. doi:10.1017/S0266467400001292

Carroll CR, Janzen DH (1973) Ecology of foraging by ants. Annu Rev Ecol Syst 4:231–257. doi:10.1146/annurev.es.04.110173.001311

Carter C, Finley W, Fry J, Jackson D, Willis L (2007) Palm oil markets and future supply. Eur J Lipid Sci Technol 109:307–314. doi:10.1002/ejlt.200600256

Carvalho KS, Vasconcelos HL (1999) Forest fragmentation in central Amazonia and its effects on litter-dwelling ants. Biol Conserv 91(2–3):151–157. doi:10.1016/S0006-3207(99)00079-8

Chazdon RL (1998) Tropical forests-log'em or leave'em. Science 281:12–13. doi:10.1126/science.281.5381.1295

Colwell RK (1997) EstimateS: statistical estimation of species richness and shared species from samples. Version 5. http://viceroy.eeb.uconn.edu/esimates

Colwell RK, Coddington JA (1994) Estimating terrestrial biodiversity through extrapolation. Philos Trans R Soc Lond Ser B 345:101–118. doi:10.1098/rstb.1994.0091

Danielsen F, Beukema H, Burgess ND, Parish F, Brühl CA, Donald PF, Murdiyarso D, Phalan B, Reijnders L, Struebig M Fitzherbert EB (2008) Biofuel plantations on forested lands: double Jeopardy for biodiversity and climate. Conserv Biol. doi:10.1111/j.1523-1739.2008.01096.x

Dauber J, Wolters V (2000) Microbial activity and functional diversity in the mounds of three different ant species. Soil Biol Biochem 32:93–99. doi:10.1016/S0038-0717(99)00135-2

Davis NE, O'Dowd DJ, Green PT, Mac Nally R (2008) Effects of an alien ant invasion on abundance, behavior, and reproductive success of endemic island birds. Conserv Biol 22(5):1165–1176. doi:10.1111/j.1523-1739.2008.00984.x

Dennis RA, Mayer J, Applegate G, Chokkalingam U, Colfer CJP, Kurniawan I, Lachowski H, Maus P, Permana RP, Ruchiat Y, Stolle F, Suyanto S, Tomich TP (2005) Fire, people and pixels: linking social science and remote sensing to understand underlying causes and impacts of fires in Indonesia. Hum Ecol 33:465–504. doi:10.1007/s10745-005-5156-z

Donald PF (2004) Biodiversity impacts of some agricultural commodity production systems. Conserv Biol 18:17–38. doi:10.1111/j.1523-1739.2004.01803.x

Drescher J, Blüthgen N, Feldhaar H (2007) Population structure and intraspecific aggression in the invasive ant species *Anoplolepis gracilipes* in Malaysian Borneo. Mol Ecol 16:1453–1465. doi:10.1111/j.1365-294X.2007.03260.x

FAOSTAT (2007) Online statistical service. FAO, Rome. Available from http://faostat.fao.org/. Accessed January 2008

Fitzherbert EB, Struebig MJ, Morel A, Danielsen F, Brühl CA, Donald PF, Phalan B (2008) How will oil palm expansion affect biodiversity? Trends Ecol Evol 23:538–545. doi:10.1016/j.tree.2008.06.012
Folgarait PJ (1998) Ant biodiversity and its relationship to ecosystem functioning: a review. Biodivers Conserv 7:1221–1224. doi:10.1023/A:1008891901953
Gunadi B, Verhoef HA (1993) The flow of nutrients in a *Pinus merkusii* forest plantation in Central Java; the contribution of soil animals. Eur J Soil Biol 29:133–139
Hölldobler B, Wilson EO (1990) The ants. Belknap Press of Harvard University Press, Cambridge MA, 733 pp
Jones CG, Lawton JH, Shachak M (1994) Organisms as ecosystem engineers. Oikos 69:373–386. doi:10.2307/3545850
Koh LP, Wilcove DS (2007) Cashing in palm oil for conservation. Nature 448:993–994. doi:10.1038/448993a
Koh LP, Wilcove DS (2008) Is oil palm agriculture really destroying tropical biodiversity? Conserv Lett 1:60–64. doi:10.1111/j.1755-263X.2008.00011.x
Laurance WF, Bierregaard RO (1996) Fragmented tropical forests. Bull Ecol Soc Am 77:34–36
Lawton JH, Bignell DE, Bolton B, Blowmers GF, Eggleton PM (1998) Biodiversity inventories, indicator taxa and effects of habitat modification in tropical rain forest. Nature 391:72–76. doi:10.1038/34166
Lester PJ, Tavite A (2004) Long-legged ants, *Anoplolepis gracilipes* (Hymenoptera: Formicidae), have invaded Tokelau, changing composition and dynamics of ant and invertebrate communities. Pac Sci 58:391–401. doi:10.1353/psc.2004.0031
MacKinnon K, Hatta G, Halim H, Mangalik A (1996). The ecology of Kalimantan—Indonesian Borneo, Vol. 3. Periplus Editions, Singapore, 802 pp
Majer JD (1993) Comparison of the arboreal ant mosaic in Ghana, Brazil, Papua New Guinea and Australia—its structure and influence on arthropod diversity. In: La Salle J, Gauld ID (eds) Hymenoptera and biodiversity. CAB International, Wallingford, pp 115–141
McMorrow J, Talip M (2001) Decline of forest area in Sabah, Malaysia: relationship to state policies, land code and land capability. Glob Environ Change 11:217–230. doi:10.1016/S0959-3780(00)00059-5
MOPB (2007) Malaysian oil palm statistics 2006, 26th edition. Malaysian Palm Oil Board. Available from http://econ.mpob.gov.my. Accessed June 2008
O'Dowd DJ, Green PT, Lake PS (2003) Invasional 'meltdown' on an oceanic island. Ecol Lett 6:812–817. doi:10.1046/j.1461-0248.2003.00512.x
Perfecto I, Vandermeer J (2007) The agricultural matrix and a future paradigm for conservation. Conserv Biol 21:274–277. doi:10.1111/j.1523-1739.2006.00582.x
Pfeiffer M, Ho CT, Teh CL (2008) Exploring arboreal ant community composition and co-occurrence patterns in plantations of oil palm *Elaeis guineensis* in Borneo and Peninsular Malaysia. Ecography 31:21–32. doi:10.1111/j.2007.0906-7590.05172.x
Schultz TR, McGlynn TP (2000) The interactions of ants with other organisms. In: Agosti D, Majer J, Alonso E, Schultz TR (eds) Ants: standard methods for measuring and monitoring biodiversity. Biological diversity handbook series. Washington DC, Smithsonian Institution Press, pp 35–44
Tilman D, Fargione J, Wolff B, D'Antonio C, Dobson AP, Howarth R, Schindler D, Schlesinger WH, Simberloff D, Swackhammer D (2001) Forecasting agriculturally driven global environmental change. Science 292:281–284. doi:10.1126/science.1057544
Torres JA (1984) Diversity and distribution of ant communities in Puerto Rico. Biotropica 16:296–303. doi:10.2307/2387938
Turner EC, Snaddon JL, Fayle TM, Foster WA (2008) Oil palm research in context: identifying the need for biodiversity assessment. PLoS One 3:e1572. doi:10.1371/journal.pone.0001572
Underwood EC, Fisher BL (2006) The role of ants in conservation monitoring: if, when, and how. Biol Conserv 132(2):166–182. doi:10.1016/j.biocon.2006.03.022
Wetterer JK (2005) Worldwide distribution and potential spread of the long-legged ant, *Anoplolepis gracilipes* (Hymenoptera: Formicidae). Sociobiology 45:1–21
Wilson EO (2000) Foreword. In: Agosti D, Majer J, Alonso E, Schultz TR (eds) Ants: standard methods for measuring and monitoring biodiversity. Biological diversity handbook series. Smithsonian Institution Press, Washington DC
Yeboua K, Ballo K (2000) Caractéristiques chimiques du sol sous palmeraie. Cahiers d'études et de recherches francophones/Agricultures 9:73–76

Biodivers Conserv (2010) 19:531–541
DOI 10.1007/s10531-009-9729-9

ORIGINAL PAPER

Effects of habitat fragmentation on the movement patterns and dispersal ability of the brown spiny rat (*Maxomys rajah*) in the Planted Forest Zone of Sarawak, Eastern Malaysia

Antony Bruce Shadbolt · Roslina Ragai

Received: 18 March 2008 / Accepted: 30 September 2009 / Published online: 16 October 2009

Abstract Brown spiny rats (*Maxomys rajah*) were translocated from continuous secondary forest to small isolated patches of remnant native forest embedded within *Acacia mangium* plantation in the Planted Forest Zone of Sarawak, East Malaysia, and fitted with tracking spools to monitor behaviours in novel environments and to record responses to a range of habitat edge features. Forest roads, large clearings and acacia plantation compartments were found to pose barriers to dispersal of brown spiny rats over short temporal scales, whereas old regenerating haul trails were readily crossed on 50% of the encounters. Downed woody debris accounted for a greater proportion of the travel route compared with brown spiny rats tracked in secondary and primary forest in Sabah, which may represent heightened predator avoidance in new environments. Provision of downed woody debris within plantation compartments may improve the dispersal ability of brown spiny rats in this modified landscape, and thus promote metapopulation dynamics and colonisation of vacant habitat patches.

Keywords Brown spiny rat · *Maxomys rajah* · Spool and line · *Acacia mangium* · Habitat fragmentation · Translocation · Reintroduction · Planted forests · Sarawak

Introduction

Habitat loss is one of the most important threats to global biodiversity (Chapman et al. 2007). A by-product of habitat loss, habitat fragmentation poses potentially dire consequences for species persistence (Fahrig and Merriam 1994; Brinkerhoff et al. 2005), with

A. B. Shadbolt (✉)
School of Forestry, University of Canterbury, Private Bag 4800, Christchurch 8020, New Zealand
e-mail: antony.shadbolt@ccc.govt.nz

R. Ragai
Grand Perfect Sdn Bhd, No. 183, Lot 3430, 1st Floor, Parkcity Commerce Square,
97000 Bintulu, Sarawak, Malaysia
e-mail: rragai81@gmail.com

the survival of many species depending on their ability to cope with such changes in their environment (Oderdonk and Chapman 2000).

Most societies are unlikely to set aside large tracts of land specifically dedicated to full protection (Salwasser et al. 1996), and accordingly just 10% of Borneo's land area is in fact designated as 'protected areas' (Stuebing 2005). Due to the continued threat of poaching, illegal logging, unauthorised shifting cultivation, storm damage and fires within these areas, and with the spatial isolation of forest reserves, wildlife managers are investigating alternatives to the reliance on traditional enforcement measures for the protection of biodiversity.

Plantation forestry is taking over much of the formerly natural forested landscape of Borneo (Padoch and Peluso 2003), which includes the Malaysian states of Sarawak and Sabah, the Indonesian state of Kalimantan and to a lesser extent Brunei Darussalam. With the suggestion that raw material supply to wood-based Malaysian industries may soon be derived from the plantation sector (Woon and Norini 2002), planted forests in Sarawak and Sabah are set to become a major feature of landscapes. Due to an increasing demand for wood fibre (Chan et al. 1998), the Sarawak Forest Department has begun to establish a fast growing *Acacia mangium* plantation in a specially designated 504,000 ha Planted Forest Zone (PFZ) in the Bintulu Division, Sarawak. While the plantation has a target of 205,000 planted hectares by 2009, it is envisioned that strategically planned and located 'conservation set-asides' will eventually comprise 35% of the PFZ landscape. Conservation set-asides will include large tracts of continuous forest, one kilometre wide river corridors, riparian buffer strips and remnant patches of native forest embedded within all plantation compartments to provide lifeboats for biodiversity (Stuebing 2005).

While these measures potentially provide benefits for wildlife conservation and management at a landscape scale, habitat fragmentation is unavoidable, with fragmentation occurring across a range of scales. Natural habitats are becoming increasingly fragmented by human activities, and in many areas only small patches remain (Bakker 2006). Small forest patches are most likely to lose some species as their increased isolation makes re-colonisation less probable in the long term (Jacquemyn et al. 2003), thus compromising metapopulation dynamics.

Forest fauna show diverse responses to edge features, and these responses can vary amongst species from preference to avoidance (Murcia 1995). Such responses to fragmentation are likely to be species specific, and changes in abundance of one species can affect forest processes including changes in plant communities and other biota (Harrington et al. 2001). While many small mammal species are able to deal with changing habitat conditions including heterogeneous and patchy environments (Wells et al. 2004), most species in South East Asia are poorly studied, and their life history patterns are often anecdotal (Wells 2005). An understanding of the relationship between animal distribution, landscape pattern and habitat types is of fundamental importance in applied ecology (Wheatley et al. 2005), and research is therefore needed to determine the real effects of fragmentation on the forest fauna occurring in modified landscapes such as the PFZ.

This study investigates the potential impacts of habitat fragmentation on the predominantly terrestrial murid, the brown spiny rat, *Maxomys rajah* (Yasuda et al. 2003; Payne and Francis 2005). Although *Maxomys* rats have been described as the most common rodents in Southeast Asian forests (Nakagawa et al. 2007a) little is known of the of the life history and ecology of the brown spiny rat, (Nakagawa et al. 2007b). Brown spiny rats may be important dispersers of seed (McConkey 2005), and in terms of their cultural use, although adult rats are not eaten, some believe that consuming new born rats could render immunity from the charm of black magic and may also act as a cure for asthma (Hin and Lading 2000).

While recent studies have investigated habitat use of this species (e.g. Wells 2005; Wells et al. 2004; Nakagawa et al. 2007a, b), none have attempted to experimentally mimic dispersal through a novel and fragmented environment. Here I use the spool and line tracking method to record the movement patterns of brown spiny rats released in unfamiliar habitat patches. This was achieved by trapping animals in secondary forest and releasing them in small patches of secondary forest separated from their assumed home range where they would have a high probability of encountering habitat edges (e.g. forest roads, haul trails, plantation compartments, riparian areas and other areas of assumed non-habitat) that may affect their dispersal ability across the landscape. Not only will such insight be of practical use to the conservation of this species within the PFZ and similar habitat types, but it will also be of interest to those involved in the evolving science of reintroduction biology.

Materials and methods

Study site

The study was conducted in the PFZ situated between longitude 113°00′ and 113°15′ East and latitude 3°15′ and 3°00′ North in the Bintulu Division, Sarawak, Eastern Malaysia over a 7-week period between 22nd June and 10th August 2007. The study area ranged in elevation from 30 to 80 m above sea level and consisted of a secondary mixed dipterocarp and kerangas forest mosaic in the process of being converted to a fast growing *A. mangium* plantation. Trapping was carried out in large continuous tracts of secondary forest at four sites; Kalabat, Samarakan (N 02°56.462′ E 113°07.603), Tubau and Kelawit. One suitable patch of secondary forest embedded within an acacia compartment was selected for animal release at each study site. Release patches were selected that had defined edges with acacia compartments, forest roads, old haul-trails and riparian areas, and that were considered small enough that subject animals would have a good chance of encountering such edges soon after release. These patches ranged in distance from the respective trapping sites from 500 m (Kalabat) to >3 km (Kelawit), and were thus considered to be unfamiliar to the subject animals. Furthermore, all release patches were separated from the sites of capture by areas of acacia compartments, secondary forest and forest road networks.

Trapping

One-hundred Sherman type 160 × 160 × 450 mm wire sprung cage traps were located in pairs at 20 m intervals either side of a 1000 m transect at each of the four trapping sites. Traps were set at ground level in association with downed woody debris (fallen tree trunks and/or branches lying at or near the forest floor) where present in order to target non arboreal, non-volant small mammal species. Traps were baited with ripe aromatic banana and checked daily for a period of seven nights for each site. Traps were not pre-baited.

Handling and spool attachment

Danfield Nylon 120/2 No.10 quilting bobbins (Texspec NZ Limited, Auckland, New Zealand) containing 240 m length lightweight nylon thread (white) measuring 40 mm × 15 mm and weighing 5 g were wrapped in plastic shrink-wrap, and then and wrapped with 12.5 mm zinc oxide adhesive plaster tape (Snowflake™, Shanghai China) (weight 0.5 g). The animals' fur was clipped short in the centre of the upper back area near the shoulders, and the spool device was attached directly to the animal with cyanoacrylic adhesive. The adhesive was applied directly to the spool casing, allowed to become tacky and held firmly against the animal for 10–20 s while the adhesive dried. Animals were not anaesthetized during any stage of their handling.

Once spools were securely in place, animals were left in half open bags while field workers retreated from the release area. Assistance was given only if animals did not leave their bags within a few minutes to avoid animals becoming entangled in the nylon thread. Where assistance was required, this was administered by gently tipping the animal from the bag on one side of a large diameter tree trunk with the fieldworker out of view on the opposite side of the trunk. Where animals showed signs of removing their spools immediately after release, a small fright was administered by a fieldworker to encourage the animal to move on. The rationale behind this strategy is discussed in more detail later in this paper.

Animal tracking

On the morning following release, the route of the animal as revealed by the path of the nylon thread was recorded by measuring step-lengths (the straight-line distance travelled between distinct changes of direction) to the nearest 10 cm and nearest degree using a 100 m tape measure and a sighting compass, respectively. Use of substrate was grouped under the four categories: G (ground); L (log); B (branch); and W (water), and recorded for the entire length of the tracked thread. Qualitative habitat variables that described vegetation class, edge type and various fine scale landscape elements encountered by the animal were also recorded along the route. The first 10 m of spool was not used in any analysis as this potentially constituted a flight response immediately post-release.

Habitat edges and linear landscape features were grouped under five categories: roads, haul trails, acacia compartments, riparian areas and open clearings. Roads were differentiated from haul trails in that they had either a mineral earth or loose chip surface, had no overhead canopy cover and were over 10 m wide. Conversely, haul-trails were much narrower, had overhead canopy closure, some leaf-litter accumulation and sporadic seedling/sapling regeneration. Where the patches adjoined acacia compartments, low definition edges were created as a result of the low tree diversity in the acacia monoculture, lower tree and canopy height and different under-storey characteristics consisting of sparse fern, exposed mineral soils and sparse, un-decomposed leaf-litter. Riparian areas included both wetland areas and small streams that resulted in a change in vegetation type and the presence of open/running water, whereas clearings were characterised by wide areas lacking overhead canopy cover and vegetated with pioneer species such as dense grasses and ferns. Responses to these features were recorded as binary responses where 1 represents a movement across the edge feature, and 0 where an animal did not cross. Where an animal did not cross the feature, the movement pattern was described as either parallel; a parallel movement along the feature then away, or away; a definite movement away from the feature. Data was analysed using the statistical functions of Microsoft Excel®.

Results

Over the course of the study 22 brown spiny rats were trapped and fitted with tracking spools. Of these, 15 were successfully tracked over a combined distance of 2554.40 m, giving an average tracked spool length of 170.30 m (42.30–245.10 m). Animals encountered various habitat edge features on 51 occasions (Table 1), translating to an edge encounter on average once every 50.60 m.

Brown spiny rats encountered roads on 11 occasions, haul roads on 14, acacia on 16, wetland/riparian areas on four occasions, and large clearings on six occasions. Of the 11 encounters with forest roads just one individual was recorded crossing a small road on a single occasion (9.09%), while the remaining ten encounters (90.01%) resulted in either parallel movement along the feature then away or 180° changes in direction away from the feature. Four individuals traversed parallel with the edge on single occasions (36.36%) and five individuals moved away in the opposite direction of the road once it was reached on six occasions (54.55%). That only one individual crossed a forest road was found to be highly significant (binomial test: $P = 0.005$, df = 10).

At old haul trails, four individuals crossed these on seven occasions (50.00%), three individuals moved parallel on five occasions (35.71%), and two individuals moved away on two occasions (14.29%). Therefore, no significant difference in their ability to cross this type of landscape feature was detected (binomial test: $P = 0.209$, df = 13).

At acacia compartments, four individuals made single excursions across the edge and into acacia forest (25.00%), while the remaining 75.00% did not cross. Of these, five individuals traversed parallel with the edge on single occasions (31.25%), and six individuals moved away from the edge on seven occasions (43.75%). That only four individuals made excursions into the acacia forest was found to be significant (binomial test: $P = 0.027$, df = 15).

Where riparian areas were encountered, one individual was tracked across the feature (25.00%); two moved parallel (50.00%) and one animal recorded moving away from the riparian edge (25.00%). The ability of brown spiny rats to cross riparian areas was not found to be significant (binomial test: $P = 0.25$, df = 3). Although the rats were not recorded entering large clearings, they traversed parallel with the edge on two occasions (33.33%), and turned away from the clearings on four occasions (66.67%). That none entered these clearings was found to be significant (binomial test: $P = 0.016$, df = 5).

Rats spent on average 85.20% (58.14–98.82%) of their time on the ground and 12.25% (1.18–42.14%) on or associated with logs. Where movement was associated with logs, step lengths were generally greater with an average length of 2.80 m (1.20–5.05 m) associated with logs compared to an average step length of 1.90 m (1.44–2.68 m) on the ground (Table 2). This difference in recorded step lengths on the two substrates was found to be highly significant (t-test: $P = 0.0015$, df = 14).

Table 1 Response to edge features encountered by brown spiny rats

	Road	Haul trail	Acacia	Riparian	Clearing	Total
CROSS	1 (1)	7 (4)	4 (4)	1 (1)	0	13
NCParallel	4 (4)	5 (3)	5 (5)	2 (2)	2 (2)	18
NCAway	6 (5)	2 (2)	7 (6)	1 (1)	4 (4)	20
NCPooled	10	7	12	3	6	38
Total	11	14	16	4	6	51

CROSS animal crossed over edge feature, *NCParallel* animal moved parallel along feature without crossing, *NCAway* animal moved away from edge feature without crossing, *NCPooled* NCParallel + NCAway. Numbers in parenthesis indicate number of individuals contributing to responses

Table 2 Tracked distances and step-lengths of brown spiny rats at the four release sites

Animal	Location	Tracked distance (m)	Dist Log (m)	% Route Log	StLGrd (m)	StLLog (m)
MRAJ1	Kalabat	210.00	88.50	42.10	2.34	4.43
MRAJ2	Kalabat	175.50	28.20	16.10	2.68	3.13
MRAJ3	Kalabat	147.10	13.50	9.20	1.92	2.70
MRAJ4	Samarakan	231.00	6.60	2.90	1.60	1.65
MRAJ5	Tubau	157.00	53.70	34.20	1.67	2.98
MRAJ6	Tubau	227.30	9.00	4.00	2.16	4.50
MRAJ7	Tubau	147.80	11.60	7.80	1.87	2.32
MRAJ8	Tubau	102.00	1.20	1.20	2.02	1.20
MRAJ9	Tubau	42.30	9.80	23.20	1.63	2.45
MRAJ10	Tubau	103.60	18.20	17.60	1.62	3.03
MRAJ11	Tubau	233.00	4.70	2.00	1.60	1.20
MRAJ12	Kelawit	211.00	10.10	4.80	2.60	5.05
MRAJ13	Kelawit	104.00	2.00	1.90	1.64	2.00
MRAJ14	Kelawit	229.30	27.80	12.10	1.68	3.48
MRAJ15	Kelawit	245.10	11.70	4.80	1.44	1.95
Average		171.10	19.78	12.25	1.90	2.80

Tracked Distance total tracked distance minus initial 10 m, *Dist Log* total distance recorded on or associated with logs; *% Route Log* percent of total route on or associated with logs, *StLGrd* average step-length on ground, and *StLLog* average step-length on or associated with logs

Discussion

Where others have recently investigated the habitat use of brown spiny rats in primary and secondary forest on Borneo and in Peninsula Malaysia (Yasuda et al. 2003; Nakagawa et al. 2007a, b; Wells 2005; Wells et al. 2004), this is the first study to specifically address the effects of fragmentation and broad-scale land use change on the species. The study drew on a small sample size ($n = 15$) which is similar to that of Wells (2005) who sampled 10 brown spiny rats in primary forest and 13 in secondary forest in Sabah, East Malaysia. These small sample sizes highlight one of the major difficulties of studying small mammals in tropical regions within limited time constraints; i.e. high species diversity but low population densities. However, by using longer tracking spools (240 m compared with 120), and modifying our release techniques as discussed below, we achieved a greater overall tracked length per animal.

Use of the spool and line method for tracking brown spiny rats over both a small spatial and temporal scale proved very successful in that it provided a very detailed picture of habitat use, however, initial trials with the species met with mixed results. Of the seven rats released at the first study site, four managed to remove their spools within the first few meters from the release point, yet the remaining three covered distances of 210.00, 175.50 and 147.10 m before the spools detached. Given that the latter three rats had obviously habituated to having the spools attached, it was felt that by giving any animal that showed signs of removing the device immediately upon release a small fright to encourage them to keep moving, may assist them to habituate to the attachment. Following these initial tracking failures we employed this strategy, and the longevity of the spool attachment improved markedly from that point on, suggesting that any discomfort experienced by that animal as a result of the spool attachment was soon forgotten.

Of all the species tracked by the authors using the method during this and previous rounds of fieldwork, brown spiny rats proved to be the most difficult to fit spools without them being removed soon after attachment. Interestingly, brown spiny rats were also the most docile of all species trapped and handled, and appeared to exhibit little fear of humans, even when the 'small fright' was administered. On subsequent fieldwork involving the attachment of radio collar transmitters to the species, the subject animals showed no obvious flight response following release, with one individual commencing to feed within 5 m of the release point and in full view of field workers. Therefore, as the main focus of the study was to assess animal responses to habitat edges, the release method described here was not considered to have affected such responses in this species, or to have compromised the study in any way.

Released rats exhibited significant negative responses to forest roads, acacia compartments and clearings with a reluctance to cross over or move into them, respectively, whereas old regenerating haul roads within the closed patch were detected but did not appear to constitute a barrier to movement for half of these edge encounters. As the brown spiny rat is associated with tall and secondary forest with sandy soils (Payne and Francis 2005; Nakagawa et al. 2007a; Maryanto et al. 2000), and is described as a primary forest species by Yasuda et al. (2003), it is understandable that the species may have a low tolerance for forest edges with their associated high light, modified vegetation class and increased litter depth—an environment intrinsically different to that of continuous forest.

Maryanto et al. (2000) suggest that litter depth in primary and secondary forests may exert major influence over the presence of this species, and it was not recorded in either *Paraserianthes* or rubber plantations during their survey. The Southeast Asian Mammal Database (Boitani et al. 2006) list the species as not occurring outside primary and disturbed lowland evergreen forest and list plantations as being unsuitable habitat. However, although the species has not been recorded in acacia in the PFZ to date, they have been recorded in exotic timber plantations in Sabah, Eastern Malaysia (Stuebing and Gasis 1989). Therefore, as some individuals did cross the edges in this study, this demonstrates that at least a proportion of the population may be able to move across these edge features and through acacia compartments in order to disperse to new habitats.

Furthermore, as this study is limited by a relatively short window of observation, dictated by the length of traceable spool, a greater number of individuals may have moved out of the patch and across the edge features after the nylon thread was exhausted. This phenomenon may be attributed to the animal exploring its new environment and possibly an initial reluctance to move from the site of release. Innes et al. (2007) found that radio tagged ship rats *(Rattus rattus)* remained within 100 m of their release point for 3 days before gradually making larger ranging movements following translocation to a forest sanctuary in Maungatautari, New Zealand. Brown spiny rats may exhibit similar behaviour when introduced to new environments, and therefore tracking methods such as radio tagging that allow observation over longer period of time will be required to assess the real barrier effect of roads and acacia.

However, if behavioural responses to habitat fragmentation such as those observed at forest roads and acacia compartments persist over a longer temporal scale, this may have effects on higher organizational levels such as populations and communities (Debinski and Holt 1999) through reduced ability of individuals to move across edge features. Where suitable habitat patches are small and distant from one another, moving between them through the acacia matrix entails traversing habitat that may be unsuitable for breeding and feeding for the species, and may also present a substantial risk of failing to locate another suitable patch to colonize (Spellerberg and Sawyer 1999). The inability to negotiate these edges may therefore prevent individuals from re-colonizing vacant habitat patches, consequently limiting meta-population dynamics and thus compromising population viability to the point where local extinctions may occur.

Local extinctions of brown spiny rats, whose diets consist mainly of fruits and invertebrates, may therefore have implications for forest structure. McConkey (2005) reported a hoard of seeds of the tree *Litsea ferruginea* (Lauraceae) formed by a single brown spiny rat in a study site in dipterocarp forest in central Borneo, indicating that this species may play a role in seed dispersal for certain tree species. Yasuda et al. (2000) also reports 'scatter hoarding' behaviour by brown spiny rats in Peninsula Malaysia, where fruit was cached on the forest floor and covered with leaves. While it is virtually unknown that any plant species may be dependent on a single animal species for its dispersal to the point that it may be endangered by the animal's extinction (Silvertown 1990), the local extinction of brown spiny rats may limit the recruitment of selected species, thereby potentially altering forest composition and structure.

Miura et al. (1997) found in Peninsula Malaysia that both the brown spiny rat and closely related red spiny rat (*M. surifer*) selected small seeds weighing less than 10 g. Although Wells (2005) found that the red spiny rat shared an ecologically similar niche to the brown spiny rat in Sabah, Payne and Francis (2005) suggest an altitudinal partitioning, with brown spiny rats being more associated with lowland sites. Furthermore, to date, red spiny rats have not been recorded from the PFZ. Therefore, if brown spiny rats were to become locally extinct in habitat patches of the PFZ, the vacant niche would unlikely be taken up by the congeneric red spiny rat.

Forty-three percent of the brown spiny rats tracked by Wells (2005) moved in the scansorial habitat at heights of less than one-meter above ground, whereas this study revealed that only one rat was found to have left the terrestrial environment (3.10 m distance) representing 6.60% of animals and just 0.12% of the pooled tracked distance for the sample. Pooled average step-lengths on or associated with logs (2.80 m) were found to be significantly longer (47.37% longer) than those on the ground (1.90 m). This may indicate selection for logs as direct movement trajectories compared with movement on the ground.

Use of downed woody debris may be attributed to a strategy which maximizes travel efficiency whereby logs are used as navigational cues (Waldien et al. 2006) and offer fast and direct routes with fewer obstacles than travelling on the ground through dense understorey vegetation (Bakker, 2006). Wells (2005) recorded brown spiny rats moving on logs for a mean 4.40%, whereas this study revealed the higher mean of 12.25%. Higher use of logs in this study compared with that of Wells (2005) may represent a heightened predator avoidance response in a new environment as a result of offering both a relatively silent travel route and/or the provision of readily accessible protective cover (Waldien et al. 2006). McAdam and Kramer (1998) suggest that where small mammals select travel routes on more complex substrates requiring focusing on nearby objects, scanning more distant fields of vision for potential predators may require the animal to cease forward movement. In contrast, logs provide a comparatively easy substrate for movement which would allow animals to scan for and detect potential predators whilst in motion.

Furthermore, under increased perceived predation threat mammals have been found to reduce normal foraging activity and increase refuging behaviour (Brinkerhoff et al. 2005) which may account for the observed increase in log use in the unfamiliar habitat patch. However, for eastern chipmunks (*Tamias striatus*), Zollner and Crane (2003) found that the use of downed woody debris by this species was a correlate of the amount of the material present in the study site, and therefore may also account for differences between the two studies.

Conclusion

Habitat edges are inextricably linked with habitat fragmentation and have potentially great implications for wildlife management and conservation (Desrochers et al. 2003). Because species may differ in their responses to local environmental factors such as habitat edge features and modified land-cover types, a landscape that appears fragmented to one species may be contiguous to another (Diffendorfer et al. 1995). The patterns of habitat use and responses to the range of edge features illustrated by this study add to our understanding of the habitat requirements of this species. Furthermore, the brown spiny rat may be a good indicator species for the utility of acacia plantation as movement habitat as to date the species has not been trapped in this forest type in the PFZ.

The study illustrates that brown spiny rats are able to detect a range of habitat edge features which they respond to in different ways. The higher use of downed woody debris recorded in this study compared to that of Wells (2005), indicates that the incorporation of this substrate in acacia compartments may aid movement and dispersal of the species through the modified landscape. Tracking of movement patterns over a longer temporal scale than that achievable with the spool and line method will give a clearer picture of the real barrier effect of habitat edges and land cover types on dispersal ability. Intensive trapping inventories carried out in small isolated habitat patches within the acacia matrix may also provide evidence for or against the utility of planted forests as a land use type conducive to dispersal and colonization of isolated habitat patches.

Acknowledgements Much thanks to Wayne Wooff and the management of Grand Perfect Conservation (GP Con) for allowing us the opportunity to work with them and their team in the Planted Forest Zone, for organizing the field expeditions, and for all the invaluable information and advice offered. We also wish to thank the GP Con team and all of the field staff for their hard work and patience in the field. Thanks also go to the Sarawak Timber Association (STA), and particularly to Barney Chan and Peter Kho for their parts in giving us the opportunity to carry out research work in Sarawak, and especially for their continued support and hospitality. Thank you to our supervisory team: Hamish Cochrane, David Norton and Robert Stuebing. Finally, we also wish to thank the University of Canterbury Scholarships Office, the New Zealand Postgraduate Study Abroad Award and to the McKelvey Fund (School of Forestry) for their generous funding, without which this ongoing research program would not be possible.

References

Bakker VJ (2006) Microhabitat features influence the movements of red squirrels (*Tamiasciurus hudsonicus*) on unfamiliar ground. J Mamm 87(1):124–130

Boitani L, Catullo G, Marzetti I, Masi M, Rulli M, Savini S (2006) The Southeast Asian mammal databank; a tool for conservation and monitoring mammal diversity in Southeast Asia. Instituto di Ecologia Applicata, Roma

Brinkerhoff RJ, Haddad NM, Orrock JL (2005) Corridors and olfactory predator cues affect small mammal behavior. J Mamm 86(4):662–669

Chan B, Kho PCS, Lee HS (1998) In: Proceedings of planted forests in Sarawak; an international conference. Forest Department Sarawak, Sarawak Timber Association, Sarawak Development Institute

Chapman CA, Naughton-Treves L, lawes MJ, Wasserman MD, Gillespie TR (2007) Population declines of colobus in Western Uganda and conservation value of forest fragments. Int J Primatol 28:513–528

Debinski DM, Holt RD (1999) A survey and overview of habitat fragmentation experiments. Conserv Biol 14(2):342–355

Desrochers A, Hanski IK, Selonen V (2003) Siberian flying squirrel responses to high- and low-contrast forest edges. Landsc Ecol 18:543–552

Diffendorfer JE, Gaines MS, Holt RD (1995) Habitat fragmentation and movement of three small mammals–Sigmodon, Microtus and Peromyscus. Ecology 76(3):827–839

Fahrig L, Merriam G (1994) Conservation of fragmented populations. Conserv Biol 8:50–59
Harrington GN, Freeman AND, Crome FHJ (2001) The effects of fragmentation of an Australian tropical rain forest on populations and assemblages of small mammals. J Trop Ecol 17:225–240
Hin HK, Lading E (2000) Development of Lanjak-Entiman wildlife sanctuary as a totally protected area, phase II; a species inventory of small mammals for the development of Lanjak-Entiman wildlife sanctuary. ITTO Workshop Proceedings
Innes J, Watts C, Fitzgerald N, Thornburrow D, Burns B, Mackay J, Speedy C (2007) Behaviour of invader ship (roof, black) rats behind pest-proof fences. Xclude It; Issue 3. Xcluder Pest Proof Fencing Company, Cambridge, New Zealand
Jacquemyn H, Butaye J, Hermy M (2003) Impacts of restored patch density and distance from natural forests on colonization success. Restor Ecol 11(4):417–423
Maryanto I, Kartono AP, Sinaga MAH (2000) Survey of mammals on different land use types. In: Gillison AN (ed) Above ground biodiversity assessment working group summary report 1996–1999: impact of different land uses on biodiversity and social indicators. ASB Working Group Report, ICRAF, Nairobi
McAdam AG, Kramer DL (1998) Vigilance as a benefit of intermittent locomotion in small mammals. Animal Behav 55:109–117
McConkey KR (2005) Influence of faeces on seed removal from gibbon droppings in a dipterocarp forest in Central Borneo. J Trop Ecol 12:117–120
Miura S, Yasuda M, Ratnam LC (1997) Who steals the fruits? Monitoring frugivory of mammals in a tropical rain forest. Malay Nat J 50:183–193
Murcia C (1995) Edge effects in fragmented forests: implications for conservation. Trends Ecol Evol 10: 58–62
Nakagawa M, Miguchi H, Sato K, Nakashizuka T (2007a) A preliminary study of two sympatric Maxomys rats in Sarawak, Malaysia: spacing patterns and population dynamics. Raffles Bull Zool 55(2):381–387
Nakagawa M, Miguchi H, Sato K, Sakai S, Nakashizuka T (2007b) Population dynamics of arboreal and terrestrial small mammals in a tropical rainforest, Sarawak, Malaysia. Raffles Bull Zool 55(2): 389–395
Oderdonk DA, Chapman CA (2000) Coping with forest fragmentation: the primates of Kibale National Park, Uganda. Int J Primatol 21(4):587–611
Padoch C, Peluso NL (2003) Borneo people and forests in transition: an introduction. In: Padoch C, Peluso NL (eds) Borneo in transition: people, forests, conservation and development. Oxford University Press, Oxford, UK
Payne J, Francis CM (2005) A field guide to the mammals of Borneo. The Sabah Society, Kota Kinabalu, Malaysia
Salwasser H, Schonewald-Cox C, Baker R (1996) The role of interagency cooperation in managing for viable populations. In: Soule ME (ed) Viable populations for conservation. Cambridge University Press, Cambridge, UK
Silvertown J (1990) Introduction to plant population ecology, 2nd edn. Longman Scientific and Technical, Essex, UK
Spellerberg IF, Sawyer JWD (1999) An introduction to applied biogeography. Cambridge University Press, Cambridge, UK
Stuebing RB (2005) Wildlife conservation in the planted forests of Sarawak: blind ambition? In: Tuen AA, Das I (eds) Wallace in Sarawak–150 years later. An international conference on biogeography and biodiversity. Institute of Biodiversity and Environmental Conservation, Universiti Malaysia, Sarawak, Kota Samarahan
Stuebing RB, Gasis J (1989) A survey of small mammals within a Sabah tree plantation in Malaysia. J Trop Ecol 5:203–214
Waldien DL, Hayes JP, Huso MMP (2006) Use of downed wood by Townsend's chipmunks (*Tamias townsendii*) in Western Oregon. J Mamm 87(3):454–460
Wells KL (2005) Impacts of rainforest logging on non-volant small mammal assemblages in Borneo. Thesis, Department of Experimental Ecology, University of Ulm, Ulm, Germany
Wells K, Pfeiffer M, Lakim MB, Linsenmair KE (2004) Use of arboreal and terrestrial space by a small mammal community in a tropical rain forest in Borneo, Malaysia. J Biogeogr 31:641–652
Wheatley M, Fisher JT, Larsen K, Litke J, Boutin S (2005) Using GIS to relate small mammal abundance and landscape structure at multiple spatial extents: the northern flying squirrel in Alberta, Canada. J Appl Ecol 42:577–586
Woon WC, Norini H (2002) Trends in Malaysian forest policy. Policy Trend Report 2002, pp 12–28
Yasuda M, Miura S, Nor Azman H (2000) Evidence for food hoarding behaviour in terrestrial rodents in Pasoh Forest Reserve, a Malaysian lowland rain forest. J Trop For Sci 12(1):164–173

Yasuda M, Ishii N, Okuda T, Nor Azman H (2003) Small mammal community: habitat preference and effects after selective logging. In: Okuda T, Manokaran N, Matsumoto Y, Niiyama K, Thomas SC, Ashton PS (eds) Pasoh: ecology of a lowland rain forest in Southeast Asia. Springer–Verlag, Tokyo, pp 533–546

Zollner PA, Crane KJ (2003) Influence of canopy closure and shrub coverage on travel along coarse woody debris by eastern chipmunks (*Tamais striatus*). Am Midl Nat 150:151–157

Biodivers Conserv (2010) 19:543–561
DOI 10.1007/s10531-009-9691-6

ORIGINAL PAPER

Non-volant small mammal community responses to fragmentation of kerangas forests in Brunei Darussalam

Joseph K. Charles · Bee Biaw Ang

Received: 26 April 2008 / Accepted: 23 June 2009 / Published online: 11 July 2009

Abstract This paper summarises the effects of kerangas forest fragmentation on the non-volant small mammal community resident in the fragments based on a long term study conducted in Brunei Darussalam. These effects are shown as eight responses—a significant reduction in the species richness and diversity in relation to size of fragments, the disappearance of species with large home ranges and specific habitat requirements, invasion of generalists/opportunists into the fragments from degraded habitats, absence of predators, a possible reduction in the activity space of generalists in fragments, erosion in the trophic structure of the small mammal community, changes in the small mammal abundance in relation to fragment size and dominance of one species over other co-occurring species in an isolated fragment. These responses are briefly explained and their significance discussed.

Keywords Small mammal communities · Kerangas forest fragmentation · Small mammal responses

Introduction

Research on disturbance of pristine forests in the Tropics have shown that drastic changes occur in the vegetation and small mammal communities of disturbed and fragmented forests, e.g., alteration of species richness and abundance, species invasions, changes in forest dynamics and trophic structure of communities, edge effects playing key roles in fragment dynamics, influence of the matrix on fragment connectivity (e.g., Bentley et al. 2000; Bierregaard et al. 1992; Chiarello 1999, 2000; Laurance 1994, 1997; Laurance et al. 2002; Turner 1996; Yahner 1988), modifications in the population dynamics of resident species and movement patterns, disappearance of specialised species, invasion of weed species (Charles and Ang 2006) etc. Most of the research efforts carried out in Southeast

J. K. Charles (✉) · B. B. Ang
Faculty of Science, Biology Department, Universiti Brunei Darussalam, Bandar Seri Begawan, Brunei Darussalam
e-mail: jcharles@fos.ubd.edu.bn

Asia have concentrated on lowland dipterocarp forests (e.g., Benneth and Dahanan 1992; Bernard 2004; Curran et al. 1999; Dahaban et al. 1996; Kemper 1988; Lynam and Billick 1999; Nakagawa et al. 2007; Stuebing and Gasis 1989; Walker and Rabinowitz 1992; Zubaid 1993, 1995; Zubaid and Rizal 1995; Zubaid and Ariffin 1997; Wells et al. 2006, 2007). The kerangas forest is one type that has been largely ignored. Brunei Darussalam has seven forest types, namely, mixed dipterocarp, montane, heath, beach, mangrove and brackish water, freshwater swamp and peat swamp (Davies and Kamariah 1999; Wong and Kamariah 1999). Heath or kerangas forests have a low uniform canopy, trees being generally shorter and smaller than in dipterocarp forests, and form on infertile, highly acidic, base-poor predominantly sandy soils. Most of the 3,455 ha of kerangas forests are found along the coastal zone of Brunei Darussalam and have been extensively cleared for residential and infrastructural development. Past woodcutting activities and repeated fires, exacerbated more recently in 1992 (Becker and Wong 1992) and 1998 by drought-induced fires have resulted in the deterioration of this vulnerable and endangered vegetation type.

Since this forest type has not been studied for effects of anthropogenic disturbance on mammal communities, a long term investigation commenced 18 years ago and is reported elsewhere (Ang 2000; Chang 2001; Charles and Ang 2006; Faisal 2007; Lim 2003; Madeeha 2005; Moss 1993). It was also important to explore whether results obtained from studies in lowland dipterocarp forests were comparable to those from kerangas forests. Based on our data collected, this paper summarises eight salient potential responses of small mammal communities to fragmentation of kerangas forests:

1. Reduction in species richness and diversity of small mammals
2. Disappearance of species with large home ranges and specific habitat requirements
3. Invasion of opportunists/generalists from disturbed habitats into the fragments
4. Absence of predators in forest fragments (as determined by cage trapping and camera trapping)
5. Reduction in the activity space of generalists in forest fragments
6. Changes in the trophic structure of small mammal communities in kerangas forest fragments
7. Changes in small mammal abundance in relation to fragment size
8. Dominance of one species over other co-occurring ones in an isolated small forest fragment

General methods

Study sites

Following conversion of most of the continuous pristine kerangas forests in Brunei Darussalam, eight fragments in the Brunei-Muara district and one continuous pristine forest in the Tutong district have been studied since 1989 for effects of habitat fragmentation on non-volant small mammals since—fragment sites A (04°58.239′N, 114°64′E; 10.8 ha), B (04°57.109′N, 114°53.857′E; 8.12 ha), C (04°57.124′N, 114°54.753′E; 4.15 ha), C’ (04°57.262′N, 114°54.977′E; 0.01 ha), D (04°57.288′N, 114°54.705′E, 3.22 ha), E (04°57.949′N, 114°54.421′E; 22.8 ha), F (04°56.929′N, 114°54.868′E; 3.03 ha), G (05°05′N, 114°58′E; 8.9 ha) and a continuous pristine forest Bang Nalud (BN; 04°39.705′N, 114°37.476′E). Site G is a disturbed kerangas forest patch surviving in the Berakas Forest Reserve (149 ha; Charles et al. 1999) which over the years has been severely affected by repeated fires from 1992 to 1997 (Becker and Wong 1992) resulting in extensive areas of *Acacia* spp. vegetation (an invasive species) and areas planted with mixed vegetation species to consolidate the soil. Bukit Patoi Forest Reserve (referred in Table 4; 04°45′N, 115°11′E) on the eastern part of Brunei Darussalam, about 65 km from the study fragments (as the crow flies) has an undisturbed kerangas forest at an elevation of 310 m that merges into a lowland dipterocarp forest.

Field methods

A live trapping technique was employed to sample all the study sites with modified collapsible cage traps designed by the first author (30 cm long × 14 cm wide × 14 cm high), placed at 10 m intervals along straight 200 metre-line transects (3 traps per point and 63 cage traps in total; referred to as index lines by Petticrew and Sadleir 1970; Read et al. 1988). Site C' which was 0.1 ha in size could only accommodate a 30 m transect line with a proportional decrease to nine traps. Trapping was carried out continuously for 7 days and 7 nights per trapping session. Each study site was sampled for five sessions along the same transect line with intervals of 2 weeks between sessions.

We had also tested earlier whether it would make any difference to small mammal community parameters if more than one transect is employed in a large fragment (Site E). Results from an investigation of three different transects in Site E showed that there was no significant difference in the abundance, species richness, species diversity, species evenness between the transects (Faizah 2006). Similar results were obtained when the test was repeated in another site Bang Nalud (a continuous forest) for three different transects.

Camera trapping was carried out at each site including Bukit Patoi Forest Reserve with SensorCamera Fieldnote 1 (Marif Co. Ltd, Iwakuni, Yamaguchi, Japan). All small mammals that were cage-trapped were marked with permanent dyes as well as with subcutaneously placed transponders (AEG). Tracking and movement studies were done by using fluorescent pigments (RADIANT COLOUR; Lemen and Freeman 1985; McShea and Gilles 1992; Mullican 1988; Mullican and Streubel 1989; Stapp et al. 1994) and by the spool-and-line method with quilting bobbins (Danfield Nylon 120/2 No. 10, Texspec NZ Limited, Auckland, New Zealand). All traps were baited with a special type of aromatic banana (pisang rasthali) as this was shown to be the most effective bait to attract small mammals, e.g., rodents, squirrels and tree shrews (Lim 1970; Charles 1996; Charles and Ang 2006). Details of all field methods employed in the long term study have been published elsewhere (Charles and Ang 2006).

Statistical analyses

Small mammal communities in all study sites were compared by examining the alpha diversity and relative abundance of the species populations concerned.

Alpha diversity of each study sites was described by looking at the number of observed species, Chao 2 species estimator, Simpson's Index and Fisher's Alpha. The mean and standard error of the indices were based on the number of species obtained in each sample for each of the sites for five samples.

All diversity calculations were performed using the software package SPECIES DIVERSITY AND RICHNESS 2.3 (Henderson and Seaby 1998). Simpson's Index and Fisher's Alpha were used as they were some of the most robust methods of comparing diversity between communities (Hayek and Buzas 1997; Southwood and Henderson 2000;

Magurran 2004). Fisher's Alpha is preferred as it is largely independent of sample efforts and gives a reliable estimate of diversity.

In each study site, species richness was calculated as the total number of species trapped using cage trapping. We also used a species richness predictor (Chao 2) to predict the number of species expected to be present in each study site. Chao 2 gives a reliable estimation of species richness even when the sample size is small (Magurran 2004).

Relative abundance of small mammal species in each study site was generated using the minimum number known alive (MNKA) method (Krebs 1966). One problem with MNKA method is it tends to underestimate population size of species (Otis et al. 1978). However, as most species caught in the study sites were quite rare with minimal encounter rate, this is the only plausible method of estimating population size.

We used one-way analysis of variance (ANOVA) to compare alpha diversity and relative abundance between study sites. When a significant variation was encountered, a post-hoc test (Bonferroni test) was used to identify groups that differed. All analyses were run using STATISTICA 6.1 (StatSoft Inc 2003).

Results

A few terms and circumstances need to be explained: 'pristine forest species' refers to small mammal species found in pristine, undisturbed forests while 'disturbed forest species' are those in disturbed, fragmented areas, e.g., secondary areas, gardens, grassland patches, plantations, riverine forests, areas that have had anthropogenic disturbance.

In Brunei, *Sundamys muelleri* is not found in pristine lowland dipterocarp and kerangas forests (from years of trapping throughout the country, Charles 1996) but only in riverine, all disturbed areas including riverine and beach forests, even homes. Payne and Francis (2005) have not qualified what they meant by 'forests' for this species. It is the common rat in forest fragments as confirmed by metric data: mean weight for adult males 235.7 ± 7.5 g ($n = 50$); adult females 205.5 ± 6.4 g ($n = 59$); subadult males (moulting) 117.8 ± 4.9 g ($n = 7$); weight range for adults 155–315 g; HF 37–49 mm; mammary formula $2 + 2$;

Rattus rattus which is usually smaller than *S. muelleri* according to Payne and Francis (2005), is similar in size to *S. muelleri* in our study areas: mean weight for adult males 238.1 ± 19.6 g ($n = 8$); adult females 248.8 ± 32.2 g ($n = 12$); weight range for adults 155–350 g; HF 35.23–40.99 mm; mammary formula $2 + 3$. These were trapped outside the fragments.

Responses of small mammal communities

Community response 1: reduction in species richness and diversity of small mammals

In the study sites (Fig. 1; C′, F, D, C, B, G, A, E) some of the pristine forest species of small mammals found in Bang Nalud—*Leopoldamys sabanus* (Long-tailed giant rat), *Maxomys rajah* (Brown spiny rat), *Maxomys whiteheadi* (Whitehead's rat), *Niviventer cremoriventer* (Dark-tailed tree rat), *Tupaia picta* (Painted treeshrew), *Tupaia gracilis* (Slender treeshrew), *Tupaia tana* (Large treeshrew), *Sundasciurus hippurus* (Horse-tailed squirrel), *Rhinosciurus laticaudatus* (Shrew-faced ground squirrel) occur together with disturbed forest species—*Sundamys muelleri* (Muller's rat), *Tupaia minor* (Lesser treeshrew), *Callosciurus notatus* (Plantain squirrel). Four fragments (C', F, C and B) have only

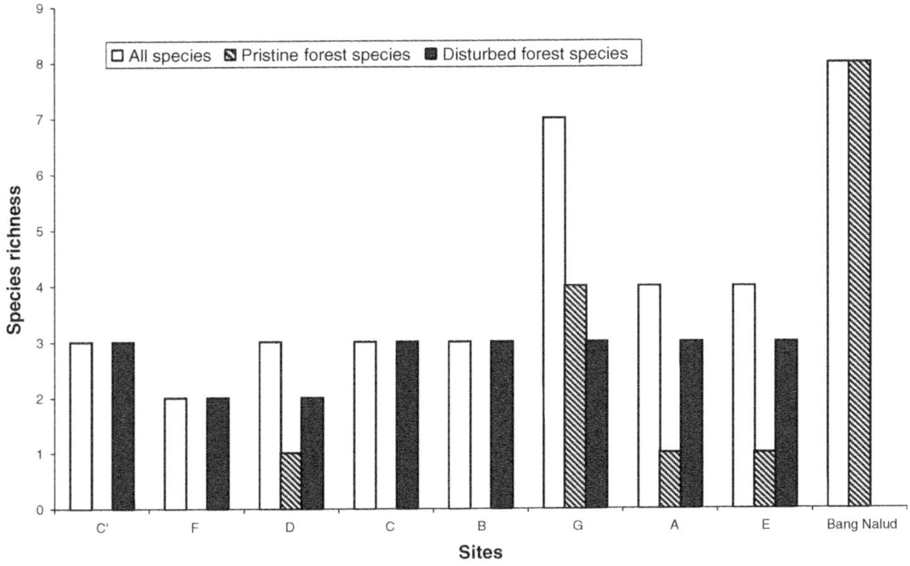

Fig. 1 Species richness of all study sites showing total number of small mammal species caught, pristine forest species only and disturbed forest species (Study sites are arranged from the smallest C′ to the largest Bang Nalud)

disturbed forest species while four others (D, G, A and E) have the disturbed and 1–4 pristine forest species. There is a statistically significant difference at the $P < 0.05$ level in the observed species richness for the study sites ($F_{(10,44)} = 12.958$, $P = 0.000$).

The species richness as determined by trap capture is similar to that as predicted by Chao 2 estimator.

Bang Nalud (the undisturbed, continuous Kerangas forest) had all eight forest species, the highest number of species compared to all study sites (ANOVA, $P < 0.05$; supported by the Chao 2 value for the study sites $F_{(10,44)} = 9.721$, $P = 0.000$). Site G, a disturbed kerangas forest patch in the larger Berakas Forest Reserve, had the highest species richness (ANOVA, $P < 0.05$) when compared to other fragments, due to the presence of four pristine forest species, including *Rhinosciurus laticaudatus* (Shrew-faced ground squirrel Fig. 3).

None of the three disturbed forest species occurred in the undisturbed forest (BN) but representatives were found in all the fragments, 2–3 species each. The presence of four pristine forest species in Site G and none in B (though they are similar in size) was probably influenced by complex vegetational structure in G (unpublished data of second author). The smallest fragment C' (0.01 ha) situated 150 m away from C shared the same species richness as C (4.15 ha) and there was evidence of these individuals moving between these two fragments.

Furthermore, fragment size has influenced species richness in that more species were found in larger fragments than smaller ones and this relationship is significant ($r = 0.792$, $n = 8$, $P = < 0.01$).

When species diversity (Simpson's Index, mean ± SE) is examined for the study sites, there is a significant reduction in the forest fragments when compared to the undisturbed forest Bang Nalud (Fig. 2; ANOVA $P < 0.05$). A similar trend is also seen when using Fisher's Alpha to analyse the data.

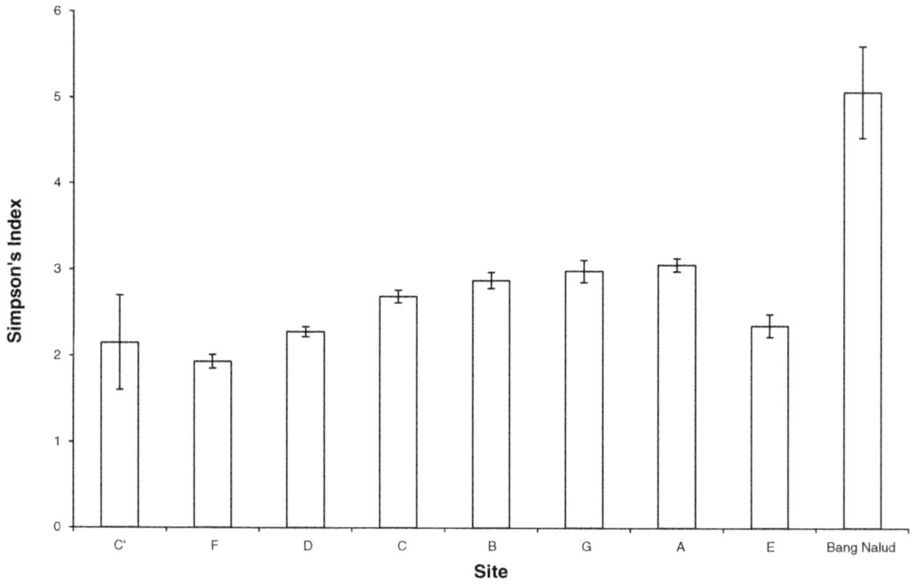

Fig. 2 Species diversity of small mammals in all study sites (Mean ± Standard Errors as *vertical bars*) (Study sites are arranged from the smallest C′ to the largest Bang Nalud)

Community response 2: disappearance of species with large home ranges and specific habitat requirements

Of the eight pristine small mammal species trapped in the continuous forest BN, *Leopoldamys sabanus*, the largest bodied rat (weight 500–700 g), also occurred in very low numbers in Site G which formerly, was a large kerangas forest. Studies have shown that it has a large home range both in kerangas and dipterocarp forests (Charles 1994, 1996; Husni and Zubaid 1998; Lim and Zubaid 1995; Moss 1993; Walker and Rabinowitz 1992; Wells et al. 2006; Zubaid and Rizal 1995). This species was absent in all other study fragments (Table 1) and its persistence at very low abundances in Site G may be precarious. A similar trend was seen in remnants of Brazilian Atlantic forests (Chiarello 2000) where bigger bodied mammals were absent from small forest fragments and higher mammal densities were found in larger fragments.

Co-occurrence of *Tupaia picta* and *T. gracilis* was seen only in undisturbed BN and Site G but not in any of the other fragments. It could be speculated that specific habitat requirements and niches for these two species do not exist in the fragments except Site G where the adjoining disturbed habitats have similar kerangas vegetation structural elements as in Site G. Hence there is a larger space for both the species to coexist.

Community response 3: invasion of opportunists/generalists from disturbed habitats into the fragments

The three disturbed forest species namely, *Sundamys muelleri* (Muller's rat), *Callosciurus notatus* (Plantain squirrel) and *Tupaia minor* (Lesser treeshrew) were found in all the

Table 1 Species caught in the different study sites using cage trapping

Species	Site A	Site B	Site C	Site C′	Site D	Site E	Site F	Site G	BN
Leopoldamys sabanus Long-tailed giant rat	–	–	–	–	–	–	–	√	√
Maxomys rajah Brown spiny rat	–	–	–	–	–	–	–	–	√
Maxomys whiteheadi Whitehead's rat	–	–	–	–	–	–	–	–	√
Niviventer cremoriventer Dark-tailed tree rat		–	–	–	√	–	–	–	√
Sundamys muelleri Muller's rat	√	√	√	√	√	√	√	√	–
Tupaia picta Painted treeshrew	√	–	–	–	–	–	–	√	√
Tupaia gracilis Slender treeshrew	–	–	–	–	–	√	–	√	√
Tupaia tana Large treeshrew	–	–	–	–	–	–	–	–	√
Tupaia minor Lesser treeshrew	√	√	√	√	–	√	–	√	–
Callosciurus notatus Plantain squirrel	√	√	√	√	√	√	√	√	–
Sundasciurus hippurus Horse-tailed squirrel	–	–	–	–	–	–	–	–	√
Rhinosciurus laticaudatus Shrew-faced ground squirrel	–	–	–	–	–	–	–	√	–
Total no. of species	4	3	3	3	3	4	2	7	8

√, Species found in the site; –, species not found in the site

fragments except Site F where *T. minor* was absent (Table 1; Fig. 3). These three species are also often trapped in disturbed habitats, e.g., gardens, orchards, urban and rural areas, vegetation close to beaches, riverine areas (Charles 1996; Charles and Ang 2006). These opportunists have also invaded fragmented kerangas which show extensive vegetational structural changes (Ang personal communication; Madeeha 2005). The changes in these habitats are probably conducive for these species to survive there. It is striking to note that none of the opportunists occur in the pristine kerangas of Bang Nalud. Perhaps another reason for their absence is the presence of predators in undisturbed kerangas.

Some fragments still retain single pristine forest species—Site A (*T. picta*), Site D (*N. cremoriventer*) & Site E (*T. gracilis*)—while Site G has four pristine forest species, three in common with the undisturbed site BN (*L. sabanus*, *T. picta*, *T. gracilis*) and one rare species *Rhinosciurus laticaudatus* (Shrew-faced ground squirrel) as yet unrecorded in BN. The latter species is extremely difficult to trap even if it is present in a habitat (Payne and Francis 1985). As explained earlier, Site G was once a large kerangas Forest Reserve and therefore it is not surprising to see a few of the pristine forest species surviving there. A follow-up study on the food requirements of the generalists/opportunists and pristine forest species would throw light on why these species occur in certain habitats.

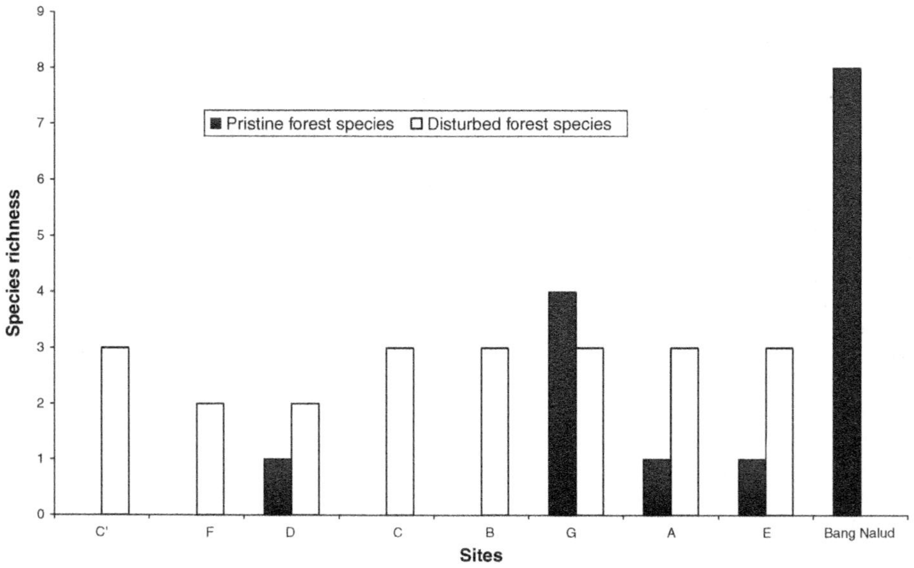

Fig. 3 Species richness of all study sites showing pristine forest species and disturbed forest species (Study sites are arranged from the smallest C′ to the largest Bang Nalud)

Community response 4: absence of predators in forest fragments (as determined by cage trapping and camera trapping)

Regular monitoring of the study sites by cage and camera trapping have revealed that six species of predators were present in the undisturbed site BN while one occurred in Site G (Table 2). It is not surprising that with one exception, the predators found in BN were generally absent from the fragments as they need large home ranges (e.g., Rajaratnam et al. 2007). Only one predator *Arctogalidia trivirgata* was recorded once in each of the

Table 2 Presence of predators in the study sites as determined by cage and camera trapping

Species	Site A	Site B	Site C	Site C'	Site D	Site E	Site F	Site G	BN
Helarctos malayanus Sun bear	–	–	–	–	–	–	–	–	√
Arctogalidia trivirgata Small-toothed palm civet	√	√	√	–	–	√	–	√	√
Paguma larvata Masked palm civet	–	–	–	–	–	–	–	–	√
Herpestes semitorquatus Collared mongoose	–	–	–	–	–	–	–	–	√
Viverra tangalunga Malay Civet	–	–	–	–	–	–	–	–	√
Felis bengalensis Leopard Cat	–	–	–	–	–	–	–	–	√
Total number of predators	1	1	1	0	0	1	0	1	6

√, Species present in the site; –, species absent in the site fragmented Sites A, B, C, and E—probably a transient as it was cage trapped or camera trapped only once over a period of a year. But in Site G it was recorded throughout the year and hence considered to be a resident. This patch of kerangas forest where this predator occurred is a part of the extensively disturbed Berakas Forest Reserve.

Earlier studies have shown that *Sundamys muelleri* Muller's rat and *Callosciurus notatus* Plantain squirrel had larger populations in fragments than in disturbed areas (Charles and Ang 2006) and this could be probably attributed to absence of predators in fragments. This therefore is an important consequence of habitat fragmentation.

Similar results are reported elsewhere. Absence of predators in small reserves was probably the main cause of the success of arboreal folivores in the Atlantic forest reserves of Brazil (Chiarello 1999). A loss of predators (Viverridae) also occurs when a rainforest is logged and habitats degraded, as shown in the lowland dipterocarp forests of Sabah, Malaysia (Wells et al. 2007).

Community response 5: reduction in the activity space of generalists in forest fragments

The space of activity of *Sundamys muelleri*, (a generalist) was studied over 3 days in three fragments of different size using fluorescent pigments to track them (Table 3).

The mean activity space for males and females in Site A (10.58 ha) was 21.84 m^2, 19.44 m^2 in Site C (4.15 ha) and 1,580.13 m^2 in Site G (8.9 ha) indicating that there is no reduction of activity space with fragment size. But this need to be explained further—though Site G is 8.9 ha of kerangas patch in the disturbed Forest Reserve, this patch merges with the disturbed vegetation with kerangas elements all around it. There was no

Table 3 The activity space (Mean ± SD) of *Sundamys muelleri* in kerangas fragments A, C and G

Fragment	Size of fragment (ha)	Animal code	Sex	Activity space (m^2)
A	10.58	Sm 1	Female	17.58
A	10.58	Sm 2	Male	26.10
Mean activity space = 21.84 m^2 ± 6.02				
C	4.15	Sm 3	Male	16.94
C	4.15	Sm 4	Female	11.98
C	4.15	Sm 5	Female	11.89
C	4.15	Sm 6	Female	47.94
C	4.15	Sm 7	Female	15.59
C	4.15	Sm 8	Female	12.28
Mean activity space = 19.44 m^2 ± 14.12				
G	8.9	Sm 9	Male	450
G	8.9	Sm 10	Female	754
G	8.9	Sm 11	Female	2,864
G	8.9	Sm 12	Female	824
G	8.9	Sm 13	Female	147
G	8.9	Sm 14	Male	68
G	8.9	Sm 15	Female	2,519
G	8.9	Sm 16	Female	5,015
Mean activity space = 1,580.13 m^2 ± 1,740.28				

clear boundary separating the Site G from the disturbed vegetation outside it. The activity spaces of all the marked females included large movements into the disturbed vegetation. Hence the small size of 8.9 ha for Site G did not seem to restrict the movements of the females within it as it was continuous with the disturbed vegetation, unlike Sites A and C where the activity spaces were within the sites. An earlier study by Moss (1993) in Site G (348 ha) before it was affected by fires, showed that four marked males and one female had a mean activity space of 730 ± 213.54 m^2 (ranging from 500 to 1,100 m^2). So it seems that as far as the animals are concerned Site G is effectively larger than 8.9 ha as their activity spaces were significantly larger than in the other two fragments (Kruskal Wallis Multiple Comparison $H = 11.76$, $P = 0.0028$). So this could be interpreted to indicate that there is a reduction of activity space with decreasing fragment size in *S. muelleri.*

The differences in the activity spaces of *S. muelleri* besides being a result of fragment size could also be a result of the reduction of habitat diversity during the process of fragmentation of the kerangas forests. A study of home range sizes of four sympatric squirrels inhabiting a lowland dipterocarp forest showed that differences in home ranges could be explained by a change of habitat characteristics rather than by density or body weight (Saiful et al. 2001).

Another interesting point relates to the activity space of six female *Sundamys muelleri* in Site G—the females moved over a wider area than males. This difference in the male and female activity patterns could have been affected by their physiological states during the time of study. This needs to be tested further.

Community response 6: changes in the trophic structure of small mammal communities in kerangas forest fragments

Based on the analyses of stomach contents of small mammals in our studies (including two undisturbed sites—BN and Bukit Patoi Forest Reserve) and from published literature (Ang et al. 1999; Emmons 1987, 1991; Harrison 1962; Langham 1983; Lekagul and McNeely 1988; Lim 1965, 1970; Medway 1978; Payne and Francis 2005; Rajaratnam et al. 2007; Wong et al. 2002) those small mammals that were cage and camera trapped in eight fragments and two undisturbed sites (BN and Bukit Patoi Forest Reserve) were classified into six categories: insectivore–frugivore–herbivore, insectivore–frugivore, carnivore, herbivore, 'invertebrate feeding' and omnivore (Table 4).

Of the six categories found in the undisturbed forests BN and Bukit Patoi, only two remain in the fragmented sites A, B, C, C′, D, E and F: insectivore–frugivore–herbivore and insectivore–frugivore. The former category is represented by four pristine forest species in the undisturbed sites but by only two generalists (*S. muelleri* and *C. notatus*) in the fragments. A change in the composition of this category in fragments has occurred (Table 4). *Rattus rattus* an omnivore not trapped in the fragments, has been caught outside three sites and may perhaps invade those fragments in the future.

Even the insectivore–frugivore category has been eroded from three or four species in the undisturbed sites to one or none in most of the fragments.

Site G is an interesting site as it a kerangas patch in a Forest Reserve that has been extensively burnt, invaded by weed plant species and rehabilitated with mixed tree plantations. Four species comprise the insectivore–frugivore–herbivore category, two being generalists as in fragments while the other two are pristine forest species common to undisturbed BN.

Table 4 Erosion of the small mammal trophic structure in kerangas forest fragments

Sites	Insectivore–Frugivore–Herbivore	Insectivore–Frugivore	Carnivore	Herbivore	Invertebrate-feeding	Omnivore
A	2[a] (*Cn*, *Sm*)	1 (*Tm*)	0	0	0	1[a, b] (*Rr*)
B	2[a] (*Cn*, *Sm*)	1 (*Tm*)	0	0	0	1[a, b] (*Rr*)
C	2[a] (*Cn*, *Sm*)	1 (*Tm*)	0	0	0	0
C′	1[a] (*Sm*)	1 (*Tm*)	0	0	0	0
D	2[a] (*Cn*, *Sm*)	0	0	0	0	1[a, b] (*Rr*)
E	2[a] (*Cn*, *Sm*)	2 (*Tm*, *Tg*)	0	0	0	0
F	2[a] (*Cn*, *Sm*)	0	0	0	0	0
G	4 (*Cn*, *Sm*, *Ls*, *Rl*)	2 (*Tm*, *Tg*)	1 (*At*)	0	0	0
Bang Nalud	4 (*Mr*, *Nc*, *Ls*, *Mw*)	3 (*Tp*, *Tt*, *Tg*)	5 (*Pl*, *At*, *Hs*, *Vt*, *Fb*)	2 (*Tm*, *Tj*)	1 (*Eg*)	1 (*Hm*)
Bt Patoi	4 (*Mr*, *Nc*, *Ls*, *Mw*)	4 (*Tp*, *Tt*, *Td*, *Tg*)	3 (*Plin*, *At*, *Vt*)	2 (*Tm*, *Tj*)	1 (*Eg*)	1 (*Ph*)

[a] Generalists, [b] outside the fragment

Erinaceidae: *Eg* = *Echinosorex gymnurus*; Tupaiidae: *Td* = *Tupaia dorsalis*, *Tg* = *Tupaia gracilis*, *Tm* = *Tupaia minor*, *Tp* = *Tupaia picta*, *Tt* = *Tupaia tana*; Sciuridae: *Cn* = *Callosciurus notatus*, *Rl* = *Rhinosciurus laticaudatus*; Muridae: *Ls* = *Leopoldamys sabanus*, *Mr* = *Maxomys rajah*, *Mw* = *Maxomys whiteheadi*, *Nc* = *Niviventer cremoriventer*, *Rr* = *Rattus rattus*, *Sm* = *Sundamys muelleri*; Ursidae: *Hm* = *Helarctos malayanus*; Viverridae: *At* = *Arctogalidia trivirgata*, *Hs* = *Herpestes semitorquatus*, *Ph* = *Paradoxurus hermaphroditus*, *Pl* = *Paguma larvata*, *Vt* = *Viverra tangalunga*, *Plin* = *Prionodon linsang*; Felidae: *Fb* = *Felis bengalensis*

The carnivore category in the fragments is totally eroded, none being resident there compared with the undisturbed site BN which has five species. Site G has one resident carnivore species *Arctogalidia trivirgata*.

The other three categories—herbivore, 'invertebrate feeding' and omnivore—are absent within the fragments but as mentioned earlier, *Rattus rattus*, an invasive generalist omnivore has been recorded outside three fragments.

The presence of all six categories of trophic structure in the undisturbed kerangas forests in BN and Bukit Patoi indicates the presence of a structurally highly complex mammal community with top predators, frugivores and herbivores while the fragments could only support a very impoverished community.

Community response 7: changes in small mammal abundance in relation to fragment size

Using MNKA to represent abundance of each of the small mammals in fragments, all the three disturbed forest species, *Sundamys muelleri*, *Callosciurus notatus* and *Tupaia minor* showed a correlation with size of area—the relationship was not significant for *S. muelleri* (Spearman $r = 0.152$, $n = 40$, $P > 0.35$); significant for *C. notatus* (Spearman $r = 0.448$, $n = 40$, $P < 0.005$, Fig. 6); and significant for *T. minor* (Spearman $r = 0.441$, $n = 40$, $P = 0.005$). Thus the abundance of *C. notatus* and *T. minor* increases when fragment size increases.

The abundance of *S. muelleri* (a ground living murid) which was high in small fragments (Sites C′, F & D), did not increase with increasing fragment size possibly because the five largest fragments (Sites A, B, C, E, G) were periodically visited by a predator

Arctogalidia trivirgata (Table 2) and predation by it might have led to low abundance. A few cage traps had dead *S. muelleri* with limbs torn from their bodies.

Community response 8: dominance of one species over other co-occurring ones in an isolated small forest fragment

A detailed 13-month study of three co-occurring species (*S. muelleri*, *C. notatus* and *T. minor* Chang 2001; Charles and Ang 2006) in a small kerangas fragment (Site C 4.12 ha in size with an isolation time of 5 years in 2000 and which may be considered a 'habitat island'), showed that *Sundamys muelleri* was the most dominant species in terms of abundances while the other two species had much lower abundance throughout the year (Fig. 4).

There was no significant correlation between abundances of *S. muelleri* and *C. notatus* (Spearman $r = 0.149$; $P = 0.627$), *S. muelleri* and *T. minor* (Spearman $r = 0.329$; $P = 0.273$) and *C. notatus* and *T. minor* (Spearman $r = -0.197$; $P = 0.518$) meaning the abundance of one had no effect on the other. The lack of competition between *S. muelleri* and *C. notatus* (though similar in body size) in this site was probably due to the former being primarily terrestrial and the latter arboreal. *T. minor*, a much smaller animal was mainly arboreal in habit.

The population density of *Sundamys muelleri* (the dominant species in Site C) declined from an August 2000 high to less than 15 animals in January 2001 (Fig. 4). Generally females were more abundant than males but their numbers kept fluctuating widely (Fig. 5) while the males, showed a downward trend in abundance (Fig. 5). The recruitment of juveniles into the population did not seem to have any effect on the general population trend for the males, females and adults as a whole (Figs. 5, 6).

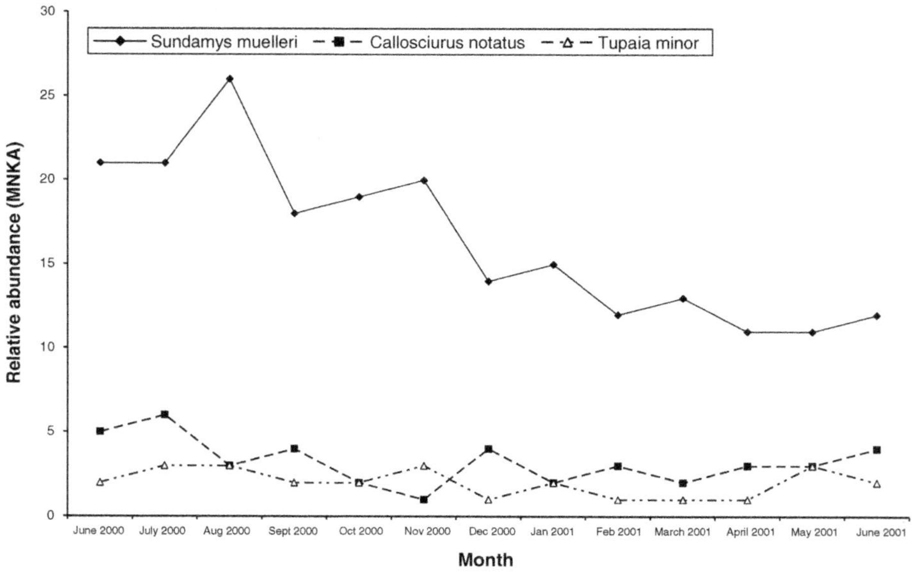

Fig. 4 Population fluctuation of Sundamys muelleri, Callosciurus notatus and Tupaia minor in Site C (4.15 ha) (Charles and Ang 2006)

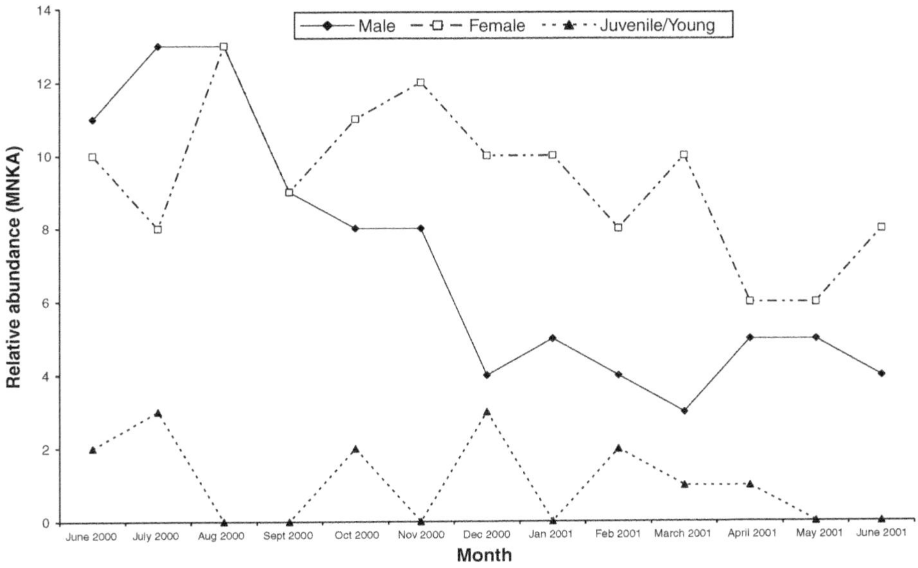

Fig. 5 Relative abundance of male, female and juvenile *Sundamys muelleri* in Site C (Charles and Ang 2006)

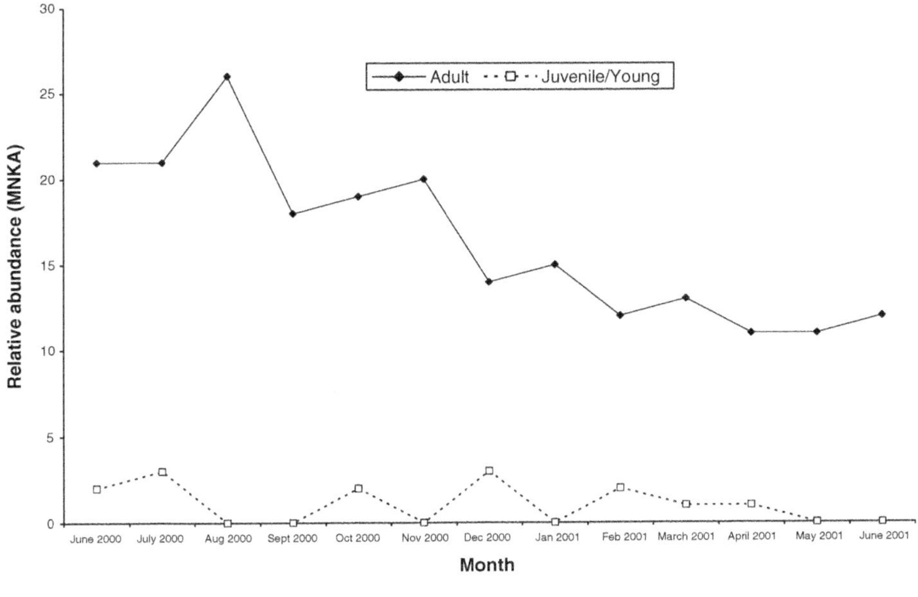

Fig. 6 Relative abundance of adult and juvenile *Sundamys muelleri* in Site C

Over 50% of *S. muelleri* males trapped every month had body injuries as a result of fighting and this high incidence of body injury could have led to the decrease in abundance. A similar trend of declining numbers was also seen among the females. The high proportion of males sustaining injuries also appeared to have changed the sex ratio in favour of the females after June 2000 (Fig. 5) even though the relationship is not significant (Spearman's $P = 0.215$).

It is probable that a high population size led to overcrowding in Site C and competition for limited resources resulted in fighting among the males and females, with many succumbing to their injuries. An ongoing study (Tee S. N. personal communication) indicates that individuals sustaining body injuries in the fragment are 'excluded' to the matrix surrounding the fragment.

Discussion

This long-term study on the effects of kerangas forest fragmentation on small mammal community dynamics has provided several insights that differ from those from disturbed lowland dipterocarp forests—a total loss of forest species in fragments, absence of predators, invasion of generalists, erosion of the trophic structure and dominance of one species in fragments.

A few studies have addressed the effects of logging on non-volant small mammals in Asian tropical forests (Zubaid and Ariffin 1997; Yasuda et al. 2003). Wells et al. (2006, 2007), in their studies of the impact of logging in rainforests of Sabah, Borneo on small mammal assemblages, showed a significant higher species richness and diversity in primary forests than in secondary forests—this was mainly due to a decrease of rare species in secondary forests. On the other hand common species of omnivorous murids or tupaiids were recorded almost equally at all forest sites. A different situation is seen in kerangas fragments—not only is there a reduction in species richness and diversity that is correlated to fragment size, there is a complete displacement and loss of native forest species from these fragments. Of the eight forest fragments studied, just three had one forest species still surviving in them while another (Site G which is actually a large disturbed kerangas forest reserve) had four forest species still present. It is habitat heterogeneity, vegetation structure and composition in the latter and in undisturbed kerangas (as shown in an ongoing current study) that favour forest species and the erosion of this heterogeneity in most kerangas fragments has encouraged the invasion of the generalists like *S. muelleri*, *C. notatus* and *T. minor*, resulting in displacement of native forest species.

The importance of habitat heterogeneity in shaping small mammal communities has been emphasised by many workers (Adler 2002; Danielson 1991; Decher and Bahian 1999; Kemper 1988; Lindenmayer and Hobbs 2004; Fox and Fox 2000; Schmid-Holmes and Drickamer 2001; Wells et al. 2007) and an experimental approach to test this idea will be carried out in all kerangas fragments in this continuing study.

Another interesting feature seen in all kerangas fragments that have been studied is the absence of large-bodied mammals that require large home ranges, e.g., *L. sabanaus*, and others that need special dietary requirements—e.g., *T. picta* and *T. gracilis*. This compounded by the absence of predators in fragments has opened up niches for the invasion of generalists like *S. muelleri* and *C. notatus* which are insect–fruit–vegetable feeders. *Rattus rattus*, an omnivore, which has not yet been recorded inside fragments was present in the surrounding matrix of some fragments (Ang 2000; Chang 2001; Lim 2003; Tee 2008) probably implying that *S. muelleri*, has probably kept *Rattus rattus* from invading the fragments.

In a study that investigated the effects of habitat fragmentation on the distribution and abundance of mammalian carnivores (Crooks 2002) in coastal southern California, six species were sensitive to habitat fragmentation and disappeared as habitats became smaller and isolated, three increased in abundance in highly fragmented sites while two were tolerant of fragmentation with no changes in their distribution and abundance.

Of the six predators recorded in the undisturbed kerangas in this study only one has been seen in kerangas fragments. This surviving predator, the Small-toothed Palm Civet *Arctogalidia trivirgata* moved from one fragment to another and is known to regularly frequent disturbed sites. This aspect of predator distribution and abundance in kerangas forest fragments needs further investigation. Predator absence has probably led to another phenomenon seen in the kerangas fragments—the dominance in every fragment of the murid, *S. muelleri* over other co-occurring species.

The changes in the trophic structure of small mammal communities in kerangas fragments is another striking feature of habitat fragmentation—six categories (insectivore–frugivore–herbivore, insectivore–frugivore, carnivore, herbivore, 'invertebrate feeding' and omnivore) present in undisturbed kerangas have been reduced to only two, namely, insectivore-frugivore-herbivore and insectivore-frugivore. The two generalists *S. muelleri* and *C. notatus* occupy the first category in the fragments while this category in the undisturbed sites has only forest species (4). A similar change has also occurred in the insectivore-frugivore category—from three to four species in the undisturbed sites to one (*T. minor* only found in fragments) or none in most of the fragments. Habitat fragmentation has led to drastic changes in the vegetation structure and composition of the kerangas fragments (unpublished data of second author) and this has probably led to the loss of food plants and other resources necessary for the survival of forest species. This is reflected in this erosion of the trophic structure of small mammal communities in kerangas fragments.

The importance of matrix and corridors cannot be over-emphasised (Andreassen et al. 1998; Åberg et al. 1995; Anderson et al. 2007; August 1983; Baker et al. 2003; Barnett et al. 1978; Bennett 1990; Burkey 1989; Chetkiewicz et al. 2006; Donald and Evans 2006; Fahrig and Merriam 1985; Gascon 1993; Gonzalez et al. 1998; Lefkovitch and Fahrig 1985; Offerman et al. 1995; Pardini et al. 2005; Saunders et al. 1991; Verboom and Van Apeldoorn 1990). The role of matrices and corridors for movement of generalists and forest small mammals in kerangas fragments and undisturbed sites and in the conservation of rare species found in some fragments is currently being investigated.

The uniqueness of the kerangas forest when compared to the lowland dipterocarp forests that dominate the forest scenario of Brunei Darussalam, lies in its vegetation structure/composition and the habitat heterogeneity and in its vulnerability to disturbance and fires. When kerangas forest is subjected to anthropogenic influences, including fragmentation, this causes a series of changes in the small mammal community, some different from that seen in disturbed dipterocarp forests.

Acknowledgments The first author (JKC) thanks Universiti Brunei Darussalam for a research grant (UBD/T/RG22) and the Biology Department, Faculty of Science for facilities and for covering all incidental expenses incurred during these studies. This biodiversity survey was a laborious one and needed field assistance whenever and wherever possible. He is deeply indebted to highly motivated students over the years since 1987 and wishes to thank them all for their co-operation, patience and perseverance: Ashri Hj Ahmad, Sarimah Abu Bakar, Munah Hj Yahya, Mahani, Norlena, Malai Salbiah, Chong Kui Kian, Marinah Chek, Rostinah, Masnah, Moss Mat Juni, Penroose Saleha, Zainal, Salwana, Norhartini, Jenny Tan, Jennifer Tan, Putri Maharani, Zamri, Foo Chuan Eng, Chong Yun Onn, co-author Ang Bee Biaw (Ang B.-B.), Albert Chang, Sim Yew Hock, Rahayu S. Sukri, Rafiah, Lim Ai Giok, Isma Fazlina, Rosman, Jonathan, Teo, Norratna, Haslina Razali, Faisal Ahmad, Hamsiah, Rosli Kasah, Faizah Idris, Madeeha Samad, Siti Aisah, Khadizah Latiff, Hadizah Latiff, Husmawati, Jocelyn, Aaron and Ravi. Others who helped were the field staff of the Biology Department and Angus our guide in Bang Nalud. Dr. D. S. Edwards' feedback on some sections of this paper was valuable. The authors also appreciate the constructive comments from two anonymous peer reviewers and other detailed corrections from the Guest Editor.

References

Åberg J, Jansson G, Swenson JE, Angelstam P (1995) The effects of matrix on the occurrence of hazel grouse (Bonasa bonasia) in isolated fragments. Oecologia 103:265–269

Adler GH (2002) Tropical tree diversity, forest structure and the demography of a frugivorous rodent, the spiny rat (*Proechimys semispinosus*). J Zool Lond 250:57–74

Anderson J, Rowcliffe JM, Cowlishaw G (2007) Does the matrix matter? A forest primate in a complex agricultural landscape. Biol Conserv 135:212–222

Andreassen HP, Hertzberg K, Ims RA (1998) Space-use responses to habitat fragmentation and connectivity in the root vole *Microtus oeconomus*. Ecol Soc Am 79(4):1223–1235

Ang BB (2000) Effects of habitat fragmentation on small mammal communities in Kerangas forests. An unpublished dissertation submitted to the Biology Department, Universiti Brunei Darussalam, 116 pp

Ang BB, Chang YE, Sim YH (1999) The use of Hutchinson Ratios on small mammal communities of KBFSC. A 3rd year field course report submitted to the Biology Department, Universiti Brunei Darussalam, 6 pp

August PV (1983) The role of habitat complexity and heterogeneity in structuring tropical mammal communities. Ecology 64(6):1495–1507

Baker PJ, Ansell RJ, Dodds PAA, Webber CE, Harris S (2003) Factors affecting the distribution of small mammals in an urban area. Mamm Rev 33(1):95–100

Barnett JL, How RA, Humphreys WF (1978) The use of habitat components by small mammals in eastern Australia. Aust J Ecol 3:277–285

Becker P, Wong M (1992) Fire damage in tropical heath forest: peculiar features and vulnerability. Brunei Mus J 7(4):110–113

Benneth EL, Dahanan Z (1992) Responses of wildlife to different types of disturbance in Sarawak, and implications for forest management. Paper presented at the International Workshop on Ecology, Conservation and Management of South-east Asian Rainforests. Kuching, Sarawak, 12–14 Oct 1992

Bennett AF (1990) Habitat corridors and the conservation of small mammals in a fragmented forest environment. Landsc Ecol 4(2/3):109–122

Bentley JM, Catterall CP, Smith GC (2000) Effects of fragmentation of Araucarian vine forest on small mammal communities. Conserv Biol 14(4):1075–1087

Bernard H (2004) Effects of selective logging on the microhabitat-use patterns of non-volant small mammals in a Bornean tropical lowland mixed dipterocarp forest. Nat Hum Act 8:1–11

Bierregaard RO Jr, Lovejoy TE, Kapos V, Dos Santos AA, Hutchings RW (1992) The biological dynamics of tropical rainforest fragments. Bioscience 42(11):859–866

Burkey TV (1989) Extinction in nature reserves: the effect of fragmentation and the importance of migration between reserve fragments. Oikos 55:75–81

Chang YE (2001) A study into the density fluctuations of a population of small mammals in a forest fragment. An unpublished dissertation submitted to the Biology Department, Universiti Brunei Darussalam

Charles JK (1994) Small mammals, pp 298–305. In: Earl of Cranbrook, Edwards DS (eds) Belalong: a tropical rainforest. Sun Tree, Singapore, 389 pp

Charles JK (1996) Small mammal diversity in riparian and dipterocarp habitats, pp 175–182. In: Edwards DS, Booth WE, Choy SC (eds) Tropical rainforest research—current issues. Kluwer, Dordrecht, 566 pp

Charles JK, Ang BB (2006) Small mammal studies in Brunei Darussalam—impact of fragmentation and development on small mammal biodiversity, pp 185–214. In: Ahmad A (ed) Brunei Darussalam: a collection of essays in conjunction with the 20th anniversary of Universiti Brunei Darussalam, vol 1, 214 pp

Charles JK, Wong TM, Nair MTR, Noralinda HI, Hussin HA (1999) Forests: a key resource to be conserved, pp 36–52. In: Wong KM, Kamariah AS (eds) Forests and trees of Brunei Darussalam, Universiti Brunei Darussalam, Forestry Department and Brunei Shell Petroleum Co. Sdn. Bhd, Brunei Darussalam, 109 pp

Chetkiewicz CB, St. Clair CC, Boyce MS (2006) Corridors for conservation: integrating pattern and process. Annu Rev Ecol Evol S 37:317–342

Chiarello AG (1999) Effects of fragmentation of the Atlantic forest on mammal communities in South-Eastern Brazil. Biol Conserv 89:71–82

Chiarello AG (2000) Density and population size of mammals in remnants of Brazilian Atlantic Forest. Conserv Biol 14(6):1649–1657

Crooks KR (2002) Relative sensitivities of mammalian carnivores to habitat fragmentation. Conserv Biol 16(2):488–502

Curran LM, Caniago I, Paoli GD, Astianti D, Kusneti M, Leighton M, Nirarita CE, Haeruman H (1999) Impact of El Nino and logging on canopy tree recruitment in Borneo. Science 286(5447):2184–2188
Dahaban Z, Nordin M, Benneth EL (1996) Immediate effects on wildlife of selective logging in a hill dipterocarp forest in Sarawak: Mammals, pp 341–346. In: Edwards DS, Booth WE, Choy SC (eds) Tropical rainforest research—current issues. Kluwer, Dordrecht, 566 pp
Danielson BJ (1991) Communities in landscape: the influence of habitat heterogeneity on the interactions between species. Am Nat 138(5):1105–1120
Davies J, Kamariah AS (1999) The rain forests of Brunei, pp 15–34. In: Wong KM, Kamariah AS (eds) Forests and trees of Brunei Darussalam, Universiti Brunei Darussalam, Forestry Department and Brunei Shell Petroleum Co. Sdn. Bhd, Brunei Darussalam, 109 pp
Decher J, Bahian LK (1999) Diversity and structure of terrestrial small mammal communities in different vegetation types on the Accra Plains of Ghana. J Zool Lond 247:395–408
Donald PF, Evans AD (2006) Habitat connectivity and matrix restoration: the wider implications of agri-environment schemes. J Appl Ecol 43:209–218
Emmons LH (1987) Comparative feeding ecology of felids in a neotropical rainforest. Behav Ecol Soc 20:271–283
Emmons LH (1991) Frugivory in treeshrews (Tupaia). Am Nat 138(3):642–649
Fahrig L, Merriam G (1985) Habitat patch connectivity and population survival. Ecology 66(6):1762–1768
Faisal HMHA (2007) Small mammal dynamics and movements in Kerangas, Mixed forest and Acacia patches in Berakas Forest Reserve. An unpublished dissertation submitted to the Biology Department, Universiti Brunei Darussalam, 64 pp
Faizah HI (2006) Assessment of structure and diversity of small mammal communities in a forest fragment by index line transects. An unpublished dissertation submitted to the Biology Department, Universiti Brunei Darussalam, 97 pp
Fox BJ, Fox MD (2000) Factors determining mammal species richness on habitat islands and interactions and guild assembly rules. Global Ecol Biogeo 9:19–37
Gascon C (1993) Breeding habitat use by Amazonian primary forest frog species at the forest edge. Biodivers Conserv 2:438–444
Gonzalez A, Lawton JH, Gilbert FS, Blackburn TM, Evans-Freke I (1998) Metapopulation dynamics, abundance, and distribution in a microecosystem. Science 281:2045–2047
Harrison JL (1962) The distribution of feeding habits among animals in a tropical rain forest. J Anim Ecol 31:53–63
Hayek LC, Buzas MA (1997) Surveying natural populations. Columbia University Press, New York, p 448
Henderson PA, Seaby RMH (1998) Species diversity and richness. Pisces Conservation, Lymington, p 23
Husni IH, Zubaid A (1998) Space use by the long-tailed giant rat (*Leopoldamys sabanus*) in the UKM forest reserve. Malay Appl Biol 27:116–118
Kemper C (1988) The mammals of Pasoh Forest Reserve, Peninsular Malaysia. Malay Nat J 42:1–19
Krebs CJ (1966) Demographic changes in fluctuating populations of *Microtus californicus*. Ecol Monogr 36(3):239–273
Langham N (1983) Distribution and ecology of small mammals in three rainforest localities of Peninsular Malaysia with particular reference to Kedah Park. Biotropica 15(3):199–206
Laurance WF (1994) Rainforest fragmentation and the structure of small mammal communities in tropical Queensland. Biol Conserv 69:23–32
Laurance WF (1997) Responses of mammals to rainforest fragmentation in Tropical Queensland: a review and synthesis. Wildl Res 24:603–612
Laurance WF, Lovejoy TE, Vasconcelos HL, Bruna EM, Didham RK, Stouffer PC, Gascon C, Bierregaard RO, Laurance SG, Sampaio E (2002) Ecosystem decay of Amazonian forest fragments: a 22 years investigation. Conserv Biol 16(3):605–618
Lefkovitch LP, Fahrig L (1985) Spatial characteristics of habitat patches and population survival. Ecol Model 30:297–308
Lekagul B, McNeely JA (1988) Mammals of Thailand. Darnsutha Press, Thailand, p 758
Lemen CA, Freeman PW (1985) Tracking mammals with fluorescent pigments: a new technique. J Mammal 66(1):134–136
Lim BL (1965) Food and weight of small animals from First Division, Sarawak. Sarawak Mus J 12:360–372
Lim BL (1970) Distribution, relative abundance, food habits and parasite patterns of giant rats (*Rattus*) in West Malaysia. J Mammal 51(4):730–740
Lim AG (2003) Small mammal community in three contiguous habitats in a forest fragment in Kampong Rimba, Brunei Muara District. An unpublished dissertation submitted to the Biology Department, Universiti Brunei Darussalam, 97 pp

Lim SP, Zubaid A (1995) Determination of habitat use by some Malaysian rats using fluorescent powder tracking. Malay Nat J 49:143–146
Lindenmayer DB, Hobbs RJ (2004) Fauna conservation in Australian plantation forests—a review. Biol Conserv 119(2004):151–168
Lynam AJ, Billick I (1999) Differential responses of small mammals to fragmentation in a Thailand tropical forest. Biol Conserv 91(2–3):191–200
Madeeha AS (2005) Small mammal diversity and community structure within three different habitat patches—Acacia plantation, mixed species plantation and disturbed kerangas fragment in the Berakas Forest Reserve, Brunei Darussalam. An unpublished dissertation submitted to the Biology Department, Universiti Brunei Darussalam, 72 pp
Magurran AE (2004) Measuring biological diversity. Blackwell, Oxford, p 260
McShea WJ, Gilles AB (1992) A comparison of traps and fluorescent powder to describe foraging for mast by *Peromyscus leucopus*. J Mammal 73(1):218–222
Medway L (1978) The wild mammals of Malaya (Peninsular Malaysia) and Singapore. Oxford University Press, Kuala Lumpur, p 131
Moss MJ (1993) Population size and movements of some small mammals in a disturbed kerangas forest. An unpublished dissertation submitted to the Biology Department, Universiti Brunei Darussalam, 84 pp
Mullican TR (1988) Radio telemetry and fluorescent pigments: a comparison of techniques. J Wildl Manag 52(4):627–631
Mullican TR, Streubel DP (1989) Tracking small mammal movements with fluorescent pigments. Am Biol Teach 51(8):497–499
Nakagawa M, Miguchi H, Sato K, Nakashizuka T (2007) A preliminary study of two sympatric Maxomys rats in Sarawak, Malaysia: spacing patterns and population dynamics. Raffles Bull Zool 55(2):381–387
Offerman HL, Dale VH, Pearson SM, Bierregaard RO Jr, O'Neill RV (1995) Effects of habitat fragmentation on neotropical fauna: current research and data availability. Environ Rev 3:191–211
Otis DL, Burnham KP, White GC, Anderson DR (1978) Statistical inference for capture data from closed populations. Wildl Monogr 62:1–135
Pardini R, De Souza SM, Braga-Neto R, Metzger JP (2005) The role of forest structure, fragment size and corridors in maintaining small mammal abundance and diversity in an Atlantic forest landscape. Biol Conserv 124:253–266
Payne J, Francis CM (1985) A field guide to the mammals of Borneo. The Sabah Society, Malaysia, 332 pp
Payne J, Francis CM (2005) A field guide to the mammals of Borneo. Malaysia, The Sabah Society (original edition published in 1985), 332 pp
Petticrew BG, Sadleir RMFS (1970) The use of index lines to estimate population numbers of deermice (*Peromyscus maniculatus*) in a forest environment in British Columbia. Can J Zool 48:385–389
Rajaratnam R, Sunquist M, Rajaratnam L, Ambu L (2007) Diet and habitat selection of the leopard cat (*Prionailurus bengalensis borneoensis*) in an agricultural landscape in Sabah, Malaysia Borneo. J Trop Ecol 23:209–217
Read VT, Malafant KWJ, Myers K (1988) A comparison of grid and index-line trapping methods for small mammal surveys. Aust Wildl Res 15:673–687
Saiful AA, Idris AH, Rashid YN, Tamura N, Hayashi F (2001) Home range size of sympatric squirrel species inhabiting a lowland dipterocarp forest in Malaysia. Biotropica 33(2):346–351
Saunders DA, Hobbs RJ, Margules CR (1991) Biological consequences of ecosystem fragmentation: a review. Conserv Biol 5(1):18–32
Schmid-Holmes S, Drickamer LC (2001) Impact of forest patch characteristics on small mammal communities: a multivariate approach. Biol Conserv 99(3):293–305
Southwood TRE, Henderson PA (2000) Ecological methods, 3rd edn. Blackwell Science, Oxford
Stapp P, Young JK, Vandewoude S, Van Horne B (1994) An evaluation of the pathological effects of fluorescent powder on deer mice (*Peromyscus maniculatus*). J Mammal 75(3):704–709
StatSoft Inc (2003) STATISTICA (data analysis software system). Version 6. http://www.statsoft.com
Stuebing RB, Gasis J (1989) A survey of small mammals within a Sabah tree plantation in Malaysia. J Trop Ecol 5:203–214
Tee SN (2008) A population study of small mammals in a kerangas forest fragment and in the matrix surrounding it. An unpublished dissertation submitted to the Biology Department, Universiti Brunei Darussalam, 81 pp
Turner IM (1996) Species loss in fragments of tropical rain forest: a review of the evidence. J Appl Ecol 33:200–209
Verboom B, Van Apeldoorn R (1990) Effects of habitat fragmentation on the red squirrel, *Sciurus vulgaris* L. Landsc Ecol 4(2/3):171–176

Walker S, Rabinowitz A (1992) The small mammal community of a dry tropical forest in central Thailand. J Trop Ecol 8:57–71
Wells K, Pfeiffer M, Lakim MB, Kalko EKV (2006) Movement trajectories and habitat partitioning of small mammals in logged and unlogged rain forests on Borneo. J Anim Ecol 75:1212–1223
Wells K, Kalko EKV, Lakim MB, Pfeiffer M (2007) Effects of rain forest logging on species richness and assemblage composition of small mammals in Southeast Asia. J Biogeo 34(6):1087–1099
Wong KM, Kamariah AS (eds) (1999) Forests and trees of Brunei Darussalam. Universiti Brunei Darussalam, Forestry Department and Brunei Shell Petroleum Co. Sdn. Bhd, Brunei Darussalam, p 109
Wong ST, Servheen C, Ambu L (2002) Food habits of Malayan sun bears in lowland tropical forests of Borneo. Ursus 13:127–136
Yahner RH (1988) Changes in wildlife communities near edges. Conserv Biol 2(4):333–339
Yasuda M, Ishii N, Okuda T, Nor Azman H (2003) Small mammal community: habitat preference and effects after selective logging. In: Okuda T, Manokaran N, Matsumoto Y, Niiyama K, Thomas SC, Ashton PS (eds) Pasoh: ecology of a lowland rain forest in Southeast Asia. Springer-Verlag, Tokyo, pp 533–546
Zubaid A (1993) A comparison of the bat fauna between a primary and fragmented secondary forest in Peninsular Malaysia. Mammalia 57(2):201–206
Zubaid A (1995) Home range of Whitehead's rat, *Maxomys whiteheadi*. Wallaceana 76:13–15
Zubaid A, Ariffin KE (1997) A comparison of small mammal abundance between a primary and disturbed lowland rain forest in Peninsular Malaysia. Malay Nat J 50:201–206
Zubaid A, Rizal M (1995) The relative abundance of rats between two forest types in Peninsular Malaysia. Malay Nat J 49:139–142

Biodivers Conserv (2010) 19:563–574
DOI 10.1007/s10531-009-9699-y

ORIGINAL PAPER

Population genetic evidence for the east–west division of the narrow-barred Spanish mackerel (*Scomberomorus commerson*, Perciformes: Teleostei) along Wallace's Line

Zohrah Haji Sulaiman · Jennifer R. Ovenden

Received: 1 April 2008 / Accepted: 22 July 2009 / Published online: 7 August 2009

Abstract Mitochondrial DNA D-loop (control) region (426-bp) was used to infer the genetic structure of Spanish mackerel (*Scomberomorus commerson*) from populations in Southeast Asia (Brunei, East and West Malaysia, Philippines, Thailand, Singapore, and China) and northern Australia (including western Timor). An east–west division along Wallace's Line was strongly supported by a significant AMOVA, with 43% of the total sequence variation partitioned among groups of populations. Phylogenetic and network analyses supported two clades: clade A and clade B. Members of clade A were found in Southeast Asia and northern Australia, but not in locations to the west (Gulf of Thailand) or north (China). Clade B was found exclusively in Southeast Asia. Genetic division along Wallace's Line suggests that co-management of *S. commerson* populations for future sustainability may not be necessary between Southeast Asian nations and Australia, however all countries should share the task of management of the species in Southeast Asia equally. More detailed genetic studies of *S. commerson* populations in the region are warranted.

Keywords Mitochondrial DNA control region · Narrow-barred Spanish mackerel · Southeast Asia · *Scomberomorus commerson* · Wallace's Line · Australia · Western Timor

Introduction

The narrow-barred Spanish mackerel (*Scomberomorus commerson*) is a large epipelagic predatory species found in waters adjacent to southern and eastern Africa, the Arabian Peninsula, India, Southeast Asia, Japan, northern Australia and the western Pacific islands

Z. H. Sulaiman (✉)
Biology and Biotechnology Programme, Faculty of Science, Universiti Brunei Darussalam, Jalan Tungku Link, Gadong 1410, Brunei Darussalam
e-mail: zohrah.sulaiman@ubd.edu.bn

J. R. Ovenden
Molecular Fisheries Laboratory, Queensland Department of Primary Industries and Fisheries, P.O. Box 6097, St. Lucia, QLD 4067, Australia

(Collette and Russo 1984). Its general distribution includes the whole of the tropical Indian and Pacific Oceans, including the seas in Southeast Asia that connect these two oceans. The region has the richest fish fauna on earth, with approximately 4,000 species or about 30% of the world's total marine fish species (Allen 2000).

Wallace's Line is a major biogeographic boundary that separates the terrestrial flora and fauna of Sundaland including Borneo, Philippines and Bali, from Sulawesi, Lombok, New Guinea and Australia. It is named after Alfred Wallace (1823–1913) who first demonstrated a major separation of terrestrial flora and fauna into Asian (west) and Australasian (east) elements (Van Oosterzee 1997). There are two geophysical processes underlying Wallace's Line (Lourie and Vincent 2004): (1) the collision of Asian and Australian tectonic elements between *c.* 15 and 3 million years ago (mya), which brought significantly different terrestrial realms into close contact (Hall 2002), and (2) the recent occurrence of lowered sea level events associated with Pleistocene glaciations (2.4 mya—10,000 years), which resulted in the formation of land-bridges between mainland Asia, Borneo and the other western Indonesian islands, and among Philippine islands (Voris 2000). Various studies have tested whether Wallace's Line or other mechanisms could explain intra-specific genetic patterns in marine species (Springer and Williams 1990; Randall 1998). The phylogeography of a marine fish species, the three-spot seahorse (*Hippocampus trimaculatus*), was analysed using mitochondrial DNA (mtDNA) sequence variation in the cytochrome *b* region and found to follow a major east–west division along Wallace's Line (Lourie and Vincent 2004). Mantis shrimp (*Haptosquilla pulchella*) populations in the Indo-West Pacific showed regional genetic differentiation in mtDNA cytochrome oxidase I sequence, concordant with ocean basins isolated during Pleistocene low sea levels (Barber et al. 2002).

Using meristic studies, three species of *Scomberomorus* (*S. guttatus, S. koreanus* and *S. lineolatus)* were found to the west of Wallace's Line whereas four species (*S. multiradiatus, S. semifasciatus, S. queenslandicus* and *S. munroi*) are only found to the east of Wallace's Line (Collette and Russo 1984). It is therefore interesting to test whether the east–west division along Wallace's Line is reflected in the population genetic structure of *S. commerson.* The distribution of this species crosses Wallace's Line and is also found to the north and south of Southeast Asian archipelago in the Indian and Pacific Oceans. *S. commerson* is one of the most important and highly valued commercial species exploited in Southeast Asia and neighbouring nations. There are no accurate catch records for this species in Southeast Asia, however the larger portion of global marine catches are pelagic species which account for 21% (17.7 million tones) of the total global catches in 2002 (FAO 2005). Understanding fisheries stock structure is an important component of sustainable long-term management (Shaklee and Currens 2003). Fish stocks are panmictic, genetically discrete populations below the species level (Shaklee et al. 1990). Mitochondrial DNA analyses have been used successfully to understand the stock structure of *S. commerson* elsewhere (Ovenden and Street 2007; Hoolihan et al. 2006; van Herwerden et al. 2006). A single genetic stock of *S. commerson* was found in the Arabian Gulf, Gulf of Oman and Arabian Sea (Hoolihan et al. 2006) and another stock was restricted to one locality in the Arabian Sea (van Herwerden et al. 2006). Ovenden and Street (2007) reported the presence of a distinct genetic stock on the east coast and a single stock on the northern and western Australian coastline. Considerable genetic differences were also suggested to exist between populations in northern Australian and western Timor that were separated from the Australian coastline by only hundreds of kilometers (Ovenden and Street 2007). There has been no molecular study to date on population genetic structure of *S. commerson* in Southeast Asian waters. The present study use mitochondrial DNA D-loop (control region) variation to test whether Wallace's Line or other biogeographical barriers have influenced the population genetic structure of *S. commerson* in Southeast Asia and neighbouring areas.

Materials and methods

Population sampling

Our study area included the South China Sea, Gulf of Thailand, East China Sea, Andaman Sea, Straits of Malacca, Sulu Sea, Timor Sea and seas surrounding northern Australia. Twenty-seven narrow-barred *S. commerson* were sampled between July 2004 to August 2005 from eight locations in Southeast Asia (PHI, CHI, THA, PEN, BRU, SAB, SAR and SIN) and 20 were sampled between March to October 1998 from three sea locations in Australia (NWA, NOA and NEA) and one location in western Timor (KUP) (Table 1; Fig. 1). In Southeast Asia, samples were obtained from wet markets. The provenance of the samples was confirmed by market interviews. In Australia, samples were obtained from commercial catches by on-board fisheries observers. Unless otherwise indicated, each sampling location was considered to be a single population. Fin clips and muscle tissue samples were preserved in 5 M NaCl with 20% dimethyl sulphoxide and stored at −80°C until subjected to DNA extraction.

Laboratory analysis

Genomic DNA was isolated from 50–100 mg of tissue using DNeasyTM tissue kits (Qiagen GmbH, Germany). The 5′ end of the control region (D-loop) was amplified as described by Ovenden et al. (2002) except that the annealing temperature was 55°C for the control region. The primers used were Pro889U20 (CCW CTA ACT CCC AAA GCT AG) and TDKD1291L21 (CCT GAA ATA GGA ACC AAA TGC). Forward and reverse PCR primers for the control region were used to sequence the purified PCR product, except where indicated. Sequences were obtained on an Applied Biosystems Inc. 3130 XL

Table 1 Collection details, number of samples (*N*), number of haplotypes (*H*), number of polymorphic sites (*P*) for each population of *Scomberomorus commerson*

	Collection location	*N*	Sea	Latitude	Longitude	*H*	*P*
1	Beijing, China (CHI)	2	East China Sea	39°54′N	116°23′E	2	6
2	Bangkok, Thailand (THA)	2	Gulf of Thailand	13°45′N	100°29′E	2	1
3	Penang, Malaysia (PEN)	2	Andaman Sea/Straits of Malacca	05°24′N	100°14′E	2	12
4	Singapore (SIN)	4	South China Sea	01°18′N	103°51′E	5	39
5	Bulacan, Philippines (PHI)	5	South China Sea/Sulu Sea	14°42′N	120°55′E	5	42
6	Kota Kinabalu, Sabah (SAB)	5	South China Sea	05°15′N	117°00′E	5	42
7	Brunei (BRU)	4	South China Sea	04°03′N	114°40′E	4	38
8	Miri, Sarawak (SAR)	3	South China Sea	04°23′N	113°58′E	5	11
9	Kupang, western Timor (KUP)	5	Inshore West Timor Sea	15°06′S	123°38′E	5	34
10	Northwest Australia (NWA)	5	Southwest Timor Sea	15°06′S	124°66′E	5	26
11	Northern Australian (NOA)	5	Arafura Sea	13°39′S	137°01′E	5	18
12	Northeast Australia (NEA)	5	Coral Sea	16°44′S	146°06′E	5	4

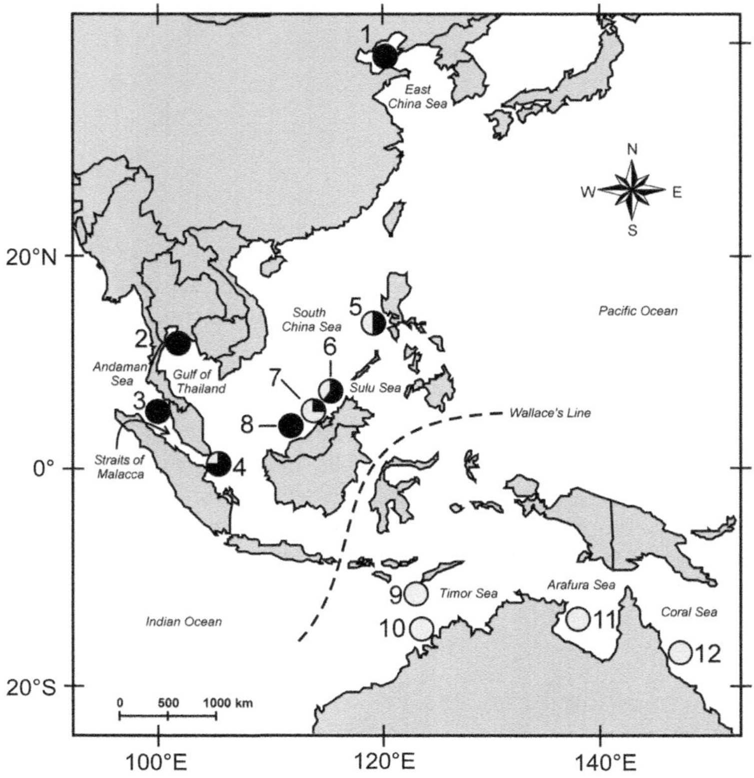

Fig. 1 Collection locations for narrow-barred Spanish mackerel (*Scomberomorus commerson*), (*1*) Beijing, China (CHI), (*2*) Bangkok, Thailand (THA), (*3*) Penang (PEN), (*4*) Singapore (SIN), (*5*) Bulacan, Philippines (PHI), (*6*) Kota Kinabalu, Sabah (SAB), (*7*) Brunei (BRU), (*8*) Miri, Sarawak (SAR), (*9*) Kupang, western Timor (KUP), (*10*) North-Western Australia (NWA), (*11*) Northern Australia (NOA) and (*12*) North-Eastern Australia (NEA). The pie charts show the distribution of clades *A* (○) and *B* (●) in each population

automated sequencer using the chain-termination method with Big Dye Terminator® chemistry. Sequence data was aligned with Sequencher v 4.8 (Gene Codes Corporation, 775 Technology Drive, Suite 100A, Ann Arbor, MI 48108, USA).

Statistical analysis

Sequence data was analysed with MacClade v 3.08 (Maddison and Maddison 1992) to identify unique haplotypes and variable sites. Phylogenetic relationships among *S. commerson* haplotypes were assessed using neighbour-joining (NJ) and maximum likelihood (ML) methods. Missing nucleotide data was excluded on a pair-wise basis. Tamura and Nei (1993) distances were used for NJ analyses. For the ML analysis, Modeltest 3.7 (Posada and Crandall 1998) was used to select the most appropriate model of DNA evolution using Akaike information criterion (Akaike 1974). The General Time Reversible Model with a gamma-distributed among site rate variation (gamma shape = 0.1226) was preferred. Neighbour-joining and ML methods were implemented in PAUP* version 4.0 b10 (Swofford 1990). Nodal support for NJ and ML analyses were assessed using non-parametric bootstrapping of 1,000 (NJ) and 100 (ML) pseudo-replicates of the original sequence alignment. A 50% majority rule consensus phylogram was constructed from NJ and ML bootstrapping runs. An intraspecific network among haplotypes was constructed using the method of Templeton et al. (1992). Haplotypes were connected into the network if their sequence varied by less than 15 single mutational steps from the remainder. The network was calculated in the software TCS v1.21 (Clement et al. 2000).

Alternate mechanisms for the genetic structure of *S. commerson* populations throughout the region were investigated using an analysis of variance framework (AMOVA, Excoffier et al. 1992). The AMOVA was based on analyses of the variance of haplotype frequencies, taking into account the number of mutations between mtDNA sequences as Tamura-Nei molecular distances (Tamura and Nei 1993), with a gamma correction of 0.12. Inter-individual and inter-populational differences were used to compute fixation indices, whose statistical significance was evaluated using a non-parametric approach involving 1,000 permutations of haplotypes, individuals and populations. Two user-defined structures were evaluated. Sampled populations were separated into groups according to their location: (1) to the east or west of Wallace's Line or (2) in waters adjacent to the Indian or Pacific Oceans (Table 1; Fig. 1). In the latter, populations 3 (PEN), 9 (KUP) and 10 (NWA) were adjacent to the Indian Ocean and the remainder was adjacent to the Pacific Ocean. Population 11 (NOA) was excluded from the analysis as being intermediate. The AMOVA method was implemented in Arlequin v 3.0 (Excoffier et al. 2005).

Results

Population and genome diversity

The mtDNA control region of the 47 *S. commerson* was highly polymorphic, with 81 variable sites across 426 aligned base pairs (Table 2). One sequence from a fish collected from the Philippines (PHI02) has been lodged on GenBank (accession no. EU526382; www.ncbi.nlm.nih.gov). Of the 47 sequences, 16 were sequenced in one direction only. Single direction sequences were missing 35–55 bases from the 5′ end of the contig of aligned sequences. Another eleven sequences had three to 40 base pairs missing from the 5′ end, however, in the first 55 bases, there was only one polymorphic site (at position 55). At the 3′ end, 15 sequences were missing between 3 and 28 bases. There were one and two base pairs indels or alignment gaps in the aligned sequences. Definitive identification of haplotypes (and hence haplotype diversity) was complicated by the pattern of missing data among sequences. Detailed comparisons revealed 43 haplotypes as each sequence was unique, except for four pairs of sequences that were assumed to be identical (CHI02 and THA01, SAB01 and SAB03, SAB02 and SAB04, PHI02 and SIN01) as they differed only by missing or ambiguous characters. Overall haplotype diversity was 0.91, as there were 43 haplotypes among the 47 sequences. The per-population nucleotide diversities were large reflecting the relatively small sample sizes and are not presented here.

Phylogenetic relationships

In the 50% consensus tree (Fig. 2) haplotypes were grouped into two main clades (A and B) with high bootstrap support. One fish from the Philippines population (PHI01) was an outlier; it did not belong to clade A or B. Clade A contained Southeast Asian, Australian

Table 2 Polymorphic sites for 47 mtDNA control region sequences from Spanish mackerel (*Scomberomorus commerson*) collected from Australia and Southeast Asia. See Fig. 3 caption for summary of sequence identity and Table 1 for collection locations. Missing data was represented as '?' and insertion or deletions (gaps) as '-'. Sequence PHI02 has GenBank accession number EU526382

Individual (Polymorphic Site Number)	55	62	76	87	88	89	94	95	96	97	98	99	107	111	112	113	115	125	128	129	155	156	167	169	172	176	177	182	187	188	190	196	200	209	211	212	213	214	215	220	223	224	225	227	230	231	232	234	241	244	247	249	252	255	256	264	268	269	272	273	274	275	276	277	282	284	293	302	310	313	314	332	334	352	353	363	381	386	388	392	407	409
NWA01	C	C	T	C	C	T	A	T	T	A	A	G	C	A	T	A	C	C	A	C	T	A	T	C	A	A	T	C	A	T	G	C	A	G	C	A	T	G	T	T	C	A	A	T	T	G	-	G	C	A	C	T	G	T	T	C	-	T	T	T	T	G	C	G	C	T	T	A	T	A	A	G	G	G	T	T	G	A	T	C	T	G
NWA02	.	.	.	.	.	.	.	.	.	.	.	.	.	.	.	.	.	.	.	T	.	.	.	.	.	.	.	.	.	.	.	.	.	A	.	.	.	.	.	.	.	.	.	.	.	.	-	.	.	.	.	.	.	.	.	.	-	.	.	.	C	.	.	.	T	.	.	.	C	.	G	.	.	A	.	.	.	.	.	.	.	.
NWA03	.	.	.	T	.	.	.	.	.	.	G	A	.	.	.	.	T	.	.	T	.	.	.	T	.	.	.	.	.	.	.	T	.	A	.	G	C	.	.	.	.	.	.	C	.	A	A	.	.	.	.	.	A	.	.	.	T	.	.	.	C	.	.	A	.	C	.	.	.	.	G	.	.	A	.	.	.	.	.	.	.	.
NWA04	.	.	.	.	.	.	.	.	.	.	.	.	.	.	.	.	.	.	G	T	.	.	.	.	.	.	.	.	.	.	.	.	.	.	.	.	.	.	.	.	.	.	.	.	.	.	-	.	.	.	.	.	.	.	.	.	-	.	.	.	C	.	.	A	T	.	.	.	C	.	G	.	.	A	.	.	.	.	.	.	?	?
NWA05	.	.	.	.	.	.	.	.	.	.	.	.	.	.	.	.	.	.	.	T	.	.	.	.	.	.	.	.	.	.	.	T	.	.	.	.	.	.	.	.	.	.	.	.	.	.	-	.	.	.	.	.	.	.	.	.	-	.	.	.	.	.	.	A	T	C	.	.	C	.	G	.	.	A	.	.	.	.	.	.	.	.
NEA01	.	.	C	T	.	C	.	.	.	.	.	.	.	.	.	.	.	T	G	T	.	.	.	T	.	.	C	.	G	C	.	.	.	A	T	.	C	.	.	C	T	.	.	.	.	.	-	.	.	.	.	.	A	.	C	T	-	.	.	.	C	.	.	A	.	.	.	.	.	.	G	.	.	A	.	.	.	.	.	T	C	.
NEA02	.	.	.	T	.	.	.	.	.	.	G	A	.	.	.	.	T	.	.	T	.	.	.	T	.	.	.	.	.	.	.	T	.	A	.	G	C	.	.	.	.	.	.	C	.	A	A	.	.	.	.	.	A	.	.	.	T	.	.	.	C	.	.	A	.	.	.	.	.	.	G	.	.	A	.	.	.	.	.	.	.	.
NEA03	.	.	C	T	.	.	.	.	.	.	.	.	.	.	.	.	.	T	G	T	.	.	.	T	.	.	.	.	.	C	.	.	.	A	.	.	C	.	.	C	T	.	.	.	.	.	-	.	.	.	.	.	A	.	C	T	-	.	.	.	C	.	.	A	.	.	.	.	C	.	G	.	.	A	.	.	.	.	.	.	C	.
NEA04	.	.	.	.	.	.	.	.	.	.	.	.	.	.	.	.	.	.	G	T	.	.	.	.	.	.	.	.	.	.	.	.	.	.	.	.	.	.	.	.	.	.	.	.	.	.	-	.	.	.	.	.	.	.	.	.	-	.	.	.	C	.	.	A	T	.	.	.	C	.	G	.	.	A	.	.	.	.	.	.	.	.
NEA05	.	.	C	T	.	.	.	.	.	.	.	.	.	.	.	.	.	T	G	T	.	.	.	T	.	.	.	.	.	C	.	.	.	A	.	.	C	.	.	C	T	.	.	.	.	.	-	.	.	.	.	.	A	.	C	T	-	.	.	.	C	.	.	A	.	.	.	.	.	.	G	.	.	A	.	.	.	.	.	.	C	.
KUP01	.	T	.	T	.	.	.	.	.	.	.	.	T	.	.	.	.	.	.	T	.	.	C	.	.	.	.	.	.	.	.	T	.	A	.	G	C	.	.	.	.	.	.	.	.	.	A	.	.	.	.	.	A	.	.	.	T	.	C	.	G	.	.	A	.	C	C	C	.	.	G	.	.	A	.	.	.	.	.	.	.	.
KUP02	.	T	.	T	.	.	.	.	.	.	.	.	.	.	.	.	.	.	.	T	.	.	C	.	.	.	.	.	.	C	.	T	.	A	.	G	C	.	.	.	.	.	.	.	.	.	A	.	.	.	.	.	A	.	.	.	T	.	C	.	G	.	.	A	.	C	C	C	.	.	G	.	.	A	.	.	.	.	.	.	.	.
KUP03	.	T	.	T	.	.	.	.	.	.	.	.	.	.	.	.	.	.	.	T	.	.	.	.	.	.	.	.	.	.	.	T	.	A	.	G	C	.	.	.	.	.	.	.	.	.	A	.	.	.	.	.	A	.	.	.	T	.	C	.	G	.	.	A	.	C	C	C	.	.	G	.	.	A	.	.	.	.	.	.	.	.
KUP04	.	T	.	T	.	.	.	.	.	.	.	.	.	.	.	.	.	.	.	T	.	.	C	.	.	.	.	.	.	.	.	T	.	A	.	G	C	.	.	.	A	.	.	.	.	.	A	.	.	.	.	.	A	.	.	.	T	.	C	.	G	.	.	A	.	C	C	C	.	.	G	.	.	A	.	.	.	.	.	.	.	.
KUP05	.	T	.	T	.	.	.	.	.	.	.	.	.	.	.	.	.	.	.	T	.	.	C	.	.	.	.	.	.	.	.	T	.	A	.	G	C	.	.	.	.	.	.	.	.	.	A	.	.	.	.	.	A	.	.	.	T	.	C	.	G	.	.	A	.	C	C	C	.	.	G	.	.	A	.	.	.	.	.	.	.	.
NOA01	.	T	.	T	.	.	.	.	.	.	.	.	.	.	.	.	T	.	.	T	.	.	.	.	.	.	.	.	.	.	.	T	.	A	.	G	C	.	.	.	.	.	G	.	.	.	A	.	.	.	.	.	A	.	.	.	T	.	C	.	C	.	.	A	.	C	.	C	.	.	G	.	.	A	.	.	.	.	.	.	.	.
NOA02	.	.	.	T	.	.	.	.	.	.	.	.	.	.	.	.	.	.	.	T	.	.	.	.	.	.	.	.	.	.	.	T	.	A	.	G	C	.	.	.	.	.	.	C	.	A	A	.	.	.	T	.	A	.	.	.	T	.	.	C	C	.	.	A	.	C	.	.	.	.	G	.	.	A	.	.	.	.	.	T	.	.
NOA03	.	T	.	T	.	.	.	.	.	.	.	.	.	.	.	.	.	.	.	T	.	.	.	.	.	.	.	.	.	.	.	T	.	A	.	G	C	.	.	.	.	.	.	.	.	.	A	.	.	.	.	.	A	.	.	.	T	.	C	.	C	.	.	A	.	C	.	C	.	.	G	.	.	A	.	.	.	.	.	.	.	.
NOA04	.	.	.	T	.	.	.	.	.	.	.	A	T	.	.	.	.	.	.	T	.	.	.	.	.	.	.	.	.	.	.	T	.	A	.	.	.	.	.	.	.	.	.	.	.	.	G	.	T	.	T	.	A	.	.	.	T	C	.	C	C	.	.	A	.	C	.	.	.	.	G	.	.	A	.	.	.	.	.	.	.	.
NOA05	.	.	.	T	.	.	.	.	.	.	.	.	.	.	.	.	.	.	.	T	.	.	.	.	.	.	.	.	.	.	.	T	.	A	.	G	C	.	.	.	.	.	.	C	.	A	A	.	.	.	T	.	A	.	.	.	T	.	.	C	C	.	.	A	.	C	.	.	.	.	G	.	.	A	.	.	.	.	C	T	.	.
PHI01	T	.	.	T	.	C	.	.	C	.	.	.	.	.	.	.	T	.	.	T	.	C	.	.	.	.	.	.	G	.	.	T	.	A	.	.	C	A	.	.	-	T	.	.	.	.	A	A	T	G	.	C	A	C	.	-	A	C	C	.	.	A	.	A	.	A	.	.	.	.	G	.	.	A	C	.	.	.	.	.	.	.
PHI02	T	.	C	.	T	.	.	.	C	.	.	.	.	.	.	.	T	.	.	T	.	C	.	T	.	.	.	.	G	.	A	T	.	A	T	.	.	A	.	.	-	T	.	.	.	.	A	A	T	.	T	.	A	.	.	-	A	C	.	C	.	A	.	A	.	A	.	C	.	.	.	.	.	A	.	.	.	.	.	T	.	.
PHI03	.	.	.	T	.	.	.	.	.	.	.	.	.	.	.	.	T	.	.	T	.	.	.	.	.	.	.	.	.	.	.	T	.	A	.	G	C	.	.	.	.	.	.	C	.	A	A	.	.	.	.	.	A	.	.	.	C	.	.	.	C	.	.	A	.	C	.	.	.	.	G	.	.	A	.	.	.	.	.	.	.	.
PHI04	T	.	.	T	.	.	.	.	.	.	.	.	.	.	.	.	.	.	.	T	.	.	.	.	.	.	.	.	.	.	.	T	.	A	.	G	C	.	.	.	.	.	.	C	.	A	A	.	.	.	.	.	A	.	.	.	T	.	.	C	C	.	.	A	.	C	.	.	.	.	G	.	.	A	.	.	.	.	.	T	.	.
PHI05	T	.	C	.	T	.	.	-	?	.	.	.	.	.	.	.	.	.	.	T	.	C	.	.	.	.	.	.	G	.	A	T	.	A	T	.	.	A	.	.	-	T	.	.	.	.	G	A	T	.	T	.	A	C	.	-	A	C	C	C	C	A	.	A	.	A	.	C	.	.	.	.	.	A	.	.	.	.	.	T	.	A
CHI01	T	.	C	.	T	.	.	.	C	.	.	.	.	.	.	.	.	.	.	T	.	C	.	.	.	.	.	.	G	.	A	T	.	A	T	.	.	A	C	.	-	T	.	.	.	.	G	A	T	.	T	.	A	C	.	-	A	C	C	C	.	A	.	A	.	A	.	C	.	.	.	.	.	A	.	.	.	.	.	T	.	A
CHI02	T	.	.	.	T	.	.	.	C	.	.	.	.	.	.	.	T	.	.	T	.	C	.	T	.	.	.	.	G	.	A	T	.	A	T	.	.	A	.	.	-	T	.	.	.	.	G	A	T	.	T	.	A	C	.	-	A	C	.	C	.	A	.	A	.	A	.	C	.	.	.	.	.	A	.	.	.	.	.	T	.	.
THA01	T	.	.	.	T	.	.	.	C	.	.	.	.	.	.	.	T	.	.	T	.	C	.	T	.	.	.	.	G	.	A	T	.	A	T	.	.	A	.	.	-	T	.	.	.	.	G	A	T	.	T	.	A	C	.	-	A	C	.	C	.	A	.	A	.	A	.	C	.	.	.	.	.	A	.	.	.	.	.	T	.	.
THA02	T	.	.	.	T	.	.	-	?	.	.	.	.	.	.	.	T	.	.	T	.	C	.	T	.	.	.	.	G	.	A	T	.	A	T	.	.	A	.	.	-	T	.	.	.	.	G	A	T	.	T	.	A	C	.	-	A	C	.	C	.	A	.	A	.	A	.	C	.	.	.	.	.	A	.	.	.	.	.	T	.	.
BRU01	.	T	.	T	.	-	.	.	.	.	.	.	.	.	.	.	.	.	.	T	.	.	C	.	.	.	.	.	.	.	.	T	.	A	.	G	C	.	.	.	.	.	.	.	.	.	A	.	.	.	.	.	A	.	.	-	T	.	C	.	G	.	.	A	.	C	C	C	.	.	G	.	.	A	.	.	.	.	.	.	.	.
BRU02	T	.	.	T	.	-	.	.	.	.	.	.	.	.	.	.	.	.	.	T	.	.	.	.	.	.	.	.	.	.	.	T	.	A	.	G	C	.	.	.	.	.	.	C	.	A	A	.	.	.	T	.	A	.	.	.	T	.	.	C	C	.	.	A	.	C	.	.	.	.	G	.	.	A	.	.	.	.	.	T	.	.
BRU03	T	.	C	-	-	.	.	.	C	.	.	.	.	.	G	G	T	.	.	T	.	C	.	.	.	.	.	.	.	.	.	T	.	A	.	.	.	A	.	.	-	T	.	.	.	.	A	A	T	G	T	.	A	.	.	-	A	C	.	C	.	A	.	A	.	A	.	C	.	.	.	.	.	A	.	C	.	.	.	T	.	.
BRU04	T	.	.	T	.	-	.	.	.	.	.	.	.	.	.	.	.	.	.	T	.	.	.	.	.	.	.	.	.	.	.	T	.	A	.	G	C	.	.	.	.	.	.	C	.	A	A	.	.	.	T	.	A	.	.	.	T	.	.	C	C	.	.	A	.	C	.	.	.	.	G	A	.	A	.	.	.	.	.	T	.	.
PEN01	T	.	C	-	.	.	.	.	C	.	.	.	.	.	.	.	T	.	.	T	.	C	.	.	G	.	.	T	G	.	.	T	.	A	T	.	.	.	.	.	-	T	G	.	.	.	G	A	T	.	.	.	A	.	.	-	A	C	C	C	.	A	.	A	.	A	.	C	.	.	.	.	.	A	.	.	.	.	.	T	.	.
PEN02	T	.	C	-	.	.	G	.	C	.	.	.	.	.	.	.	T	.	.	T	.	C	.	T	.	.	.	.	G	.	A	T	T	A	T	.	.	A	.	.	-	T	.	.	.	.	A	A	T	.	T	.	A	.	.	-	A	C	.	C	.	A	.	A	.	A	.	C	.	.	.	.	.	A	.	.	.	.	.	T	.	.
SAB01	?	T	.	T	.	.	.	.	.	.	.	.	.	.	.	.	.	.	.	T	.	.	C	.	.	.	.	.	.	.	.	T	.	A	.	G	C	.	.	.	.	.	.	.	.	.	A	.	.	.	.	.	A	C	.	.	T	.	C	.	-	A	.	A	.	C	C	C	.	.	G	.	.	A	.	.	.	.	.	.	.	.
SAB02	?	.	C	.	T	.	.	.	C	.	.	.	.	.	.	.	.	.	.	T	.	C	.	.	.	.	.	.	G	.	A	T	.	A	T	.	.	A	.	.	-	T	.	.	.	.	G	A	T	.	T	.	A	C	.	-	A	C	C	C	.	A	.	A	.	A	.	C	.	.	.	.	.	A	.	.	.	.	.	T	.	A
SAB03	?	T	.	T	.	.	.	.	.	.	.	.	.	.	.	.	.	.	.	T	.	.	C	.	.	.	.	.	.	.	.	T	.	A	.	G	C	.	.	.	.	.	.	.	.	.	A	.	.	.	.	.	A	C	.	.	T	.	C	.	-	A	.	A	.	C	C	C	.	.	G	.	.	A	.	.	.	.	.	.	.	.
SAB04	?	.	C	.	T	.	.	.	C	.	.	.	.	.	.	.	.	.	.	T	.	C	.	.	.	.	.	.	G	.	A	T	.	A	T	.	.	A	.	.	-	T	.	.	.	.	G	A	T	.	T	.	A	C	.	-	A	C	C	C	.	A	.	A	.	A	.	C	.	.	.	.	.	A	.	.	.	.	.	T	.	A
SAB05	?	.	.	T	T	.	.	.	C	G	.	A	.	G	.	.	T	.	.	T	C	C	.	T	.	G	.	.	G	.	A	T	.	A	T	.	.	A	.	.	-	T	.	.	.	.	A	A	T	G	T	.	A	.	.	-	A	C	.	C	.	A	.	A	.	A	.	C	.	.	.	.	.	A	.	.	.	.	.	T	.	.
SAR01	?	.	.	-	.	.	.	.	C	.	.	.	.	.	.	.	T	.	.	T	.	C	.	T	.	.	.	.	G	.	A	T	.	A	T	.	.	A	.	.	-	T	.	.	.	.	G	A	T	.	T	.	A	C	.	-	A	C	.	C	.	A	.	A	.	A	.	C	.	.	.	.	.	A	.	.	.	.	.	T	.	.
SAR02	?	.	.	-	.	.	.	.	C	.	.	.	.	.	.	.	T	.	.	T	.	C	.	.	?	.	.	.	G	.	A	T	.	A	T	.	.	A	.	.	-	T	.	.	C	.	A	A	T	.	T	.	A	C	.	-	A	.	.	C	.	A	.	A	.	A	.	C	.	.	.	.	.	A	.	.	.	.	.	T	.	.
SAR03	?	.	C	-	.	.	.	.	C	.	.	.	.	.	.	.	T	.	.	T	.	C	.	.	.	.	.	T	G	.	.	T	.	A	T	.	.	.	.	.	-	T	.	.	.	.	G	A	T	.	.	.	A	.	.	-	A	C	C	C	.	A	.	A	.	A	.	C	.	.	.	.	.	A	.	.	.	.	.	T	.	.
SIN01	?	.	C	.	T	.	.	.	C	.	.	.	.	.	.	.	T	.	.	T	.	C	.	T	.	.	.	.	G	.	A	T	.	A	T	.	.	A	.	.	-	T	.	.	.	.	A	A	T	.	T	.	A	.	.	-	A	C	.	C	.	A	.	A	.	A	.	C	.	.	.	.	.	A	.	.	.	.	.	T	.	.
SIN02	?	.	.	-	.	.	.	.	C	.	.	.	.	.	.	.	T	.	.	T	.	C	.	.	.	.	.	.	G	.	A	T	.	A	T	.	.	A	.	.	-	T	.	.	.	.	G	A	T	.	T	.	A	C	.	-	A	C	.	C	.	A	.	A	.	A	.	C	.	.	.	.	.	A	.	.	.	.	.	T	.	.
SIN03	?	.	C	.	T	.	.	.	C	.	.	.	.	.	.	.	T	.	.	T	.	C	.	T	.	.	.	.	G	.	A	T	.	A	T	.	.	A	.	.	-	T	.	.	.	.	A	A	T	.	T	.	A	.	.	-	A	C	.	C	.	A	.	A	.	A	.	C	.	C	.	.	A	A	.	.	A	C	.	T	.	.
SIN04	?	.	.	T	.	-	.	.	.	.	.	.	.	.	.	.	.	.	.	T	.	.	C	.	.	.	.	.	.	.	.	T	.	A	.	G	C	.	.	.	.	.	.	.	.	.	A	.	.	.	.	.	A	.	.	.	T	.	C	.	G	.	T	A	.	C	C	C	.	.	G	.	.	A	.	.	.	.	.	.	.	.

and Timorese sequences, whereas clade B consisted only of the Southeast Asian sequences that were collected to the west of Wallace's Line. The frequencies of clades A and B in each population are shown in Fig. 1. In the eight Southeast Asian populations, the frequency of clade B varied from 100% (CHI, THA, PEN and SAR) to 25% (BRU). The three Australian populations as well as the western Timorese were fixed for clade A sequences.

There were three well supported (greater than 81% NJ bootstrap support) sub-clades within clade A. Sub-clade NWA (Fig. 2) contained four of five sequences from NWA and one sequence from NEA. Sub-clade NEA contained three sequences from the Northeast Australia only. Sub-clade KUP contained all five sequences from western Timor as well as two sequences from NOA and four sequences collected to the north-west of Kupang (BRU, SAB, SAR and SIN). There was little sub-clade structure in clade B.

Haplotype networks (Fig. 3) supported the topology of the phylogenetic tree in Fig. 2. Clade A corresponded to two networks, which had NEA03 and KUP05 as ancestral haplotypes. The latter included samples collected from Southeast Asia (BRU, SAB, SIN and PHI) and samples from NOA and KUP. Southeast Asian samples appeared in a third network that had CHI02 as their putative ancestor. This network represented populations from CHI, THA, PEN and SAR and was identical in composition to clade B in Fig. 2. There were few unresolved or circular branches on the networks suggesting that the data was generally free from convergent evolution or evolutionary reversals. As with the NJ and ML phylogenetic analyses, sample PHI01 was more than 15 mutational steps away from the remaining samples and was not part of any network.

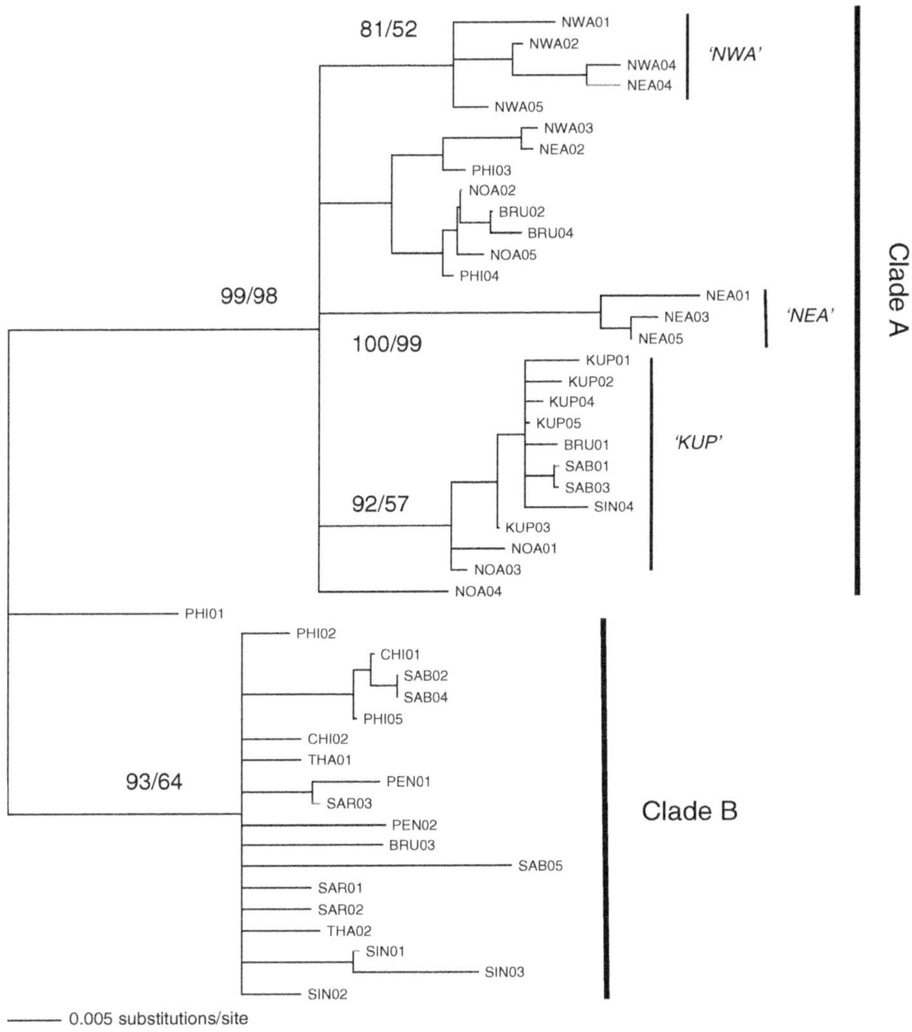

Fig. 2 A mid-point rooted phylogenetic tree (50% consensus) showing relationship between mtDNA control region sequences from Southeast Asian, Australian and western Timor Spanish mackerel. Nodal support is displayed as NJ/ML bootstrap probabilities. Haplotypes are labelled according to Table 1

Analysis of molecular variance (AMOVA)

Wallace's Line was strongly implicated as a biogeographical boundary for Spanish mackerel populations. An analysis of molecular variance (AMOVA) attributed 43% of the sequence diversity in the mtDNA control region to variance among populations collected to the west or east of Wallace's Line (Table 3). An alternative mechanism (Indian compared to Pacific Ocean origin) was not supported. Analysis of molecular variance failed to detect significant sequence diversity as a consequence of collection locations being adjacent to those Oceans (Table 3).

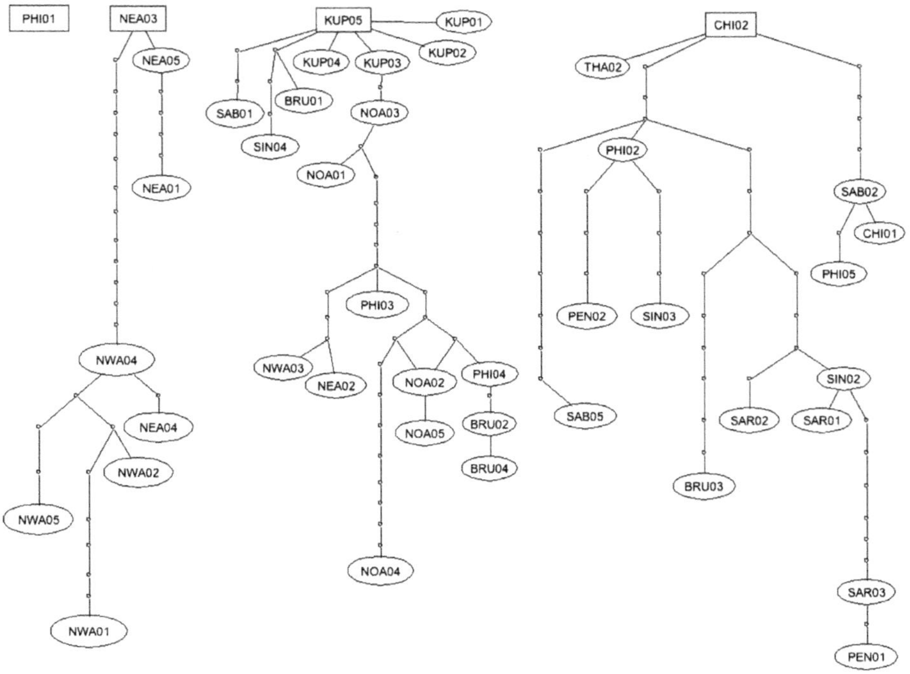

Fig. 3 Fifteen-step connection limit haplotype network for Spanish mackerel (*Scomberomorus commerson*) control region mtDNA sequences from Southeast Asia, western Timor and Australia (see Table 1 for labels). Each indicated step (*circle*) represents single nucleotide difference. The *smallest circles* represent inferred unsampled haplotypes. Sequences CHI02 and THA01, SAB01 and SAB03, SAB02 and SAB04, PHI02 and SIN01 were assumed to be identical as they differed only by missing or ambiguous characters. One of each pair is represented here. Sequences in boxes had the highest outgroup probability and were assumed to be ancestral. One sequence (PHI01) was more than fifteen mutational steps away from the remainder

Table 3 Results of AMOVA for two user-defined population groupings of Spanish mackerel (*Scomberomorus commerson*) in Southeast Asia and northern Australia

User-defined population structure consisting of two groups of populations	Fixation indices*		
	ϕ_{ST}	ϕ_{SC}	ϕ_{CT}
Separated by the Wallace Line	0.57	0.24	0.44
Adjacent to the Indian and Pacific oceans	0.49	0.44	ns

ns not significant

* ϕ-values were significant ($P < 0.01$)

Discussion

Division of Indo-Pacific populations of *S. commerson* along the Wallace's Line in the present study was strongly supported by AMOVA of mtDNA D-loop sequence data. Likewise, populations of co-distributed mantis shrimp (*Haptosquilla pulchella*) and seahorse (*Hippocampus trimaculatus*) were observed to follow east–west or north–south divisions along Wallace's Line (Barber et al. 2000; Lourie and Vincent 2004). However, the seahorse samples from the east of Wallace's Line, when analysed separately, were consistent with an Indian versus Pacific Ocean separation (Lourie and Vincent 2004). The genetic differentiation of starfish (*Acanthaster planci*) and giant clam (*Tridacna* species) was also influenced by the separation of Indian and Pacific Oceans at times of lowered sea level (Benzie 1998). Other studies on marine fishes also supported the present finding that genetic differences occur between Australia and Southeast Asia. Similar results have been obtained from Australasian goldband snapper (*Pristipomoides multidens*; Ovenden et al. 2002, 2004) and red snapper (*Lutjanus erythropterus* and *L. malabaricus*; Elliott 1996; Salini et al. 2006). The role of Wallace's Line in delimiting *S. commerson* populations needs to be further tested by an analysis of mtDNA haplotypes of fish sampled from the islands of Sulawesi and Flores, as populations immediately south and west of Wallace's Line were not sampled in this study. The results presented here would predict that Clade A haplotypes would be more frequent in Sulawesi and Flores locations. Likewise, further sampling and mtDNA analyses are necessary to confirm the temporal stability of the observed pattern. The study assumed that patterns of population subdivision in a relatively long-lived fish such as *S. commerson* (10–15 years; McPherson 1992) would be stable through time.

The mechanism driving genetic subdivision below the species level in the marine environment along Wallace's Line is unknown, but is likely to include past sea level fluctuations. Geographic isolation at times of low-sea level stands during the Pleistocene may have restricted gene flow between populations causing population differentiation in Southeast Asian and Australasian *S. commerson*. At 120 m Below Present Level (BPL), the bulk of the Sunda and Sahul shelves were largely exposed and formed massive lowland connections; Sumatra, Java and Borneo were connected by the exposed Sunda Shelf (Voris 2000). The arrangement remained very similar at 100 and 75 m BPL. There were six events (35%) associated with low-sea level (75–120 m BPL) during the past 0.25 million years (Chappel and Shackleton 1986).

As narrow-barred Spanish mackerel sampled from Southeast Asia had both clade A and B haplotypes, whereas fish to the east of Wallace's Line in Australia were characterized by clade A only, it is tempting to speculate that Australian mackerel may be descended from clade A mackerel from Southeast Asia that moved into possibly vacant Australian habitat in response to sea level fluctuations. Ovenden and Street (2007) proposed movement from the west to the east between Kupang and northwest Australian *S. commerson* populations. Reverse movement was suspected not to occur as the nucleotide diversity of 434 bp of the mtDNA control region for the Kupang population (0.5%) was significantly lower in the Australian populations (3–6%), which would not be consistent with the Kupang population receiving immigrants from the more diverse Australian populations (Ovenden and Street 2007). A study using parasites as indicators of *S. commerson* movement between Australia and Kupang arrived at the same conclusion (Lester et al. 2001). However, the hypothesis of unidirectional colonization cannot be tested with this data-set but may represent an interesting area to investigate in the future with larger numbers of samples and more comprehensive genetic markers.

Nucleotide diversity, although variable, was especially high (>10%) in a few of the Southeast Asian *S. commerson* populations to the west of Wallace's Line. Lourie and Vincent (2004) observed high haplotype but low nucleotide diversity to the west of Wallace's Line in sea-horses. The nucleotide diversity of *S. commerson* sampled from the Gulf of Oman, Arabian Gulf and Arabian Sea was between 0.035 and 0.042 (Hoolihan et al. 2006) and from Australia was 0.005 to 0.06 (Ovenden and Street 2007). Haplotype diversity in this study approached 1.0, which was higher than previous studies reported for

S. commerson sampled from Arabian Peninsula area (0.6; Hoolihan et al. 2006) and Australia (0.8; Ovenden and Street 2007).

Finer phylogenetic relationships were suggested to exist in clade A, but not in clade B in this study. A previous study on *S. commerson* using allozymes supported the existence of a distinct stock from the east coast of Australia (Shaklee et al. 1990). Results obtained from mitochondrial DNA control region sequence of *S. commerson* (Ovenden and Street 2007) confirmed this and the east coast stock corresponds here to the subclade 'NEA' within clade A that contains sequences from Northeast Australia. The ancestors of this distinct east coast clade were thought to originate from Spanish mackerel that were geographically isolated on the eastern Australian coast during past intervals of low sea levels when a land bridge connected Cape York to southern Papua New Guinea (Jones and Torgensen 1988; Ovenden and Street 2007). Members of subclade 'NWA' within clade A may also have a shared evolutionary history, however their mixed geographic origin (NWA and NEA) make this less likely.

A single sample (PHI01) was an outlier to clade A and clade B. This fish, or its maternal ancestor, may represent an immigrant of the species into Southeast Asia. Such an observation may not be an isolated incidence as outliers were also reported in *S. commerson* population from Torres Strait (Ovenden and Street 2007). Obtaining more sequences from the control region and from other mtDNA gene regions may confirm the phylogenetic status of sample PHI01. GenBank was searched for sequences similar to PHI01, but no matches were found. It may be useful to repeat the search in the future as GenBank is being constantly updated. Comparison among the sequences of additional samples collected from the Philippines would also assist in clarifying this anomaly.

While this study was not designed to delineate fisheries stocks using genetic markers, some recommendations about the management of the shared narrow-barred Spanish mackerel resource in south-east Asia are appropriate. This study has inferred the presence of two *S. commerson* stocks within the Indo-Pacific: (1) a Southeast Asian stock to the west of Wallace's Line and (2) and Australasian stock to the east of the Line. An additional distinct stock of *S. commerson* has already been identified on the east coast of Australia (Shaklee et al. 1990; Ovenden and Street 2007); the latter study also contrasts results across genetic marker systems (maternally inherited mitochondrial DNA and biparentally inherited microsatellite loci). Overfishing in combination with low gene flow can deplete fisheries resources. We suggest that Southeast Asian nations need to consider cooperative management of *S. commerson* and endorse further, more detailed studies of stock structure in the Andaman and South China Seas and adjacent regions. Australia appears to be the custodian of the northern Australian stocks of *S. commerson* and should manage them accordingly in conjunction with Papua New Guinea and western Indonesian authorities.

Acknowledgments We thank Geoff McPherson and Charles Bryce for collecting Australian samples and Raewyn Street for lab assistance. Damien Broderick provided valuable assistance in the preparation of the manuscript. Collection of Australian data was partly funded by the Australian Fishery Development Corporation. The Southeast Asian Spanish mackerel project was funded by Universiti Brunei Darussalam through research grant number UBD/PNC2/2/RG/1(18).

References

Akaike V (1974) A new look at the statistical model identification. IEEE Trans Automat Contr 19:716–723

Allen G (2000) Marine fishes of Southeast Asia. Periplus Editions (HK) Limited, Singapore, p 291

Barber PH, Palumbi SR, Erdmann MV, Moosa MK (2000) A marine Wallace's Line? Nature 406:692–693

Barber PH, Palumbi SR, Erdmann MV, Moosa MK (2002) Sharp genetic breaks among populations of *Haptosquilla pulchella* (Stomatopoda) indicate limits to larval transport: patterns, causes, and consequences. Mol Ecol 11:659–674

Benzie JAH (1998) Genetic structure of marine organisms and SE Asian biogeography. In: Hall R, Holloway JD (eds) Biogeography and geological evolution of SE Asia. Backhuys Publishers, Leiden, pp 197–209

Chappel J, Shackleton NJ (1986) Oxygen isotopes and sea level. Nature 324:137–140

Clement M, Posanda D, Crandall KA (2000) TCS: a computer program to estimate gene genealogies. Mol Ecol 9:1657–1659

Collette BB, Russo JL (1984) Morphology, systematics and biology of the Spanish mackerel (*Scomberomorus,* Scombridae). Fish Bull 82:545–692

Elliott N (1996) Allozyme and mitochondrial DNA analysis of the tropical saddle-tail sea perch, *Lutjanus malabaricus* (Schneider), from Australian waters. Mar Freshw Res 47:869–876

Excoffier L, Smouse PE, Quattro JM (1992) Analysis of molecular variance inferred from metric distances among mtDNA haplotypes: application to human mitochondrial DNA restriction data. Genetics 131:479–491

Excoffier L, Laval G, Schneider S (2005) Arlequin ver.3.0: An integrated software package for population genetics data analysis. Evol Bioinform Online 1:47–50

FAO (2005) Review of the state of world marine fishery resources. FAO Fisheries Technical paper No. 457. Rome, Italy

Hall R (2002) Cenozoic geological and plate tectonic evolution of SE Asia and the SW Pacific: computer-based reconstructions and animations. J Asian Earth Sci 20:353–434

Hoolihan JP, Anandh P, van Herwerden L (2006) Mitochondrial DNA analyses of narrow-barred Spanish mackerel (*Scomberomorus commerson*) suggest a single genetic stock in the ROPME area (Arabian Gulf. Gulf of Oman and Arabian Sea). ICES J Mar Sci 63:1066–1074

Jones MR, Torgensen T (1988) Late quaternary evolution of Lake Carpentaria on the Australia-New Guinea continental shelf. Aust J Earth Sci 35:313–324

Lester RJG, Thompson C, Moss H, Barker SC (2001) Movement and stock structure of narrow-barred Spanish mackerel as indicated by parasites. J Fish Biol 59:833–842

Lourie SA, Vincent ACJ (2004) A marine fish follows Wallace's Line: phylogeography of the three-spot seahorse (*Hippocampus trimaculatus*, Syngathidae, Teleostei) in Southeast Asia. J Biogeogr 31:1975–1985

Maddison WP, Maddison DR (1992) MacClade: analysis of phylogeny and character evolution, 3.0 edn. Sinauer Associates, Sunderland, MA

McPherson GR (1992) Age and growth of narrow-barred Spanish mackerel (*Scomberomorus commerson* Lacepede, 1800) in north-eastern Queensland waters. Aust J Mar Freshwat Res 43:1269–1282

Ovenden JR, Street R (2007) Genetic population structure of Spanish mackerel. In: Buckworth RC, Newman SJ, Ovenden JR, Lester RJG, McPherson GR (eds) The stock structure of Northern and Western Australian Spanish mackerel. Fishery Report 88 (FDRC98/159). Northern Territory Government, Australia, pp 93–143

Ovenden JR, Lloyd J, Newman SJ, Keenan CP, Slater LS (2002) Spatial genetic subdivision between northern Australian and Southeast Asian populations of *Pristipomoides multidens*: a tropical marine reef fish species. Fish Res 59:57–69

Ovenden JR, Lloyd J, Newman SJ, Keenan CP, Slater LS (2004) Pronounced genetic population structure in a potentially vagile fish species (*Pristipomoides multidens*, Teleostei; Perciformes; Lutjanidae) from East Indies triangle. Mol Ecol 13:1991–1999

Posada D, Crandall KA (1998) MODELTEST: testing the model of DNA substitution. Bioinformatics 14:817–818

Randall JE (1998) Zoogeography of shore fishes of the Indo-Pacific region. Zool Stud 37:227–268

Salini JP, Ovenden JR, Street R, Pendrey R, Ngurah H (2006) Genetic population structure of red snappers (*Lutjanus malabaricus* Bloch & Schneider, 1801 and *Lutjanus erythropterus* Bloch, 1790) in central and eastern Indonesia and northern Australia. J Fish Biol 68(Supplement B):217–234

Shaklee JB, Currens KP (2003) Genetic stock identification and risk assessment. In: Hallerman EM (ed) Population genetics-principles and applications for fisheries scientists. American Fisheries Society, Bethesda, pp 291–328

Shaklee JB, Phelps SR, Salini J (1990) Analysis of fish stock structure and mixed-stock fisheries by the electrophoretic characterization of allelic isozymes. In: Whitmore DH (ed) Application of electrophoretic and isoelectric focusing techniques in fisheries management. CRC Press, Boca Raton, pp 173–196

Springer VG, Williams JT (1990) Widely distributed Pacific plate endemics and lowered sea-level. Bull Mar Sci 47:631–640

Swofford DL (1990) PAUP*: Phylogenetic Analysis Using Parsimony, v 4.0, Sinauer Associates. Champagn, Illinois

Tamura K, Nei M (1993) Estimation of the number of nucleotide substitutions in the control region of mitochondrial DNA in humans and chimpanzees. Mol Biol Evol 10:512–526

Templeton AR, Crandall KA, Sing CF (1992) A cladistic analysis of phenotypic associations with haplotypes inferred from restriction endonuclease mapping and DNA sequence data. III. Cladogram estimation. Genetics 132:619–633

van Herwerden L, Mcllwain J, Al-Oufi H, Al-Amry W, Reyes A (2006) Development and application of microsatellite markers for *Scomberomorus commerson* (Perciformes; Teleostei) to a population genetic study of Arabian Peninsula stocks. Fish Res 79:258–266

Van Oosterzee P (1997) Where worlds collide: the Wallace Line. Cornell University Press, New York

Voris HK (2000) Maps of Plesitocene sea-levels in South East Asia: shorelines, river systems, time durations. J Biogeogr 27:1153–1167

Biodivers Conserv (2010) 19:575–586
DOI 10.1007/s10531-009-9744-x

ORIGINAL PAPER

Case studies on decapod crustaceans from the Philippines reveal deep, steep underwater slopes as prime habitats for 'rare' species

Jose Christopher E. Mendoza · Tohru Naruse · Swee-Hee Tan · Tin-Yam Chan · Bertrand Richer de Forges · Peter K. L. Ng

Received: 21 July 2009 / Accepted: 27 October 2009 / Published online: 15 November 2009

Abstract Relatively few studies have been done to define or assess rarity in the marine environment. Published studies have focused on shallow-water and intertidal habitats, and the available information appears to reflect the same pattern observed in terrestrial environments, i.e., that there are many rare species and few common species in any one given area. However, our studies of the abundance of new and/or supposedly rare taxa of decapod crustaceans from the deep, steep slopes of the island of Balicasag, in the central Philippines, have raised questions on how rarity should be defined in marine invertebrates. Examples of such supposedly rare species of crabs and lobsters (Crustacea: Decapoda) are presented here. That these animals come from deep, steep slopes, a relatively under-studied habitat, highlights the major gaps in current knowledge of marine biodiversity that are in part due to the inadequacy of both traditional and high technology sampling methodologies and the limited habitat types that the former can target. Low-technology, artisanal tangle nets have proved to be an optimal capture technique for deep-water decapod crustaceans on deep, steep slopes; many new taxa have been discovered and, in other cases, perceptions of rarity and endemicity have been corrected.

J. C. E. Mendoza · T. Naruse
Department of Biological Sciences, National University of Singapore,
14 Science Drive 4, Singapore 117543, Singapore

S.-H. Tan · P. K. L. Ng
Raffles Museum of Biodiversity Research, National University of Singapore,
6 Science Drive 2, Level 3, Singapore 117546, Singapore

T.-Y. Chan
Institute of Marine Biology, National Taiwan Ocean University, Keelung 202, Taiwan, ROC

B. Richer de Forges
Muséum national d'Histoire naturelle, Paris (UMR SAE),
5, rue Félix Franchette, 98800 Nouméa, Nouvelle-Calédonie

P. K. L. Ng (✉)
Tropical Marine Science Institute, National University of Singapore,
14 Kent Ridge Road S, Singapore 119223, Singapore
e-mail: peterng@nus.edu.sg

Keywords Crustacea · Decapoda · Balicasag · Rarity · Marine biodiversity · Deep · Steep slopes

Introduction

Most studies on species rarity have been conducted in the terrestrial environment, for which there have been many reviews (e.g. Gaston 1994). Rarity can occur at several levels. Rare species may be considered as such due to restrictions in geographical distribution (limited range), habitat distribution (few suitable habitats), and/or population size (genuinely low numbers) (Rabinowitz 1981; Pitman et al. 1999; Ricklefs 2000). However, rarity may also be erroneously attributed due to sampling limitations. For instance, some 'rare' species may actually be geographical edge species, which are rare in the study area but common outside it, or methodological edge species, which are rare because they are not readily sampled by the method used (Longino et al. 2002). They may also be 'tourists' or transient, non-resident species (Novotny and Basset 2000; Longino et al. 2002; Magurran and Henderson 2003).

In the marine environment, however, very few studies have been done to define and/or understand rarity. The general lack of quantitative data on various aspects of marine invertebrate diversity as well as their relative abundance has impeded the proper assessment of 'rare' marine species (Chapman 1999). Molluscs have probably been more systematically studied than most other marine invertebrate groups (Chapman and Underwood 1996; Bouchet et al. 2002; Benkendorff and Przeslawski 2008) and these reports have shown that rare species (uniques and duplicates) probably do exist, at least in the intertidal and shallow (diving depth) subtidal zones, and that a greater proportion of species assemblages consists of rare species. Soft-sediment benthic communities have also been studied in the same manner, with similar observations being made (Grassle and Maciolek 1992; Gray 1994, 2002; Schlacher et al. 1998; Ellingsen et al. 2007).

Deep, steep slopes are near-vertical underwater walls, typically with a hard substratum, and rising up from depths greater than 100 m. They are an example of what has been popularly labeled as "neglected habitats" (Dennis and Aldhous 2004; Ng 2006), so called because they have been largely untouched by biodiversity studies due to either of two reasons: their seemingly depauperate state (coral rubble beds, caves and peat swamps) or their relative inaccessibility to conventional sampling methods (deep, steep slopes). Deep, steep slopes, in particular, cannot be effectively sampled by ship-deployed trawls or dredges, which are normally effective only on flat, horizontal or gently-sloping substrates of less than 20° incline (see Andrae 2002; Grehan 2003) and/or relatively soft or less rocky surfaces.

Furthermore, although scuba divers can do some intensive sampling on the upper reaches of the slope, this method is only effective above the maximum diving depth ($\sim$60–120 m), leaving the deeper portions of the slope largely unexplored. This is even with recent advances in deep-diving technology that allow scuba divers to descend to much deeper habitats (see Pyle 2000).

In this paper we review some of the discoveries made of the many interesting decapod crustaceans collected by an artisanal fishing method (tangle net fishing) currently being used in Balicasag (see Ng et al. 2009). Some of these crustaceans have been recently described from material from Balicasag whereas the rest have been known from elsewhere but from a limited number ($\sim$10) of individuals.

Methods

Study site: Balicasag is a small, remote, coralline island in the Bohol Sea, about 8 km southwest of the island of Panglao and the larger island of Bohol in the central Philippines. Balicasag is owned by the Philippine Coast Guard, which maintains a lighthouse there. The island is about 700 m in diameter (<30 ha in surface area), and is surrounded by a narrow coral shelf and steep drop-offs (down to about 750 m) particularly near the northwestern coast (Christie et al. 2002; Ng et al. 2009). The quality of the island's shallow marine resources declined in the 1970s and 1980s due to destructive fishing practices. In 1985, an 8-ha, no-take marine sanctuary was established along the southwestern coast as part of a community project implemented by Silliman University and financed by the United States Agency for International Development. Since 1990, the management of this sanctuary has been the responsibility of the Philippine Tourism Authority, which has also built a dive resort there (Christie et al. 2002). A small community of approximately 800 residents (Christie et al. 2002) depends heavily on the reef's resources for food as well as for cottage industries such as shell-craft and the souvenir trade.

Sampling: Tangle net fishing is an artisanal method practiced in the island of Balicasag and other nearby islands in the Bohol Sea. The method targets gastropods, whose processed shells fuel a lucrative souvenir and collectors' trade on Balicasag and nearby islands, although other marine organisms—including crustaceans—are also captured. A detailed description of this method is provided by Ng et al. (2009). All the specimens studied in this report were collected by the fishermen of Balicasag between December 2000 and December 2004. Specimens were typically preserved by the fishermen in 10% formalin, and were later transferred into 70% ethanol. Large specimens were dried after soaking in 10% formalin or 70% alcohol by the fishermen, presumably in preparation for sale as souvenirs to visiting tourists. These specimens are now deposited in the crustacean reference collections of the Philippine National Museum, Manila (NMCR); Muséum national d'Histoire naturelle, Paris (MNHN); National Taiwan Ocean University, Keelung (NTOU); University of San Carlos, Cebu (USC); and Raffles Museum of Biodiversity Research, National University of Singapore (RMBR).

Case studies: results

To demonstrate our point that tropical deep, steep slopes are actually species-rich habitats that have not been well-studied by past sampling methods, we have put together a number of case studies of decapod crustaceans (crabs and lobsters) based on work the authors and their colleagues have done on Balicasag. These case studies highlight examples of taxa which were: (1) previously considered rare prior to their discovery in Balicasag (cases 1–5, 7–13, 15); (2) previously considered endemic to a certain region prior to their discovery in Balicasag (case 14); or (3) simply new and never before seen, having been collected solely from the deep, steep slopes of Balicasag and solely by tangle nets (case 6) (Fig. 1).

Case 1. *Acanthodromia margarita* (Alcock) (family Dynomenidae Ortmann) (Fig. 2b). Prior to the Balicasag tangle netting operations, this species was only known from three specimens from the Andaman Sea and Japan, all from areas which have been relatively well-sampled and well-studied. It was a well known 'rare' species. By 2004, 19 specimens had been collected from Balicasag alone over a period of three years (McLay and Ng 2005).

Case 2. *Paradynomene* Sakai (family Dynomenidae Ortmann). This genus was established to accommodate one unusual species, *Paradynomene tuberculata* Sakai (Fig. 2a),

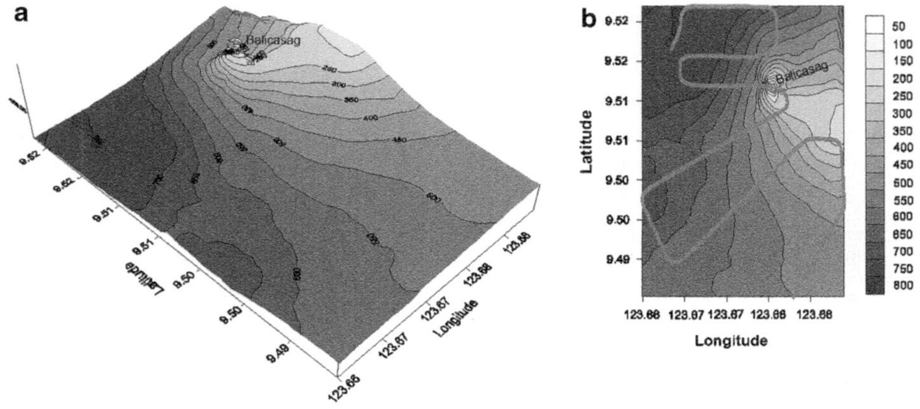

Fig. 1 Bathymetric map of area surrounding Balicasag (modified from Richer de Forges et al. 2009)—**a** 3D map, southwest view; **b** 2D map—showing the location of the deep, steep slope on the western side. Red line shows course followed by research vessel, MV DA-BFAR, while taking soundings off the seafloor. Adjacent contour lines have a depth difference of 50 m

which was, for many years, represented by only one male and one female specimen from Japan. McLay (1999) subsequently assembled the known material of *P. tuberculata* in his revision of the family Dynomenidae, and could only list 18 specimens from around the Indo-West Pacific, and continued to recognize the genus as monotypic. Later, from Balicasag, a dozen individuals of this supposedly rare genus were collected, and this allowed its taxonomy to be resolved. As it turned out, there were at least two distinct species of *Paradynomene* in the Balicasag collection. This allowed McLay and Ng (2004) to recognize one new species from Balicasag (*P. quasimodo*, with three specimens, Fig. 2c) as well as four other new species across the Indo-West Pacific, from all but two of the 18 specimens earlier listed by McLay (1999). Since then, more specimens of both species (*P. quasimodo* and *P. tuberculata*) have been collected in Balicasag.

Case 3. Family Homolidae De Haan. The surveys around the island of Balicasag have yielded several important new records for the Philippines, as well as two new species: *Yaldwynopsis saguili* and *Latreillopsis mariveneae* (Richer de Forges and Ng 2007a). One of the new records, *Homolomannia occlusa* Guinot and Richer de Forges (Fig. 2e) merits further discussion. This species was known previously from only three specimens; two from Madagascar and one from Taiwan. The catch from Balicasag yielded over 20 specimens (Richer de Forges and Ng 2007b). The family Homolidae is represented by 14 genera, 56 species in the Indo-West Pacific (Ng et al. 2008b). Surprisingly, seven genera and 11 species are found in Balicasag, which represents the highest concentration of homolids known to date in such a small area (cf. Richer de Forges and Ng 2007b).

Case 4. *Patulambrus petalophorus* (Alcock) and *Rhinolambrus cybelis* (Alcock) (family Parthenopidae MacLeay). *P. petalophorus* (Fig. 2g) was previously known from only three specimens, but 17 more specimens were subsequently collected from Balicasag alone by tangle netting (Tan 2004). The situation for *R. cybelis* (Fig. 2f) is even more dramatic. This species was previously known from only six specimens obtained from various parts of the Indo-West Pacific. From Balicasag, over 500 specimens were collected by tangle nets, with many more which were not preserved and were discarded by the fishermen (Tan 2004).

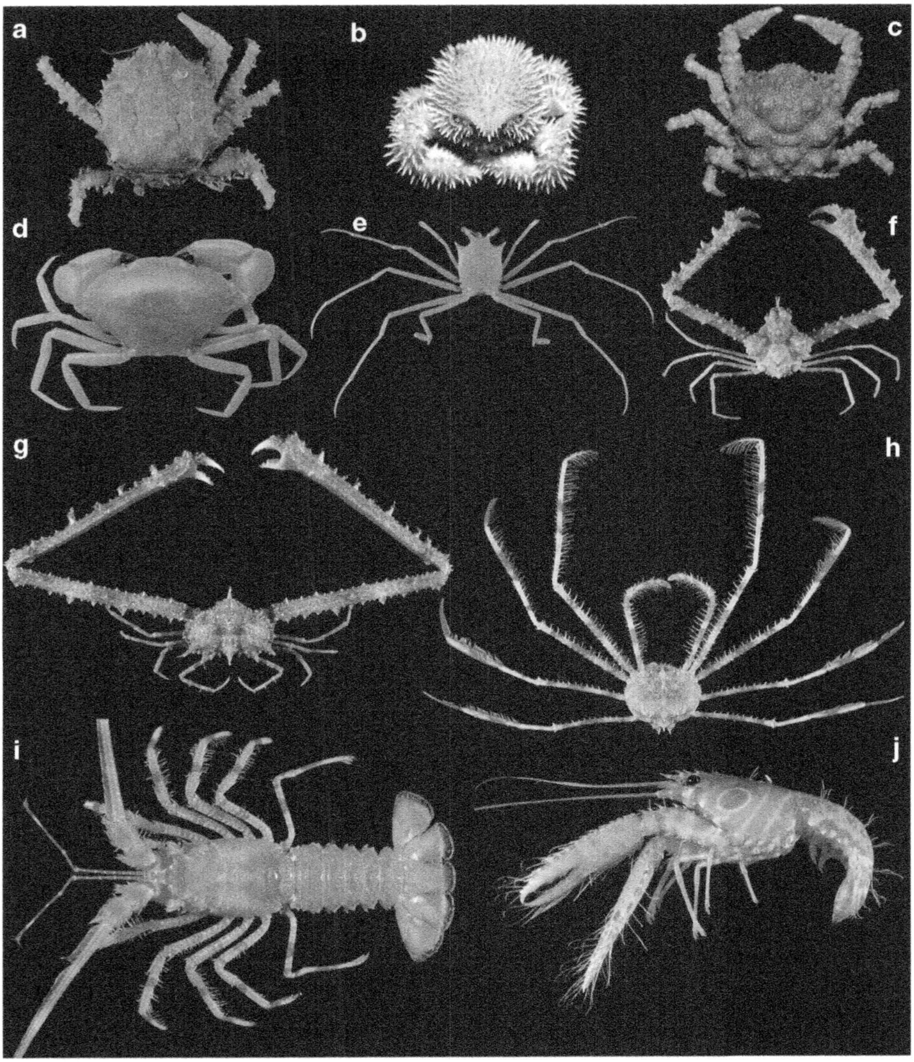

Fig. 2 Examples of decapod crustaceans from Balicasag, live colouration: **a** *Paradynomene tuberculata* Sakai; **b** *Acanthodromia margarita* (Alcock); **c** *Paradynomene quasimodo* McLay and Ng; **d** *Euryozius camachoi* Ng and Liao; **e** *Homolomannia occlusa* Guinot and Richer de Forges; **f** *Rhinolambrus cybelis* (Alcock); **g** *Patulambrus petalophorus* (Alcock); **h** *Cyrtomaia murrayi* Miers; **i** *Palinustus waguensis* Kubo; **j** *Enoplometopus crosnieri* Chan and Yu

Case 5. *Cyrtomaia* Miers (family Inachidae MacLeay). *Cyrtomaia murrayi* Miers (Fig. 2h) and *C. horrida* Rathbun are spider crab species which were considered relatively rare because only a few specimens could be obtained by trawling during past surveys, even though they have been known to science for over a century. Richer de Forges and Ng (2007a), however, reported large numbers of these species (*C. murrayi*, $n = 76$; *C. horrida*, $n = 23$) collected by deep-set tangle nets on the steep slopes of Balicasag. It is important to state that their numbers are based only on museum-preserved specimens and discounts the many dozens more of each species which were discarded (due to the broken condition of the crabs) or kept by the fishermen (larger specimens are dried and sold as souvenirs). As they were preferentially caught by tangle nets rather than trawls, Richer de Forges and Ng (2007a) hypothesized that these two species prefer microhabitats associated with deep, steep slopes.

Case 6. *Kasagia arbastoi* Richer de Forges and Ng (family Majidae Samouelle). *K. arbastoi* was described as a new genus and new species of spider crab by Richer de Forges and Ng (2007c) after examining three individuals collected by deep-set tangle nets on the steep slopes of Balicasag. Subsequently, seven more individuals have been found among previously unsorted lots which had been collected around the same time as the types. This spider crab species has not been collected elsewhere or by any other method thus far.

Case 7. *Ladomedaeus* Števčić (family Xanthidae MacLeay). The genus *Ladomedaeus* was established for a rare Japanese species, *Medaeus serratus* Sakai known only from 12 type specimens, all of which were collected from Sagami Bay (see Manuel-Santos and Ng 2007). In Balicasag, several hundred individuals of a related species, *L. fungillus* Manuel-Santos and Ng have been collected since 2000. These crabs were observed to be associated with glass sponges which were often collected by the tangle nets set on the western coast of the island. Trawls around the island and nearby Panglao also brought up large numbers of these crabs, together with the glass sponges. It may be possible that *L. serratus* is also associated with hexactinellid sponges, and the patchy distribution of these sponges could be the reason for the apparent rarity of *L. serratus*.

Case 8. *Euryozius* Miers (family Pseudoziidae Alcock). This rarely reported genus used to be represented by only three Atlantic species, *E. bouvieri* (A. Milne-Edwards), *E. pagalu* Manning & Holthuis, and *E. sanguineus* (Linnaeus), and two Indo-West Pacific species, *E. canorus* (Rathbun) and *E. danielae* Davie. The Indo-West Pacific species, *E. canorus* and *E. danielae*, were each known only from the single female holotype specimen, and no new material has been collected since they were described (see Ng and Liao 2002). In 2002, Ng and Liao described a new species, *Euryozius camachoi* (Fig. 2a), from five specimens collected by deep-set tangle nets in Balicasag. Subsequently, about a dozen such individuals have been collected (not all preserved) from the same vicinity, all by tangle nets.

Case 9. *Euclosia scitula* Galil (family Leucosiidae Samouelle). This very distinctive species was described from four specimens provided to Galil (2003) from tangle nets in Balicasag for her revision of the genus, and one collected from Thailand in the 1960s. In the collections from Balicasag are hundreds of specimens of this species. It has turned out to be one of the most common large leucosiid species from the deep steep slopes of the island.

Case 10. *Mursia* Desmarest (family Calappidae De Haan). *Mursia* is a relatively large genus, with 28 known species, of which 17 have been described only over the last 20 years (see Galil and Takeda 2004; Galil and Ng 2009). Most are deep-water species and are known on the basis of only few specimens. Galil and Takeda (2004) described four species (*M. baconaua*, *M. buwaya*, *M. diwata* and *M. mameleu*), most from good series of specimens from Balicasag, and all collected from tangle nets. All these species have since been supplemented by very extensive material from tangle-net collections made at Balicasag between 2000 and 2004.

Case 11. *Parathranites* Miers (family Portunidae Rafinesque). Crosnier (2002) revised the deep-water genus *Parathranites*, and described six new species from the Indo-West Pacific. Two of these, *P. granosus* and *P. tuberogranosus*, described from the Philippines, merit discussion. For *P. granosus*, Crosnier (2002) had 14 specimens from the Admiralty Islands (Papua New Guinea), Indonesia and the Philippines which had been collected by trawling or dredging from five separate cruises between 1884 and 1991. However, the tangle nets of Balicasag provided an additional 31 specimens for his revision. This one locality provided two-thirds of his sample size for this new species! For *P. tuberogranosus*, Crosnier (2002) only had three specimens from two cruises, one to the Philippines (1976) and another to Indonesia (1991). From the tangle nets of Balicasag, 28 specimens were obtained in the space of only three years (2000–2002). Since 2002, many more specimens of *P. granosus* and *P. tuberogranosus* have been collected from the same area by the same methods.

Case 12. *Mathildella rubra* Ng and Ho (family Mathildellidae Karasawa and Kato). The genus *Mathildella* Guinot and Richer de Forges is a small genus with only five known species, most of which have been described from a handful of specimens. One notable exception is the new species *M. rubra* Ng and Ho 2003, which was described from 45 specimens, all obtained by tangle nets on the deep cliffs off Balicasag. More specimens have been obtained since then by the same method. This species has not yet been obtained by trawling, dredging or other methods.

Case 13. The lobster genus *Palinustus* A. Milne-Edwards (family Palinuridae Latreille) is generally believed to be rare or uncommon (Chan and Yu 1995). *Palinustus unicornutus* Berry was previously known from only eight specimens from scattered localities in the Indo-West Pacific. Seven additional individuals have since been collected by tangle net from the deeper parts of the drop-offs in Balicasag. Another *Palinustus* species, *P. waguensis* Kubo (Fig. 2i), is also found in Balicasag. Although *P. waguensis* is the best known species of the genus, it is still regarded as uncommon, with only a few specimens collected each year from each locality. In Balicasag, however, it appears to be quite common, with 105 specimens collected thus far, and with much more material of this 'common' species not retained and preserved (authors' obs.). Fishermen eat the larger specimens while smaller ones are sold dried as souvenirs.

Case 14. *Epistocavea mururoa* Davie (family Xanthidae MacLeay). Davie (1993) established *Epistocavea*, with one species (*E. mururoa*) on the basis of 17 specimens collected by traps set in deep waters surrounding the islands and atolls of French Polynesia. This species has been considered as endemic to the Tuamotu Archipelago (see Poupin 1996). Subsequently, however, *E. mururoa* was collected in large numbers ($\sim$50) in Balicasag solely with the use of tangle nets (Mendoza and Ng in press), extending the range of the species by almost 10000 km across the Pacific Ocean. In Balicasag, as in French Polynesia (where there is no continental shelf), the depth and the underwater terrain had prevented the effective use of conventional methods such as scuba, trawls and dredges. Hence all specimens of this species were collected either by trap or by tangle net.

Case 15. *Enoplometopus crosnieri* Chan and Yu, and *E. macrodontus* Chan and Ng (family Enoplometopidae de Saint Laurent). *E. crosnieri* (Fig. 2j) was once regarded as one of the rarest species in the genus with only 14 specimens known from French Polynesia, Australia and Taiwan. In Balicasag, as well as in nearby Panglao Island, this species is frequently brought up by tangle nets such that there are now 17 preserved specimens of *E. crosnieri* from this area. One species, *E. macrodontus*, described as new from Balicasag (on the basis of nine specimens), has interestingly, also been found in the Indian Ocean (see Chan and Ng 2008).

Discussion

Deep, steep slopes, such as those found in Balicasag, are inaccessible to the primary methods used for sampling marine biodiversity today. The deeper profiles cannot be reached by scuba divers, whose maximum effective depth, even with the most advanced technology (see Pyle 2000), does not go much beyond 120 m. Furthermore, at the deeper limits of such dives, sampling time is very short, so intensive sampling is difficult to achieve. The steep angle of the slope and hard substrate make it impossible for ship-towed trawls and dredges to be used effectively. Traps have also been utilized, but with limited success, and in all cases, only work for larger species (see Richer de Forges and Ng 2007b). Remotely operated vehicles have limited capability to capture active or cryptic species and, like deep-sea submersibles, are expensive to build and operate and also require extensive logistical support (Pyle 2000). This failure of conventional methods to capture certain species in their preferred habitats may be the main reason for such species' apparent rarity, a clear example of a methodological edge species as defined by Longino et al. (2002). Clearly, the use of tangle nets can fill in the information gaps left by these conventional methods, as evidenced by the disparity in species captured (Fig. 3), as well as the number of new species discovered and the number of rare species being reassigned to relatively more common status (cf. Ng et al. 2009).

It has been discussed elsewhere (e.g. Bouchet 2006) that our knowledge of marine biodiversity is extremely poor. For brachyuran crabs, between the years 1991 and 2000, 810 new species were formally named; with about 60 new genera and species named every year since 2000 (Ng et al. 2008a, b). Excluding freshwater crabs, which represent about one-sixth of all known crab species (Cumberlidge and Ng 2009), researchers are still discovering, on average, 50 new marine brachyuran taxa every year globally (Ng et al. 2008a)! Figure 3 contrasts the species richness of brachyuran crabs collected from slopes

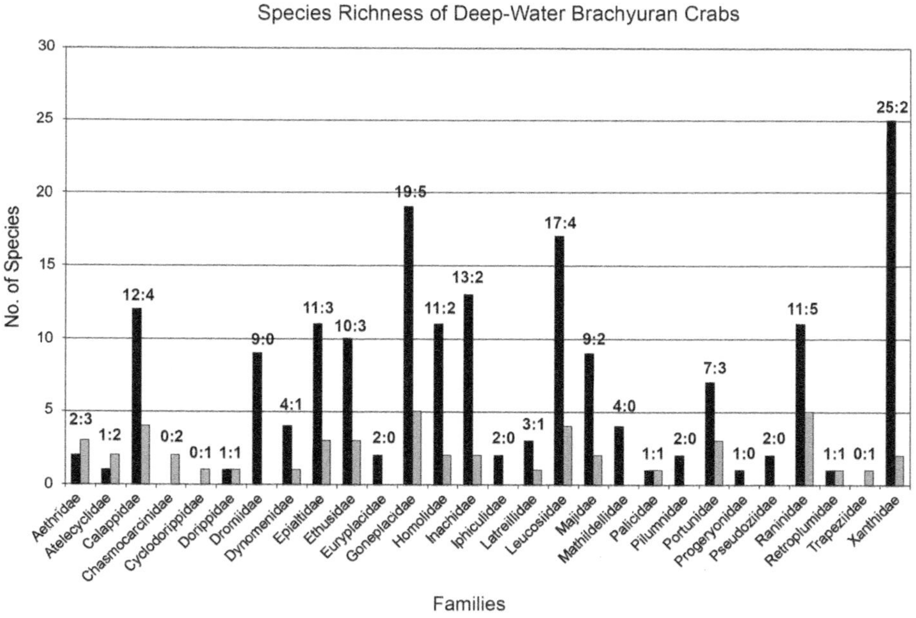

Fig. 3 Comparison of the brachyuran species richness in samples collected by tangle nets from the deep, steep slopes off Balicasag during the years 2000–2004 (in *black*), and samples collected by trawl and dredge from the relatively flatter areas around Balicasag during the PANGLAO 2004 and 2005 expeditions (in *gray*). Species richness values (tangle net: trawl and dredge) for each family are shown above each bar. Values were determined by direct examination and counting of specimens deposited at ZRC

of Balicasag versus the flatter bottoms surrounding the island. There clearly are more species and families represented in the slope samples than in the bottom samples. While the datasets are not completely comparable because it was not possible to standardize the sampling protocols, the pattern is nevertheless still very insightful. It is important to bear in mind that the slope datasets in Fig. 3 have all been obtained from a relatively short but almost continuous period between 2000 and 2004; whereas the bottom datasets are derived from the collections of the PANGLAO 2004 and 2005 expeditions. It is not possible to compare these datasets with those from the 19th and 20th century expeditions to the Philippines because all of them only used trawls and/or dredges, and none were focused on just one relatively small area. For the Panglao and Balicasag datasets, notable are the patterns seen in the Calappidae, Dromiidae, Dynomenidae, Epialtidae, Ethusidae, Goneplacidae, Homolidae, Inachidae, Iphiculidae, Latreilliidae, Leucosiidae, Mathildellidae, Portunidae, Raninidae and Xanthidae, in which slope species are more than double those known from flat substrates (Fig. 3). While the number of species of Xanthidae in Balicasag is notable for being high with some 25 species, considering the size of this family which has some 641 species (Ng et al. 2008a), this is to be expected. In any case, the slopes only had 0.04% of the world xanthid fauna. More interesting are the results for much smaller families like the Homolidae and Goneplacidae, with 11 out of 66 species, and 19 out of 72 species, respectively. What this translates to is that the small island of Balicasag has 17% of the world's fauna of Homolidae and 26% of the known species of Goneplacidae!

The present discoveries lead to several questions. Key is whether there are likely to be proportionately fewer rare species than currently acknowledged? And is this a general trend for marine invertebrate (or even vertebrate) diversity as a whole? Would increasing numbers of "rare species" be shifted to the "common" category; even as we discover new "rare ones"? The datasets from Balicasag certainly suggest this, although it is not really possible to draw generic conclusions based on just this series of studies and from one area. While several species mentioned in this paper have been erroneously considered as rare, we do not discount the possible occurrence of truly rare organisms in the deep, steep slopes of Balicasag, as the discovery of some new taxa (e.g., *Y. saguili* and *L. mariveneae*, see Richer de Forges and Ng 2007b) in relatively small numbers suggests.

Endemicity of both rare and locally common species in the marine environment has been shown to occur despite the openness and inter-connectivity of the different marine regions of the world (Richer de Forges et al. 2000; Roberts et al. 2002; Carpenter and Springer 2005). Nevertheless, it is rather premature to establish the endemicity and/or true rarity of marine animals, decapod crustaceans for example, without having exhaustively sampled all the possible habitats they could occupy. The above examples involving the crab *Epistocavea* and lobster *Enoplometopus* are cases in point (see Mendoza and Ng in press; Chan and Ng 2008). Both involve substantial range extensions of species as well as substantial increases in sample sizes. Subsequent studies on rarity must take these limitations of sampling into account in order for us to have a clearer picture of the marine community structure.

Regarding the use of tangle nets, we can now clearly advocate their controlled use as scientific tools—they allow scientists to have a valuable insight into a poorly known habitat. On the other hand, we do not support attempts to increase the commercial scale of their use beyond that seen in Balicasag at present. We are aware that similar fishing methods are practiced in other neighbouring islands (Siquijor and Pamilacan) but none are to the extent seen in Balicasag. The expansion of commercial tangle-netting will bring about the usual suite of environmental problems. In addition to overfishing and the sharp depletion of stocks, the damage to the habitat, lack of time for recovery and destruction of commercially unimportant species (bycatch); there is also the equally serious issue of "ghost-fishing". The latter terms refers to destruction caused by discarded and lost nets which will continue to kill and destroy all manner of marine life for many years. Thus far, there have been no specific studies to measure the impact of tangle net fishing on the health of the ecosystems in which they are deployed and until such studies are done we recommend a cautionary approach in their commercial use. While science has clearly benefited from this exercise, the negative environmental impacts cannot be over-emphasised. We see parallels in the commercial deep-sea trawling operations in northern Taiwan that have been going on for several decades. These operations have yielded large numbers of new and rare deep-sea crustacean and fish species from the 1990s to the present (see Chan et al. 2009), but clearly the habitats have also been impacted, with species diversity decreasing and stocks dropping.

Acknowledgements The study has been aided and facilitated by many colleagues and friends; notably the fishermen of Balicasag, especially Jose Arbasto; Lawrence Liao and Danilo Largo (USC), Father Florante Camacho and his staff (Holy Name University in Bohol); Ludivina Labe (National Fisheries Research and Development Institute, NFRDI); and Marivene Manuel-Santos (NMCR). The PANGLAO 2004 and 2005 Expeditions were organized by Philippe Bouchet (MNHN) with USC, NFRDI, the Bureau of Fisheries and Aquatic Resources, the National University of Singapore, and the Taiwan National Ocean University; and supported by TOTAL Foundation, the French Ministry of Foreign Affairs and the ASEAN Regional Centre for Biodiversity and Conservation. We thank Joelle Lai, Lin Chia-Wei, Dwi Listyo Rahayu and Noel Saguil for their kind help. Important suggestions from two anonymous reviewers have also helped improve the paper. This study was partially supported by ARF Grant No. R-154-000-334-112 from the National University of Singapore.

References

Andrae D (2002) Seamount Fisheries—myths and realities. Fishing Boat World, September 2002:27–28

Benkendorff K, Przeslawski R (2008) Multiple measures are necessary to assess rarity in macro-molluscs: a case study from southeastern Australia. Biodivers Conserv 17:2455–2478

Bouchet P (2006) Chapter 2. The magnitude of marine biodiversity. In: Duarte CM (ed) The exploration of marine biodiversity. Scientific and technological challenges. Fundación BBVA, pp 33–64

Bouchet P, Lozouet P, Maestrati P, Heros V (2002) Assessing the magnitude of species richness in tropical marine environments: exceptionally high numbers of molluscs at a New Caledonia site. Biol J Linn Soc 75:421–436

Carpenter KE, Springer VG (2005) The center of the center of marine shore fish biodiversity: the Philippine Islands. Environ Biol Fish 72:467–480

Chan T-Y, Ng PKL (2008) *Enoplometopus* A. Milne-Edwards, 1862 (Crustacea: Decapoda: Nephropoidea) from the Philippines, with description of one new species and a revised key to the genus. Bull Mar Sci 83(2):347–365

Chan T-Y, Yu H-P (1995) The rare lobster genus *Palinustus* A. Milne-Edwards, 1880 (Decapoda: Palinuridae), with description of a new species. J Crustac Biol 15(2):376–394

Chan T-Y, Ng PKL, Ahyong ST, Tan SH (2009) Crustacean fauna of Taiwan: Brachyuran crabs, vol I: carcinology in Taiwan and Dromiacea, Raninoida, Cyclodrippoida. National Taiwan Ocean University, Keelung, Taiwan, pp i–vii, 1–198

Chapman MG (1999) Are there adequate data to assess how well theories of rarity apply to marine invertebrates? Biodivers Conserv 8:1295–1318

Chapman MG, Underwood AJ (1996) Experiments on effects of sampling biota under intertidal and shallow subtidal boulders. J Exp Mar Biol Ecol 207:103–126

Christie P, White A, Deguit E (2002) Starting point or solution? Community-based marine protected areas in the Philippines. J Environ Manage 66:441–454

Crosnier A (2002) Révision du genre *Parathranites* Miers, 1886 (Crustacea, Brachyura, Portunidae). Zoosystema 24(4):799–825

Cumberlidge N, Ng PKL (2009) Systematics, evolution, and biogeography of freshwater crabs. In: Martin JW, Crandall KA, Felder DL (eds) Crustacean issues 18: decapod crustacean phylogenetics. CRC Press, pp 491–508
Davie PJF (1993) Deepwater xanthid crabs from French Polynesia (Crustacea, Decapoda, Xanthoidea). Bulletin du Muséum national d'Histoire Naturelle, Paris, (4) 14(2):501–561
Dennis C, Aldhous P (2004) A tragedy with many players. Nature 430:396–398
Ellingsen KE, Hewitt JE, Thrush SF (2007) Rare species, habitat diversity and functional redundancy in marine benthos. J Sea Res 58:291–301
Galil BS (2003) Contribution to the knowledge of Leucosiidae II. *Euclosia* gen. nov. (Crustacea: Brachyura). Zoologische Mededelingen 77(20):331–347
Galil BS, Ng PKL (2009) Calappoidea and Leucosioidea (Crustacea: Decapoda: Brachyura) from Luzon, Philippines, with descriptions of two new species of *Mursia*. Zootaxa 2085:45–60
Galil B, Takeda M (2004) On a collection of *Mursia* (Crustacea, Decapoda, Brachyura, Calappidae) from Balicasag Island, Philippines. Bull Natl Sci Mus Tokyo (A) 30(1):23–35
Gaston KJ (1994) Rarity. Chapman & Hall, London, p 220
Grassle JF, Maciolek NJ (1992) Deep-sea species richness: regional and local diversity estimates from quantitative bottom samples. Am Nat 139(2):313–341
Gray JS (1994) Is the deep sea really so diverse? Species diversity from the Norwegian continental shelf. Mar Ecol Prog Ser 112:205–209
Gray JS (2002) Species richness of marine soft sediments. Mar Ecol Prog Ser 244:285–297
Grehan A (2003) Deep-water corals off the wests coast of Ireland. In: Grehan AJ, Long RL, Deegan R, Cinneide MO (eds) The Irish Coral Task Force and Atlantic Coral Ecosystem Study (ACES). Report on two deep-water coral conservation stakeholder workshops held in Galway in 2000 and 2002. Section B-Workshop Papers. Marine, Environment and Health Series No. 11, pp 26–32
Longino JT, Coddington J, Colwell RK (2002) The ant fauna of a tropical rainforest: estimating species richness three different ways. Ecology 83(3):689–702
Magurran AE, Henderson PA (2003) Explaining the excess of rare species in natural species abundance distributions. Nature 422:715–716
Manuel-Santos MR, Ng PKL (2007) On the genus *Ladomedaeus* Števčić, 2005, from the Philippines and Japan, and the status of the Ladomedaeidae Števčić, 2005 (Decapoda: Brachyura: Xanthoidea). Raffles Bull Zool 16:177–185
McLay CL (1999) Crustacea: decapoda: revision of the family Dynomenidae. In: Crosnier A (ed) Résultats des Campagnes MUSORSTOM. vol 20. Mémoires du Muséum National d'Histoire Naturelle, Paris 180:427–569
McLay CL, Ng PKL (2004) A taxonomic revision of the genus *Paradynomene* Sakai, 1963 (Crustacea: Decapoda: Brachyura: Dynomenidae). Zootaxa 657:1–24
McLay CL, Ng PKL (2005) On a collection of Dromiidae and Dynomenidae from the Philippines, with description of a new species of *Hirsutodynomene* McLay, 1999 (Crustacea: Decapoda: Brachyura). Zootaxa 1029:1–30
Mendoza JCE, Ng PKL (in press) The euxanthine crabs (Crustacea: Decapoda: Brachyura: Xanthidae) of the Philippines. Raffles Bull Zool
Ng PKL (2006) La biodiversité et la rareté des espèces: etude de cas deux habitats "cryptiques" (Biodiversity and species rarity: a case study from two "cryptic" habitats). Forum Biodiversité des Ecosystèmes Coralliens (Coral Reef Ecosystem Biodiversity Forum), BIODEC, Institut de recherches pour le développement (IRD), Nouméa, New Caledonia, 30 October–4 November 2006. Livre des Résumés (Abstract Book), pp 36–39
Ng PKL, Ho P-H (2003) *Mathildella rubra*, a new species of deep-water carcinoplacine crab (Decapoda, Brachyura) from the Philippines. Crustaceana 76(3):333–342
Ng PKL, Liao LM (2002) On a new species of *Euryozius* Miers, 1886 (Crustacea: Decapoda: Brachyura: Pseudoziidae) from the Philippines, with notes on the taxonomy of the genus. Proc Biol Soc Wash 115(3):585–593
Ng PKL, Guinot D, Davie PJF (2008a) How many living crab species are there? Update on the Brachyuran Crab Project. In: Abstracts, advances in decapod crustacean phylogenetics. A symposium during the Society of Integrative and Comparative Biology (SICB) annual meeting, San Antonio, Texas, USA, January 2–6, 2008, p 73
Ng PKL, Guinot D, Davie PJF (2008b) Systema Brachyurorum: Part I. An annotated checklist of extant brachyuran crabs of the world. Raffles Bull Zool 17:1–286
Ng PKL, Mendoza JCE, Manuel-Santos MR (2009) Tangle net fishing, an indigenous method used in Balicasag Island, central Philippines. Raffles Bull Zool 20:39–46

Novotny V, Basset Y (2000) Rare species in communities of tropical insect herbivores: pondering the mystery of singletons. Oikos 89:564–572
Pitman NCA, Terborgh J, Silman MR, Nuñez V P (1999) Tree species distributions in an Upper Amazonian forest. Ecology 80(8):2651–2661
Poupin J (1996) Crustacea Decapoda of French Polynesia (Astacidea, Palinuridea, Anomura, Brachyura). Atoll Res Bull 442:1–114
Pyle RL (2000) Assessing undiscovered fish biodiversity on deep coral reefs using advanced self-contained diving technology. Mar Technol Soc J 34(4):82–91
Rabinowitz D (1981) Seven forms of rarity. In: Synge H (ed) The biological aspects of rare plant conservation. Wiley, Chichester, pp 205–217
Richer de Forges B, Ng PKL (2007a) Notes on the deep-sea spider crabs of the genus *Cyrtomaia* Miers, 1886, from the Philippines (Crustacea: Decapoda: Brachyura: Majidae), with description of a new species. Raffles Bull Zool 16:55–65
Richer de Forges B, Ng PKL (2007b) New records and new species of Homolidae De Haan, 1839, from the Philippines and French Polynesia (Crustacea: Decapoda: Brachyura). Raffles Bull Zool 16:29–45
Richer de Forges B, Ng PKL (2007c) On a new genus and new species of deep-water spider crab from the Philippines (Crustacea, Decapoda, Brachyura, Majidae). Zootaxa 1644:59–68
Richer de Forges B, Koslow JA, Poore GCB (2000) Diversity and endemism of the benthic seamount fauna in the southwest Pacific. Nature 405:944–947
Richer de Forges B, Tan SH, Bouchet P, Ng PKL, Chan T-Y, Saguil N (2009) PANGLAO 2005—survey of the deep-water benthic fauna of the Bohol Sea and adjacent waters. Raffles Bull Zool (Suppl No 20):21–38
Ricklefs RE (2000) Rarity and diversity in Amazonian forest trees. Trends Ecol Evol 15(3):83–84
Roberts CM, McClean CJ, Veron JEN, Hawkins JP, Allen GR, McAllister DE, Mittermeier CG, Schueler FW, Spalding M, Wells F, Vynne C, Werner TB (2002) Marine biodiversity hotspots and conservation priorities for tropical reefs. Science 295:1280–1284
Schlacher TA, Newell P, Clavier J, Schlacher-Hoenlinger MA, Chevillon C, Britton J (1998) Soft-sediment benthic community structure in a coral reef lagoon–the prominence of spatial heterogeneity and 'spot endemism'. Mar Ecol Prog Ser 174:159–174
Tan SH (2004) A systematic revision of the Parthenopidae (Crustacea: Decapoda: Brachyura), 2 vols. Unpublished PhD thesis, National University of Singapore, 729 pp

Biodivers Conserv (2010) 19:587–600
DOI 10.1007/s10531-009-9733-0

ORIGINAL PAPER

The integration of biodiversity conservation with oil and gas exploration in sensitive tropical environments

Claude-Henri Chaîneau · Jacques Miné · Suripno

Received: 7 March 2008 / Accepted: 3 October 2009 / Published online: 22 October 2009

Abstract This paper describes the approach of TOTAL Exploration & Production concerning the protection of biodiversity during oil and gas exploration and production activities, particularly in sensitive environments such as tropical islands. This approach was developed according to three stages: knowing biodiversity, protecting biodiversity and, going further, by contributing to scientific research. This three step approach is embedded in an environment protection policy and an associated set of actions already implemented for many years. *Knowledge of the biodiversity* is an indispensable preliminary to evaluating the sensitivity of a site which is likely to be impacted by an industrial activity. TOTAL E&P Indonésie (TI) has been operating for over 30 years in the East part of Borneo Island in areas of increasing sensitivity with respect to mangrove ecosystems; for example the Mahakam delta which includes the Tambora, Handil and Tunu fields. This area was initially covered by dense mangrove vegetation. Regular biodiversity surveys in the delta, especially for fishes, birds and benthos, have allowed TI to witness the progressive changes occurring in the delta. *Reducing significant impacts and conserving biodiversity* is considered as an integral part of sustainable development. With regard to the Mahakam delta, in the last decades there has been a marked increase of local population activities and deforestation due to shrimp-culture development. TI has made significant efforts with environmental management actions to protect the mangrove ecosystem through several environmental programs such as minimizing land clearing and reinstating cleared areas. The programs have involved local communities and other stakeholders such as NGO's and governmental bodies. *Contributing to scientific knowledge of biodiversity* is of mutual interest to both science and industry. The TOTAL Group, through its *Foundation for Biodiversity and the Sea*, has supported the establishment of a comprehensive inventory of the marine biodiversity of the island of Panglao, situated

C.-H. Chaîneau (✉) · J. Miné
TOTAL S.A., DGEP/HSE/ENV, 2 place Jean Millier, La Défense 6, Paris la Défense
Cédex 92078, France
e-mail: claude-henri.chaineau@total.com

Suripno
TOTAL E&P Indonésie, DSE/ENV, Jalan Yos Sudarso, P.O. Box 606, East Kalimantan,
Balikpapan 76123, Indonesia

southwest of Bohol, Philippines. The Panglao Marine Biodiversity project was driven in collaboration between the Muséum National d'Histoire Naturelle, Paris, and the University of San Carlos, Cebu City, Philippines. Recently, another major biodiversity inventory has been also supported by the TOTAL Foundation: the Santo 2006 biodiversity project. On this island, located in the Vanuatu archipelago, the research program has focused on four themes: marine, forest, karstic environments and alien species. Furthermore, TOTAL in partnership with IFREMER, a French oceanographic research centre, has also developed a comprehensive biodiversity research and development program dedicated to Deep Sea ecosystems. Even if initially focused off Western Africa, this emerging issue is also of concern in most of the tropical areas in the world where deep offshore environments are present.

Keywords Biodiversity · Tropics · Policy · Oil and Gas · Mangrove · Deep-sea ecosystems

Introduction

As a producer of energy from natural resources, TOTAL Exploration & Production learned early on how to operate with respect for the environment and with specific care for areas that are sensitive from a biodiversity viewpoint. The company devotes its undivided attention to the protection of sensitive areas such as tropical islands. Protection of biodiversity is one of the targets of the Convention on Biological Diversity, which was a major achievement of United Nation Conference held in Rio in 2002 (Siebenhüner 2007). Protection of biodiversity is part of the corporate social responsibility. By implementing the commitment to safeguard biodiversity as a component of sustainable development the company is optimizing its activities and consolidating its performance over the long term. This determination concerning biodiversity protection is reflected in a series of different actions already implemented for many years: (1) Establishment of the *Foundation for Biodiversity and the Sea* in 1992; (2) Establishment of the Biodiversity Policy validated by the executive committee of the company in January 2005; (3) Writing of a roadmap defining the company position, drawn up after consultation with the World Conservation Union; (4) Issuing an internal practical guide expressing the company's commitment to enforce the group biodiversity policy.

The commitment to biodiversity protection is applied at every stage of an exploration and production project, from the seismic acquisition, through the drilling and production phase until the end of the field's life. To achieve protection of the environment and obviously of biodiversity, the company has developed a systematic and constant approach based on the following three stages: (1) knowing biodiversity, (2) protecting biodiversity and (3) contributing to scientific/biodiversity research.

Two classical project tools are used to achieve the different stages: (1) The Environmental Baseline Study (EBS), and (2) the Environmental Impact Assessment (EIA). The Environmental Baseline Study determines a *reference status* for many environmental parameters and especially those concerning biodiversity. The EBS includes, most of the time, a biological survey allowing a thorough inventory of species present at the site. During this study, independent expertise is sought from biological scientists (ecologists, zoologists and botanists) in order to assess the ecological sensitivity of the site likely to be impacted by the industrial activity. A preliminary evaluation of the potential impacts is also assessed. Following the EBS, the Environmental Impact Assessment determines the potential impacts of the project on the natural and human environment and defines the mitigation measures and solutions required to eliminate or, if not possible, to reduce the impacts. Specific attention is paid to evaluate the residual impact on biodiversity and specific protection measures are implemented if necessary.

This paper aims to highlight the three steps process: knowing the biodiversity, protecting the biodiversity and the promotion and support of biodiversity research.

Knowing the biodiversity: the example of the Mahakam delta in Kalimantan (Indonesia)

TOTAL E&P Indonésie (TI), a subsidiary of TOTAL E&P, has been operating in the Mahakam contract area for more than 30 years as contractor of BPMIGAS (Executive Indonesian Agency for Upstream Oil and Gas Business). Several oil and gas fields operated by TI are located mainly in the Mahakam delta i.e. Handil, Tambora and Tunu. Other fields are also operated offshore, i.e. Bekapai, Peciko and Sissi-Nubi. The Mahakam delta was initially covered by a dense mangrove vegetation but, in recent years, shrimp pond development has spread over more than 70% of the delta area with consequent major loss of mangrove habitat (Fig. 1) (Suripno et al. 2004; Van Zwieten et al. 2006). Aquaculture activities have also resulted in an increasing number of people working and/or living in the area. This situation has made the Mahakam delta more sensitive to oil and gas operations in terms of social and environmental issues. Hence the company has had to face more challenges in managing its activities in such a sensitive social environment and mangrove ecosystem.

The Mahakam delta, a highly sensitive area, as are many other deltas, is located at the mouth of the Mahakam river on the East coast of Kalimantan, Indonesia, along the Makassar Strait. The delta plain stretches over an area of 1,500 km^2, including the distributary and interdistributary channels. Around 1,300 km^2 are covered by land. It was originally covered by dense mangrove, including *Avicennia* sp., *Sonneratia* sp., *Rhizophora* sp., *Nypa nypa*, and as well as fresh water mangroves (MacKinnon et al. 1996; Suripno et al. 2004; INRR 2006). In 1986, *Nypa nypa* dominated the mangrove vegetation of the delta (Figs. 1, 2). Salinity conditions range from fresh in the upstream, brackish in

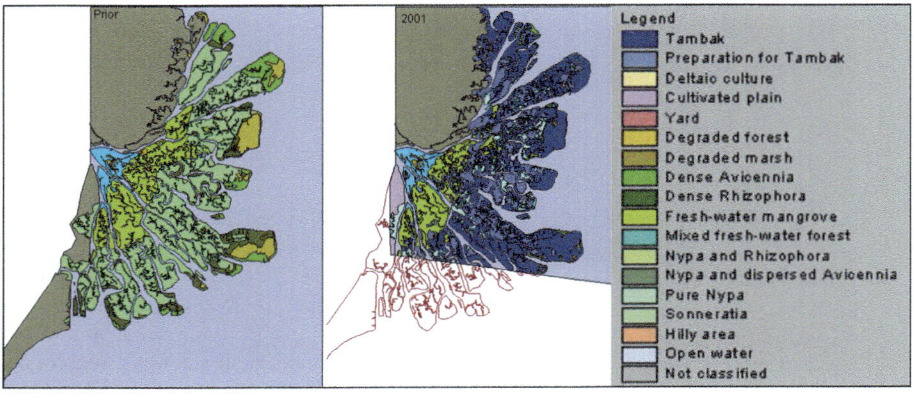

Fig. 1 Integrated GIS maps demonstrating a major biodiversity loss due to shrimp farming over time (prior: 1986; part of the south of the Mahakam was not covered by the Spot satellite in 2001)

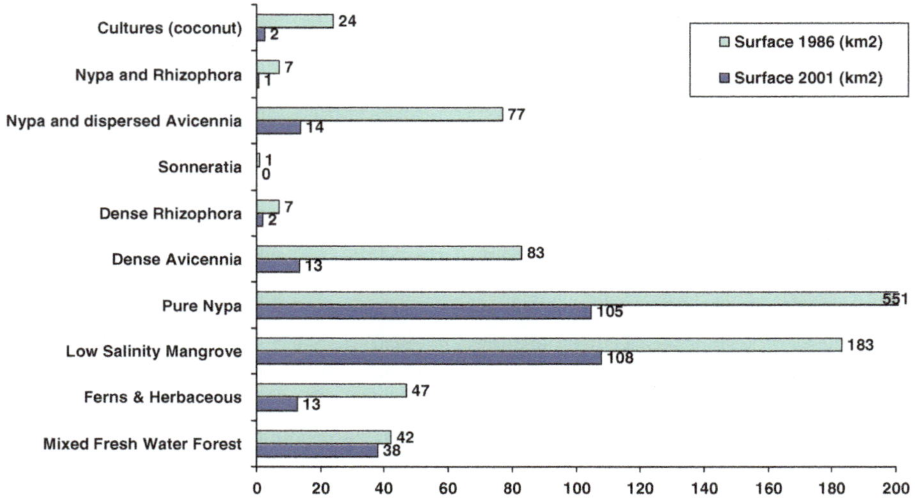

Fig. 2 Variation of vegetation cover and types between 1986 and 2001

the intertidal zone and saline in the outer part of the delta. The mangrove ecosystem provides many benefits to the environment (MacKinnon et al. 1996) such as:

- Stabilization and protection of shorelines and river banks,
- Filtering and trapping water borne pollutants released from large cities and industrial activities upstream,
- Providing a nursery and feeding ground for a wide spectrum of animals such as numerous species of fishes, prawns, crabs, molluscs, mammals including marine mammals (Kreb and Budiono 2005) and reptiles,
- Providing habitat for a large number of animals,
- Providing nesting site for sea and shore birds,
- And, to a lesser degree, providing resources for tourism and recreation.

The vegetation in the Mahakam delta consists of four main vegetal communities (MacKinnon et al. 1996). From the external part of the delta (seashore) going upstream, these are:

- Coastal mangrove fringe, dominated by the *Avicennia* sp., *Rhizophora* sp., *Sonneratia* sp., and *Bruguiera* sp,
- *Nypa* plain: *Nypa nypa* is a monospecific community which spreads over the oldest stabilized ground. They grow in central part of the delta with variable salinity conditions,
- Low-salinity mangrove, i.e. the transition zone between the *Nypa* plain and the rain forest. The low-salinity mangrove is dominated by *Heritiera* sp. It grows in areas which are weakly influenced by marine saline waters,
- Fresh water forest: located at the apex of the delta and consisting of highly diversified vegetation dominated by *Oncosperma* sp., *Pandanus* sp. and *Acanthus* sp.

In the last decade, the mangrove tract in the delta has been increasingly cleared due to conversion to shrimp ponds for aquaculture. The rapid destruction of mangrove vegetation peaked in 1997 when the East Asian economic crisis hit Indonesia. As a result, more than

73% of mangrove has been converted into shrimp ponds to date (Van Zwieten et al. 2006). Pond areas range from less than 2 ha to up to 50 ha, the average shrimp pond size being about 10 ha (Bourgeois et al. 2002). Most of the shrimp ponds are located in the external part of the delta where the salinity is high. However, in recent years many shrimp ponds were also constructed in the low salinity area, i.e. the central part of the delta. Shrimp culture and fishing activities in the delta attract many people to invest and work in the area. Some of them are residing in the area while others work in the delta but reside outside. The population growth rate in the Mahakam delta between 1996 and 2001 was relatively high (7–8%), compared with the national population growth rate in 2000 (3%) (Bourgeois et al. 2002; Suripno et al. 2004; INRR 2006). These rapid changes impacted the mangrove ecosystem by degrading its ecological functioning. Several problems have arisen such as degradation of water quality, loss of vegetation surface area, increasing coastal erosion, loss of biological habitats, soil losses, soil quality impoverishment, decrease of shrimp and fish productivity, increasing occurrence of shrimp diseases. This deterioration also resulted in social and economic impacts such as failures of shrimp pond harvests and declining fish catches. These social and economic impacts have also affected TOTAL E&P Indonésie operations in the area. The massive reduction of the area available to natural life (Fig. 2), has affected the ecology of the indigenous biota. It is now common to observe some wild animals in the vicinity of camps and installations.

Over the last 10 years, TI has been developing a geographical information system dedicated to its own operations and additionally to environmental and social issues. The environmental objectives were to monitor the variations in vegetation cover of the delta with satellite images (Fig. 1), then to gain a better knowledge of the delta status and to have sufficient information for taking decisions in case of any crisis. For that purpose, TI has established various partnerships with international research centres, local universities, and international engineering companies. They have also made use of the existing scientific studies covering Borneo Island. Periodic surveys, with the support of scientists, have been, and still are, conducted regularly in the delta to update knowledge of the local biodiversity, especially fishes, birds and marine benthic organisms. The discoveries and results of the studies have been made available through several environmental baseline reports. These studies have been presented to the Indonesian authorities and constituted an integrated part of the whole environmental management system implemented by the company and which was certified (ISO 14001) in 2005 for its operations. EBS surveys are conducted for each planned activity (seismic, drilling, pipeline installation, oil and gas producing platforms installation, dredging…). The systematic use of aerial photography and on site baseline surveys allows the identification of sensitive areas that need to be protected. The implementation of the environmental impact assessment studies then allows the development of mitigation measures in order to protect these areas and more specifically the local biodiversity.

Protecting the biodiversity: the example of the TOTAL E&P Indonésie initiatives in the Mahakam delta

In parallel to the above studies, TI has launched sustainable development programs centralized in an organization called the Sustainable Development Steering Committee. The committee consists of four main sections, each corresponding to the following: Community development, Mahakam delta, Environment, and Protection of land assets. Each section of the committee has defined its objectives and programs in relation to institutional aspects, regulatory framework, law enforcement and community awareness improvement. As part of the work with the different stakeholders, including local and national authorities, NGOs, Pongawas (the richest owners of shrimp ponds), several initiatives have been implemented through operational programs dedicated to the preservation of the mangrove ecosystem.

To this end, the subsidiary steering committee has implemented some actions as part of the programs mentioned above. These include minimization of land clearance, implementation of an offset policy and associated actions, such as replanting programs, mangrove rehabilitation and sustainable programs of shrimp pond farming (Suripno et al. 2004; INRR 2006).

Minimizing land clearing

Land clearing is minimized by initiating a proper assessment prior to implementing any new development project. Project locations and/or extension alternatives are assessed in order to select the most acceptable alternative which has minimum effects on the environment. Some examples are the installation of new pipelines in existing right of ways (ROW), extension of existing oil and gas gathering facilities instead of installing additional new ones, optimization of land clearing in accordance with safety and environmental aspects, horizontal drilling to lay down pipelines that avoid vegetation removal, etc.

Offset policy

When land clearance is unavoidable, an offset policy has been put into place since 2002. It mainly consists in purchasing degraded lands, or non degraded areas in order to compensate for the clearance even if rehabilitation is going to be conducted after field operations. The ratio, usually 2 preserved for 1 cleared, allows the creation "protected" areas which can be considered to some degree as representing the native environment.

Mangrove reinstatement in cleared and eroded areas

In addition, TI has initiated a mangrove replanting program (Fig. 3). To date more than 3 million mangrove trees have been planted on previously cleared areas, mainly pipeline right of ways. Since the launching of the program in year 2000, mangrove replanting has covered more than 200 km of ROW and adjacent purchased ponds which represents more than 600 ha. This operation is performed in cooperation with local villagers who provide the seedlings and perform the planting as well as part of the monitoring of the planted trees. In cooperation with the Environment Ministry (BAPEDALDA) and local NGOs, a specific program of mangrove replanting was conducted on eroded areas along the coastal fringe of the delta. It was carried out by local villagers, local students, BAPEDALDA and TI.

Sustainable shrimp farming: pilot pond test

One of the community development programs is to provide shrimp farmers with information on environmentally friendly and sustainable methods for the management of shrimp culture, in particular through the silvo-fishery method. This method allows the reinstatement of the mangrove trees in ponds. The aim is to strike a balance between natural nutrient input, voluntary supplements (from the prawn feed) and nutrient consumption in the ponds by the natural microbiota and the planted trees. The training sessions

Fig. 3 **a** Mangrove plantation on a pipeline right of way, **b** plants being monitored after 3 months and **c** 3 years

were provided to more than 400 participants at several Sub-Districts, in cooperation with governmental bodies and NGOs. To support shrimp pond training programs, a pilot project of environmentally sustainable shrimp ponds has been launched to provide a good example to local communities as well as to promote awareness of mangrove ecosystem sustainability.

Hydrocarbon exploration and production activities in a sensitive social and environmental mangrove ecosystem require the minimization of environmental impacts from the operations as well as efforts to improve the sustainability of the mangrove ecosystem. The establishment of effective organization and the actions taken to minimize the impacts to the mangrove ecosystem and to manage the sensitive social context have allowed the promotion of stakeholder awareness of the environmental issues in the Mahakam delta. The actions taken have also shown positive results in the betterment of communication with the local communities as well as in solving disputes and claims. Recently, in 2007, TI has decided to again take the initiative and has earmarked two million dollars over 5 years to promote sustainable management of the Mahakam Delta. The company has organized a symposium on the risks posed by unsustainable shrimp farming and destruction of the mangrove ecosystem and it is training the villagers in more environmentally friendly methods for shrimp farming.

Going further through scientific surveys and research programs

Partnering to develop biodiversity scientific projects is of mutual interest to science and industry. Science contributes to a better knowledge of the biodiversity and its evolution, which is necessary for a better evaluation of the potential impact of industrial activity on the environment and for a better identification of the mitigation measures to eliminate or, if not possible, reduce this impact. Industry contributes to scientific research by way of logistics and financing. As an example, oil and gas exploration occurs often in isolated areas where the scientific knowledge is relatively poor: rain forest, deep offshore, deserts, arctic environments and tropical islands. The availability of heavy logistic support, provided by the oil and gas industry, can be shared with scientists in order to give them access to such isolated environments.

Some examples of partnership or financial support to scientific projects linked to the inventory/protection of tropical biodiversity are presented below to illustrate the involvement and philosophy of TOTAL: the Panglao project in the island of Panglao (Philippines), the Espiritu Santo project in Vanuatu and the BioZaïre project dedicated to deep offshore environments.

The 2004 Panglao marine biodiversity project

The objective of the international Panglao Marine Biodiversity project was to evaluate the marine species richness around the island of Panglao, situated southwest of Bohol in the central Philippines. It was carried out in collaboration with the Muséum National d'Histoire Naturelle, Paris, France (principal investigator Dr Philippe Bouchet), the National University of Singapore (principal investigator Dr Peter Ng) and the University of San Carlos, Cebu City, Philippines (principal investigator Dr Danilo Largo). The project was supported by the *TOTAL Foundation for Biodiversity and the Sea*, the French Ministry of Foreign Affairs and the ASEAN Regional Centre for Biodiversity Conservation. The project faced three main challenges and objectives: (1) to evaluate biodiversity in, reputedly the richest marine eco-region, (2) to select a study area with maximal habitat diversity (from mangrove to corals) and (3) to focus on two of the most diverse marine taxa: molluscs and decapod crustacea.

Numerous species were observed alive, collected and photographed during the survey. Species description is still underway but it is estimated that several hundred new species have been discovered (Anker and Jeng 2006; Anker et al. 2006; Mitsuhashi and Chan 2008; Reid et al. 2008; Chan et al. 2009). Twice as many marine species have been found in this area (15,000 ha) than in all the Mediterranean Sea. It demonstrates the extraordinary richness of a tropical coastal environment. This program was not only an innovative research program but it was also a workshop for local students and scientists with a view to capacity building and establishing specific measures to protect such a sensitive environment, one of the biologically richest ecosystems on Earth.

The 2006 Santo biodiversity project

In the wake of the Panglao project, *TOTAL Foundation for Biodiversity and the Sea* supported the Santo 2006 expedition. Espiritu Santo island is the largest island in the Vanuatu archipelago; it is also the highest island with its summit at 1,879 m. Considering its location (tropical island), size, relief and geology, Santo island was an excellent place to undertake a comprehensive assessment of the biodiversity of a tropical island. It was also an excellent representative for the two richest and most sensitive ecosystems on the Earth: tropical forest and coral reefs. As it was under-explored, numerous biological discoveries were expected. The project was jointly coordinated by the French Museum of Natural History (MNHN), the French Centre for the Research and Development (IRD) and an NGO (ProNatura International) with the support from different sources, in addition to the TOTAL Foundation: The Stavros Niarchios Foundation, the French Ministry of Foreign Affairs, the French Ministry for the Research, the European Distributed Institute of Taxonomy, Sodexho, Veolia-Environment Foundation. The TOTAL Foundation supported all marine operations.

The scientific program focused on four major "themes": (1) "marine biodiversity" including deep-sea ecosytems and coral reefs, (2) "forests, mountains, rivers", (3) "karst" and (4) "fallows and alien species". The programme on "marine biodiversity" required heavy logistics with the use of an oceanographic vessel. The objective was to undertake a comprehensive assessment of the island's marine biodiversity with specific studies dedicated to molluscs, crustaceans, echinoderms and algae. The objective of the "forest, mountains and rivers" study was to establish an inventory of the onshore biodiversity and to sample all forest strata. A large-scale canopy access system (including a canopy-glider) was used to reach the top of the trees. "Karstic" environments, either caves or surface formations, are known to host specific fauna and flora and the presence of these geological formations in the eastern part of the island provided an excellent opportunity to study ecosystems linked to this specific environment. The last theme "fallows and aliens" was dedicated to the study of the anthropogenic impact of cultivated lands and the introduction of alien/invasive species.

A comprehensive inventory for the different themes was achieved in four months. Over 10,000 species of plants, animals and fungi were collected. Among these species, 1,000 could be "new". The taxonomic identification is ongoing and, due to the species richness, may take many years (Bouchet et al. 2009).

The BioZaïre research program

Knowledge of the deep-sea environments is relatively new, due to extreme operational and environmental conditions (e.g. high pressures, low temperatures around 2–4°C, fluids containing chemical compounds and the absence of light). Until the mid 19th century, nobody imagined that "life" might exist at depths greater than 600 m. In fact, this "abyssal" ecosystem, covering vast surface areas (307 million km^2), represents a major reservoir of biodiversity on the planet. For the offshore oil and gas industry, the lack of knowledge of ecosystems in the deep sea environment became apparent, as detailed below, when the first environmental surveys were carried out for the purposes of exploration and development projects in the vicinity of the Angolan deep-offshore at water depths greater than 1,000 m.

As classically performed by oil and gas companies, two types of environmental studies are generally undertaken, as already explained: Environmental Baseline Studies (EBS) and Environmental Impact Assessments (EIA). Although Environmental Baseline Studies were carried out for these deep offshore developments, they consisted merely of a description of the fauna, using methodologies already applied in shallower domains and well-suited to such contexts. Even though specific methodologies and protocols were adapted to this specific deep-sea environment (Sutton et al. 2004), such EBS's were not enough. The company needed a broader, regional vision, hence a research programme was initiated in partnership with IFREMER, a French oceanographic research centre. In this respect, the

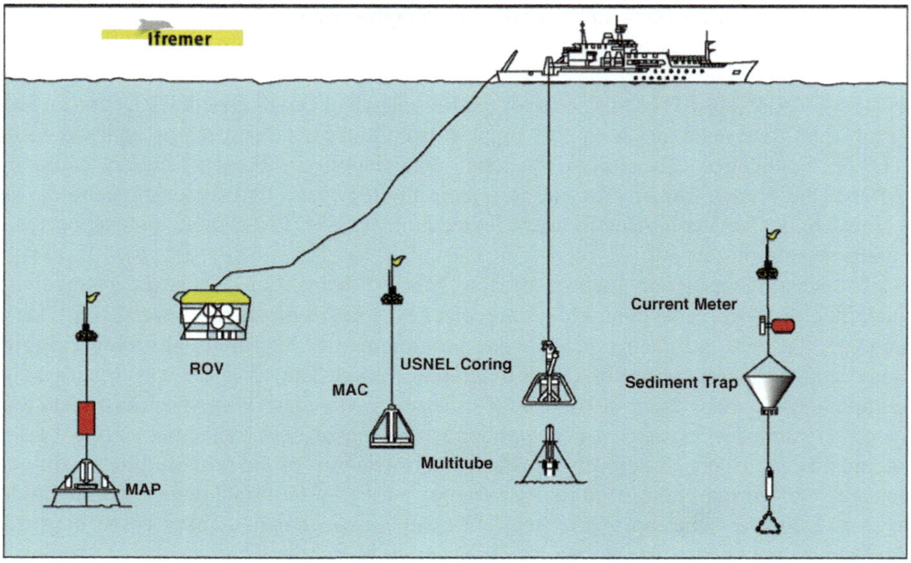

Fig. 4 Sampling methodology used during the oceanographic campaigns of BioZaïre program

main objectives of the "BioZaïre" programme, so named after its location near the Zaïre River channel, were (1) to obtain a representative description of the benthic populations at depths ranging from 600 m to 4,000 m, based on a suitable and reproducible sampling methodology and, (2) to get a clear understanding of the structure and behaviour of the ecosystems in order to assess the potential impact of the oil and gas exploration/production activity on them.

The research centres involved in the "BioZaïre" programme were: the Deep Sea Environment Department of IFREMER, its university partners, the French National Centre for Scientific Research (CNRS), the Max-Planck Institute and the Alfred-Wegener Institute (Germany), Akvamiljø in Stavanger (Norway) and the TOTAL technical and research centres in Stavanger and in Pau (France). Such studies required heavy logistics: four main oceanographic cruises have been organised on board IFREMER vessels with important equipment for sampling, measuring and observing, such as remotely controlled submersible vehicle (ROV Victor 6000), current meters, sediment traps, respirometers, multitube corers, etc. (Fig. 4).

Two benthic ecosystems were studied during the survey: one a detritus-based ecosystem, particularly influenced by the continental sediment from the Zaïre river and a second ecosystem, chemosynthetic-based, and closely linked to methane-bearing cold fluid emissions in specific areas called pockmarks due to the typical topography of these structures on the sea bed. The detritus-based ecosystem is reliant on a source related to the photosynthetic process. Its fauna is clearly adapted to use the chlorophyllian production settling from the sea surface and/or the continental sediments transported by the Zaïre river. Even when small-sized species dominate the ocean floor, large megafauna are also present and some fishes have developed anatomical and physiological adaptations in order to feed and reproduce (Fig. 5). The chemosynthetic-based ecosystem is associated with fluids or methane-rich gases rising up to the water–sediment interface (cold seepages). It comprises communities developed at the apex of active pockmarks, forming "oases of

Fig. 5 Example of a detritivorous fish with anatomical adaptation of its jaws for capture of food (*Source*: IFREMER)

Fig. 6 Dense fauna in the hollow of a pockmark: mussels, vestimentiferan tubeworms, shrimps and purple sea cucumbers observed at 3,160 m water depth, Gabon (*Source*: IFREMER)

life". Both diversity and biomass are very high (Figs. 6, 7, 8): communities consist of large fields of deep sea clams and mussels, clumps of giant tube worms (Vestimentifera) more than two metres high, clusters of shrimps, blue holothurians, and carpets of bacteria. These communities are dominated by species living in symbiosis with methanotrophic (methane oxidizing) or thiotrophic (sulphur oxidizing) bacteria. The first results of this pluridisciplinary environmental program have been presented by Sibuet et al. (2002), Savoye et al. (2003) and Andersen et al. (2004). Further recent discoveries of similar communities have been made in Nigerian offshore waters 1,200 m deep (Cauquil and Adamy 2008).

Conclusion

Biodiversity preservation, considered as an integral part of sustainable development, is today a universal concern. It requires involvement from all actors at all levels. In particular,

Fig. 7 "Bush of vestimentiferan tube worms" sampled at 3,160 m water depth, Gabon (*Source*: IFREMER)

Fig. 8 A shrimp on a vestimentiferan tubeworm that has been colonized by sedentary organisms (small sea anemones) at a water depth of 3,160 m, Gabon (*Source*: IFREMER)

it implies the participation of the administrative authorities, local communities, industry representatives and, obviously, the scientific community.

Partnership between the oil and gas exploration industry and the scientific community is in the mutual interest of each party. Science contributes to industry knowledge of the environment which then allows a better protection of the biodiversity around the industrial sites (through implementation of mitigation measures). On the other hand, the industry contributes to scientific research in different ways: (1) setting up scientific partnerships, (2) sharing scientific information from the different Environmental Baseline Studies, (3) facilitating access to remote areas where the industry is operating and (4) funding biodiversity Research & Development programs.

Acknowledgments The authors thank TOTAL E&P Borneo B.V. for its support, Michel Grandprat from TOTAL S.A. and Joelle Galeron from IFREMER who kindly checked the article. The authors are grateful to the reviewers and to the editor for their comments.

References

Andersen AC, Hourdez S, Marie B, Jollivet D, Lallier FH, Sibuet M (2004) *Escarpia southwardae* sp. nov., a new species of vestimentiferan tubeworm (Annelida, Siboglinidae) from West African cold seeps. Can J Zool 82:980–999

Anker A, Jeng MS (2006) *Richalpheus palmeri*, N GEN,N.SP. an infaunal alpheid shrimp from the Philippines, with redescription of *Amphibetaeous jousseaumei* (Coutière, 1896) (Decapoda: Caridea). J Crust Biol 26:379–391

Anker A, Poddoubtchenko D, Jeng MS (2006) *Acanthanas pusillus*, new genus, new species, a miniature alpheid shrimp with spiny eyes from the Philippines (Crustacea: Decapoda). Raffles Bull Zool 54:341–348

Bouchet P, Le Guyader H, Pascal O (eds) (2009) Natural history of Santo. Patrimoines naturels 69. Coéd./Copubl. IRD éditions.

Bourgeois R, Gouyon A, Jésus F, Levang P, Langeraar W, Rahmadani F, Sudiono E, Sulistani B (2002) Socioeconomic and institutional analysis of Mahakam delta stakeholders. Final Report to TotalFinaElf, contract no 501125/DKI/204, 108 pp

Cauquil E, Adamy J (2008) Seabed imagery and chemosynthetic communities: examples from deep offshore West Africa. Paper OTC 19352 presented at the 2008 Offshore Technology Conference, Houston, Texas, USA, 5–8 May 2008

Chan TY, Ho CK, Li CP, Chu KH (2009) Origin and diversification of the clawed lobster genus *Metanephrops* (Crustacea: Decapoda: Nephropidae). Mol Phylogenet Evol 50:411–422

INRR (2006) Mahakam Delta in space and time: ecosystem, resources, and management. Bogor: BPMigas, TOTAL, 190 pp

Kreb D, Budiono (2005) Cetacean diversity and habitat preferences in tropical waters of East Kalimantan, Indonesia. Raffles Bull Zool 53:149–155

MacKinnon K, Hatta G, Halim H, Mangalik A (1996) The ecology of Kalimantan. The ecology of Indonesia series, vol III, s.1. Periplus Editions (HK) Ltd., Singapore, xxiv, 802 pp

Mitsuhashi M, Chan TY (2008) Pontoniine shrimps of the genus *Apopontonia* Bruce, 1976 (Crustacea: Decapoda: Palaemonidae) from the Philippines PANGLAO 2004 expedition, with a description of one new species. Raffles Bull Zool 19:27–38

Reid DG, Dyal P, Lozouet P, Glaubrecht M, Williams ST (2008) Mudwhelks and mangroves: the evolutionary history of an ecological association (Gastropoda: Potamididae). Mol Phylogenet Evol 47:680–699

Savoye B, Sibuet M, Morash A, Miné J (2003) Zaïango-BioZaïre—exploring the depths. Energies-Technology special issue-winter (2002–2003), pp 24–33

Sibuet M, Galéron J, Khripounoff A, Menot L, Olu-Leroy K, Durrieu J, Miné J (2002) Deep-sea ecosystems on the equatorial African margin: First results of a pluridisciplinary environmental programme and discovery of chemosynthetic based ecosystem. Paper SPE 78875 presented at the SPE international conference on health, safety and environment in oil and gas exploration and production, Kuala Lumpur, Malaysia, 20–22 March 2002

Siebenhüner B (2007) Administrator of global biodiversity: the secretariat of the convention on biological diversity. Biodivers Conserv 16:259–274

Suripno, Najib M, Asmaradewi G (2004) Management of oil & gas exploration and production in sensitive social environment and mangrove ecosystem. Paper SPE 86578 presented at the SPE international conference on health, safety, and environment in oil and gas exploration and production, Calgary, Alberta, Canada, 29–31 March 2004

Sutton C, Miné J, Canovas S (2004) Environmental baseline surveys in deep waters: the lesson learned offshore West Africa. Paper SPE 86775 presented at the SPE international conference on health, safety, and environment in oil and gas exploration and production, Calgary, Alberta, Canada, 29–31 March 2004

Van Zwieten PA, Sidik AS, Noryadi, Suyatna I, Abdunnur (2006) Aquatic food production in the coastal zone: data-based perceptions on the trade-off between mariculture and fisheries production of the Mahakam delta and estuary, East Kalimantan. In: Hoanh CT, Tuong TP, Gowing JW, Hardy B (eds) Environment and livelihoods in tropical coastal zones. CAB International, Indonesia, pp 219–236

GPSR Compliance

The European Union's (EU) General Product Safety Regulation (GPSR) is a set of rules that requires consumer products to be safe and our obligations to ensure this.

If you have any concerns about our products, you can contact us on ProductSafety@springernature.com

In case Publisher is established outside the EU, the EU authorized representative is:

Springer Nature Customer Service Center GmbH
Europaplatz 3
69115 Heidelberg, Germany

Batch number: 10755466

Printed by Printforce, the Netherlands